Universitext

Springer
New York
Berlin
Heidelberg
Barcelona
Hong Kong
London
Milan
Paris
Singapore
Tokyo

Universitext

Editors (North America): S. Axler, F.W. Gehring, and K.A. Ribet

Aksoy/Khamsi: Nonstandard Methods in Fixed Point Theory
Andersson: Topics in Complex Analysis
Aupetit: A Primer on Spectral Theory
Bachman/Narici/Beckenstein: Fourier and Wavelet Analysis
Bădescu: Algebraic Surfaces
Balakrishnan/Ranganathan: A Textbook of Graph Theory
Balser: Formal Power Series and Linear Systems of Meromorphic Ordinary
  Differential Equations
Bapat: Linear Algebra and Linear Models (2nd ed.)
Berberian: Fundamentals of Real Analysis
Booss/Bleecker: Topology and Analysis
Borkar: Probability Theory: An Advanced Course
Böttcher/Silbermann: Introduction to Large Truncated Toeplitz Matrices
Carleson/Gamelin: Complex Dynamics
Cecil: Lie Sphere Geometry: With Applications to Submanifolds
Chae: Lebesgue Integration (2nd ed.)
Charlap: Bieberbach Groups and Flat Manifolds
Chern: Complex Manifolds Without Potential Theory
Cohn: A Classical Invitation to Algebraic Numbers and Class Fields
Curtis: Abstract Linear Algebra
Curtis: Matrix Groups
DiBenedetto: Degenerate Parabolic Equations
Dimca: Singularities and Topology of Hypersurfaces
Edwards: A Formal Background to Mathematics I a/b
Edwards: A Formal Background to Mathematics II a/b
Farenick: Algebras of Linear Transformations
Foulds: Graph Theory Applications
Friedman: Algebraic Surfaces and Holomorphic Vector Bundles
Fuhrmann: A Polynomial Approach to Linear Algebra
Gardiner: A First Course in Group Theory
Gårding/Tambour: Algebra for Computer Science
Goldblatt: Orthogonality and Spacetime Geometry
Gustafson/Rao: Numerical Range: The Field of Values of Linear Operators
  and Matrices
Hahn: Quadratic Algebras, Clifford Algebras, and Arithmetic Witt Groups
Heinonen: Lectures on Analysis on Metric Spaces
Holmgren: A First Course in Discrete Dynamical Systems
Howe/Tan: Non-Abelian Harmonic Analysis: Applications of $SL(2, R)$
Howes: Modern Analysis and Topology
Hsieh/Sibuya: Basic Theory of Ordinary Differential Equations
Humi/Miller: Second Course in Ordinary Differential Equations
Hurwitz/Kritikos: Lectures on Number Theory
Jennings: Modern Geometry with Applications
Jones/Morris/Pearson: Abstract Algebra and Famous Impossibilities
Kannan/Krueger: Advanced Analysis
Kelly/Matthews: The Non-Euclidean Hyperbolic Plane
Kostrikin: Introduction to Algebra

(continued after index)

Lucian Bădescu

Algebraic Surfaces

Translated by Vladimir Maşek

With 15 Illustrations

 Springer

Lucian Bădescu
Institute of Mathematics
Romanian Academy
PO Box 1-764
Bucharest 70700
Romania
Lucian.Badescu@imar.ro

Translator
Vladimir Maşek
Department of Mathematics
Washington University in St. Louis
Box 1146
St. Louis, MO 63130
USA
vmasek@math.wustl.edu

Mathematics Subject Classification (2000): 14-01 14Jxx, 14C17

Library of Congress Cataloging-in-Publication Data
Badescu, Lucian.
    Algebraic surfaces / Lucian Badescu.
        p. cm. —(Universitext)
    Includes bibliographical references and index.
    ISBN 978-1-4419-3149-8
    1. Surfaces, Algebraic. I. Title. II. Series.
QA571 .B32 2001
516.3′52—dc21   00-059551

Printed on acid-free paper.

Photocomposed copy prepared by Vladimir Maşek.

Printed in the United States of America.

9 8 7 6 5 4 3 2 1

Springer-Verlag   New York Berlin Heidelberg
A member of BertelsmannSpringer Science+Business Media GmbH

Foreword to the English Version

The main differences between the Romanian original and the English translation are the additions of exercise sets at the end of each chapter, a chapter on the Zariski decomposition of divisors and applications (Chapter 14), and an appendix listing further reading (Chapter 15). These new chapters contain (among other things) some developments in the theory of surfaces that are subsequent to the publication of the Romanian edition in 1981. The bibliography has also been updated. The titles added to the bibliography in the English edition are marked with an asterisk, for example [BPV*].

I would like to thank Vladimir Maşek for the excellent translation and for suggesting several slight changes that have resulted in a clearer presentation of the material.

Lucian Bădescu

Preface

The aim of this book is to present certain fundamental facts in the theory of algebraic surfaces, defined over an algebraically closed field k of arbitrary characteristic. The book is based on a series of talks given by the author in the Algebraic Geometry seminar at the Faculty of Mathematics, University of Bucharest.

The main goal is the classification of nonsingular projective surfaces (also called simply *surfaces*). In the context of *complex* algebraic varieties, the classification was obtained by Enriques and Castelnuovo. Around 1960, Kodaira [Kod1, Kod2] revived and simplified the classification of complex algebraic surfaces and extended it to the case of compact analytic surfaces. The problem of classifying surfaces in arbitrary characteristic remained open. The first step in this direction was the purely algebraic proof (valid in arbitrary characteristic), due to Zariski [Zar1, Zar2], of Castelnuovo's criterion of rationality. Then Mumford [Mum3, Mum4] introduced several new ideas, and the classification of surfaces in positive characteristic became possible. Finally, Bombieri and Mumford [BM1, BM2] completed the classification of surfaces in arbitrary characteristic. Their result was the following:

> The same types of surfaces that exist in the case when k is the complex field arise in the general case, if one sets aside certain pathologies that arise only in characteristic 2 or 3.

It is interesting to note that the new ideas and techniques introduced by Bombieri and Mumford clarified and simplified many of the classical arguments.

Our main goal is to present the methods introduced by Bombieri and Mumford regarding the classification of surfaces. The classification itself begins in Chapter 7. We had to introduce a number of other facts (interesting in themselves) that do not belong to classification proper, but do play an important role in classification, for example: the Nakai–Moishezon criterion of ampleness, the theory of normal singularities of surfaces (with special reference to rational singularities), and the relationship between Grothendieck–Serre duality and the study of singularities.

At the same time, along with the classification of surfaces we prove a remarkable result of Zariski and Mumford, which says that the canonical ring of a surface is a finitely generated k-algebra. In Chapter 12 we classify completely the minimal models of rational and ruled surfaces.

In the later stages of writing this material, we became aware of an excellent book about classification of surfaces by A. Beauville [Bea] in which some of the original ideas of Enriques and Castelnuovo are presented, via Kodaira's work referred to earlier and the Shafarevich seminar [Sha1].

The book is intended for readers who already have a certain amount of knowledge of algebraic geometry, such as the theory of algebraic curves, cohomology of coherent algebraic sheaves, the theory of Picard and Hilbert schemes of surfaces, general facts about Abelian varieties, elements of algebraic topology and étale topology (Betti numbers, Chern classes, etc.), and the Riemann–Roch theorem for surfaces (including Noether's formula). Starting from these facts, I tried to present the material in a unified manner and with complete proofs.

Finally, I would like to thank N. Manolache for his careful reading of the manuscript and his useful remarks.

Lucian Bădescu

Contents

Foreword to the English Version   v

Preface   vii

Conventions and Notation   xi

1   Cohomological Intersection Theory and the Nakai–Moishezon Criterion of Ampleness   1

2   The Hodge Index Theorem and the Structure of the Intersection Matrix of a Fiber   17

3   Criteria of Contractability and Rational Singularities   23

4   Properties of Rational Singularities   53

5   Noether's Formula, the Picard Scheme, the Albanese Variety, and Plurigenera   69

6   Existence of Minimal Models   81

7   Morphisms from a Surface to a Curve. Elliptic and Quasielliptic Fibrations   87

8   Canonical Dimension of an Elliptic or Quasielliptic Fibration   111

9 The Classification Theorem According to Canonical Dimension   123

10 Surfaces with Canonical Dimension Zero (char($\Bbbk$) $\neq$ 2, 3)   137

11 Ruled Surfaces. The Noether–Tsen Criterion   165

12 Minimal Models of Ruled Surfaces   181

13 Characterization of Ruled and Rational Surfaces   195

14 Zariski Decomposition and Applications   215

15 Appendix: Further Reading   245

References   247

Index   256

Conventions and Notation

Throughout this book we fix an algebraically closed field $\Bbbk$ of arbitrary characteristic. We use the standard notation from [EGA] (= [GD]). By *algebraic variety over* $\Bbbk$ we mean an integral algebraic $\Bbbk$-scheme.

If X is a complete algebraic variety and L is an invertible $\mathcal{O}_X$-module, we denote by $|L|$ the set of all effective Cartier divisors on X such that $\mathcal{O}_X(D) \cong L$. This set (called the *linear system associated to* L) is identified naturally with the projective space $\mathbb{P}(H^0(L))$ associated to the vector space $H^0(L)$ of global sections of L (cf. [Mum1, Lecture 10]).

Now assume that X is a nonsingular complete variety. Then the Cartier divisors on X coincide with the Weil divisors on X. We denote by $\mathrm{Div}(X)$ the set of all divisors on X. If D_1 and $D_2 \in \mathrm{Div}(X)$, then by $D_1 \sim D_2$ we denote the fact that the divisor D_1 is linearly equivalent to the divisor D_2. Let L_1 and L_2 be two invertible $\mathcal{O}_X$-modules. Then by $L_1 \approx L_2$ (resp. $L_1 \equiv L_2$) we denote the fact that L_1 is algebraically (resp. numerically) equivalent to L_2 (see Chapter 5 for the definition of these concepts). We denote by $\mathrm{Pic}(X)$ the Picard group of X (the group of all isomorphism classes of invertible $\mathcal{O}_X$-modules, with operation induced by tensor product), and by $\mathrm{Pic}^0(X)$ (resp. $\mathrm{Pic}^\tau(X)$) the subgroup of $\mathrm{Pic}(X)$ consisting of all isomorphism classes $[L] \in \mathrm{Pic}(X)$ with $L \approx \mathcal{O}(X)$ (resp. $L \equiv \mathcal{O}(X)$). Then we put $\mathrm{NS}(X) = \mathrm{Pic}(X)/\mathrm{Pic}^0(X)$ (the Néron–Severi group of X) and $\mathrm{Num}(X) = \mathrm{Pic}(X)/\mathrm{Pic}^\tau(X)$.

Other nonstandard notation and terminology will be explained later in the book.

Conventions and Notation

1

Cohomological Intersection Theory and the Nakai–Moishezon Criterion of Ampleness

In this chapter we present, following Kleiman [Kle], the cohomological theory of intersection, the Nakai–Moishezon criterion of ampleness for divisors, and some of its more important consequences. Furthermore, we prove that every nonsingular complete surface is projective.

Theorem 1.1 (Snapper). *Let F be a coherent $\mathcal{O}_V$-module on the complete algebraic $\Bbbk$-scheme V ($\Bbbk$ being an algebraically closed field of arbitrary characteristic, fixed once and for all), and let $L_1, \ldots, L_t$ ($t \geq 0$) be t invertible $\mathcal{O}_V$-modules. Then the function*

$$f_F(n_1, \ldots, n_t) = \chi(F \otimes L_1^{n_1} \otimes \cdots \otimes L_t^{n_t})$$

is a numerical polynomial in $n_1, \ldots, n_t$, of degree $\leq s = \dim(\operatorname{Supp}(F))$.

1.2. Let $\mathbb{Q}[n_1, \ldots, n_t]$ be the polynomial ring in variables $n_1, \ldots, n_t$ over the rational numbers. It is clear that the polynomials of the form $\binom{n_1 + i_1}{i_1}$, $\ldots, \binom{n_t + i_t}{i_t}$, $i_k \geq 0$, are a basis of $\mathbb{Q}[n_1, \ldots, n_t]$ over $\mathbb{Q}$. If $f \in \mathbb{Q}[n_1, \ldots, n_t]$, then f is called a *numerical polynomial* if for every $(n_1, \ldots, n_t) \in \mathbb{Z}^t$ the value of f at $(n_1, \ldots, n_t)$ is an integer. If we express f as

$$f = \sum_{i_k \geq 0} a_{i_1, \ldots, i_t} \binom{n_1 + i_1}{i_1} \cdots \binom{n_t + i_t}{i_t}, \quad a_{i_1, \ldots, i_t} \in \mathbb{Q},$$

then we have the following simple criterion: f is a numerical polynomial if and only if $a_{i_1, \ldots, i_t} \in \mathbb{Z}$ for all multi-indices $i_1, \ldots, i_t$.

Proof of Theorem 1.1. We will prove the theorem by induction on s, the cases $s = -1$ (i.e., $F = 0$) and $s = 0$ being trivial. Therefore we assume that $s \geq 1$. Replacing V with $\mathrm{Supp}(F)$ (with the closed subscheme structure given by the annihilator ideal $\mathrm{Ann}(F) \subset \mathcal{O}_V$) we may assume that $V = \mathrm{Supp}(F)$. Then let $\mathcal{L}$ be the category of all coherent $\mathcal{O}_V$-modules and let $\mathcal{L}'$ be the subclass of $\mathcal{L}$ consisting of those sheaves that satisfy the conclusion of the theorem. According to the inductive hypothesis and the additivity of the Euler–Poincaré characteristic, we see immediately that if

$$0 \longrightarrow F_1 \longrightarrow F_2 \longrightarrow F_3 \longrightarrow 0$$

is an exact sequence in $\mathcal{L}$ such that two of these sheaves are in $\mathcal{L}'$, then the third one is also in $\mathcal{L}'$. To show that $\mathcal{L}' = \mathcal{L}$, we use the *Dévissage lemma* [EGA III, (3.1.2)]. Using this lemma, it suffices to show that *if $X \subset V$ is a closed integral subscheme of V, then $\mathcal{O}_X \in \mathcal{L}'$.*

This last statement is proved by induction on t, the case $t = 0$ being trivial. Again, we may assume that V is X. Since X is an integral algebraic $\Bbbk$-scheme, there is a Cartier divisor D on X such that $L_1 \cong \mathcal{O}_X(D)$. Let

$$J = \mathcal{O}_X(-D) \cap \mathcal{O}_X, \quad I = J \cdot \mathcal{O}_X(D) = J \otimes L_1 \subset \mathcal{O}_X,$$
$$G = \mathcal{O}_X/J, \quad \text{and} \quad H = (\mathcal{O}_X/I) \otimes L_1^{-1}.$$

Then we have exact sequences

$$0 \longrightarrow J \otimes L_1^{n_1} \otimes \cdots \otimes L_t^{n_t} \longrightarrow L_1^{n_1} \otimes \cdots \otimes L_t^{n_t} \longrightarrow G \otimes L_1^{n_1} \otimes \cdots \otimes L_t^{n_t} \longrightarrow 0$$
$$\|$$
$$0 \longrightarrow I \otimes L_1^{n_1-1} \otimes \cdots \otimes L_t^{n_t} \longrightarrow L_1^{n_1-1} \otimes \cdots \otimes L_t^{n_t} \longrightarrow H \otimes L_1^{n_1} \otimes \cdots \otimes L_t^{n_t} \longrightarrow 0$$

and therefore we get the equality

$$\chi(L_1^{n_1} \otimes \cdots \otimes L_t^{n_t}) - \chi(L_1^{n_1-1} \otimes \cdots \otimes L_t^{n_t})$$
$$= \chi(G \otimes L_1^{n_1} \otimes \cdots \otimes L_t^{n_t}) - \chi(H \otimes L_1^{n_1} \otimes \cdots \otimes L_t^{n_t}).$$

But $\mathrm{Supp}(G)$ and $\mathrm{Supp}(H)$ are strictly contained in X, and therefore by the inductive hypothesis (on s) G and H are in $\mathcal{L}'$, whence the right-hand side of the equality is a numerical polynomial of degree less than s. Since by the inductive hypothesis (on t) the function $\chi(L_2^{n_2} \otimes \cdots \otimes L_t^{n_t})$ is a numerical polynomial of degree less than or equal to s, the claim (and, with it, Theorem 1.1) follows. $\square$

Remarks 1.3. (a) If V is a nonsingular complete algebraic curve and L is an invertible sheaf, then the fact that $\chi(L^n)$ is a numerical polynomial follows also from the Riemann–Roch theorem, namely, $\chi(L^n) = n \deg(L) + \chi(\mathcal{O}_V)$.

(b) If V is a nonsingular projective surface and L_1, L_2 are two invertible $\mathcal{O}_V$-modules, then from the Riemann–Roch theorem for surfaces we get the following explicit form of the numerical polynomial $\chi(L_1^{n_1} \otimes L_2^{n_2})$:

$$\chi(L_1^{n_1} \otimes L_2^{n_2}) = \tfrac{1}{2}[n_1^2 \cdot (L_1^{\cdot 2}) + 2 n_1 n_2 \cdot (L_1 \cdot L_2) + n_2^2 \cdot (L_2^{\cdot 2})]$$
$$- \tfrac{1}{2} \cdot [n_1 \cdot (L_1 \cdot \omega_V) + n_2 \cdot (L_2 \cdot \omega_V)] + \chi(\mathcal{O}_V),$$

where $(\cdot)$ is the classical intersection number [Mum1] and ω_V is the canonical sheaf of V. In particular, the classical intersection number $(L_1 \cdot L_2)$ arises as the coefficient of $n_1 n_2$ in the quadratic numerical polynomial $\chi(L_1^{n_1} \otimes L_2^{n_2})$.

These observations suggest the following general definition of intersection number.

Definition 1.4. Let $L_1, \ldots, L_t$ $(t \geq 0)$ be t invertible sheaves on the complete algebraic $\Bbbk$-scheme V, and let F be a coherent $\mathcal{O}_V$-module such that $\dim \operatorname{Supp}(F) \leq t$. The *intersection number of* $L_1, \ldots, L_t$ *with* F, notation: $(L_1 \cdots L_t \cdot F) = (L_1 \cdots L_t \cdot F)_V$, is by definition the coefficient of the monomial $n_1 n_2 \cdots n_t$ in the numerical polynomial (of degree $\leq t$) $\chi(F \otimes L_1^{n_1} \otimes \cdots \otimes L_t^{n_t})$.

From definition 1.4 and (1.2) we see that $(L_1 \cdots L_t \cdot F)$ is always an integer. Moreover, we have the following lemma.

Lemma 1.5. $(L_1 \cdots L_t \cdot F) = 0$ *if* $\dim \operatorname{Supp}(F) < t$, *and* $(F) = \dim H^0(F)$ *if* $\dim \operatorname{Supp}(F) = t = 0$.

Lemma 1.6. $(L_1 \cdots L_t \cdot F)$ *is a symmetric multilinear form in* $L_1, \ldots, L_t$.

Proof. Only linearity is nontrivial. Let M and N be two invertible $\mathcal{O}_V$-modules. We must calculate $(M \otimes N^{-1} \cdot L_2 \cdots L_t \cdot F)$. We have

$$\chi(F \otimes M^m \otimes N^{-n} \otimes L_2^{n_2} \otimes \cdots \otimes L_t^{n_t}) = (M \cdot L_2 \cdots L_t \cdot F) \cdot m n_2 \cdots n_t$$
$$- (N \cdot L_2 \cdots L_t \cdot F) \cdot n n_2 \cdots n_t + \cdots,$$

putting $n = 0$ and then $m = 0$. But now, if we put $m = n = n_1$, we have

$$\chi(F \otimes (M \otimes N^{-1})^{n_1} \otimes L_2^{n_2} \otimes \cdots \otimes L_t^{n_t})$$
$$= [(M \cdot L_2 \cdots L_t \cdot F) - (N \cdot L_2 \cdots L_t \cdot F)] \cdot n_1 n_2 \cdots n_t + \cdots,$$

and linearity follows. $\qquad\square$

Notation 1.7. If $L_1 = L_2 = \cdots = L_t = L$, we will write $(L^t \cdot F)$ instead of $(L \cdots L \cdot F)$. If $F = \mathcal{O}_W$ with W a closed subscheme of V, we will write $(L_1 \cdots L_t \cdot W)$ instead of $(L_1 \cdots L_t \cdot \mathcal{O}_W)$, and if $W = V$, we will write $(L_1 \cdots L_t)$ instead of $(L_1 \cdots L_t \cdot V)$. In particular, if $L_1 = \cdots = L_r = L$, $L_{r+1} = \cdots = L_t = L'$ and $F = \mathcal{O}_W$, we will write simply $(L^r \cdot L'^{(t-r)} \cdot W)$ instead of $(\underbrace{L \cdots L}_{r \text{ times}} \cdot \underbrace{L' \cdots L'}_{(t-r) \text{ times}} \cdot W)$.

Lemma 1.8. *If* $0 \longrightarrow F' \longrightarrow F \longrightarrow F'' \longrightarrow 0$ *is an exact sequence of coherent* $\mathcal{O}_V$-*modules such that* $\dim \operatorname{Supp}(F) \leq t$, *then*

$$(L_1 \cdots L_t \cdot F) = (L_1 \cdots L_t \cdot F') + (L_1 \cdots L_t \cdot F'').$$

The proof follows immediately from the definition of the intersection number and the additivity of the Euler–Poincaré characteristic.

Lemma 1.9. *If $D \in |L_1|$ and $D \cap \mathrm{Ass}(F) = \varnothing$, then putting $F_D = F \otimes \mathcal{O}_D$ we have $(L_1 \cdots L_t \cdot F) = (L_2 \cdots L_t \cdot F_D)$. In particular, if $\dim(V) \leq t$ then $(L_1 \cdots L_t) = (L_2 \cdots L_t \cdot D)$.*

Proof. Since $D \cap \mathrm{Ass}(F) = \varnothing$, we have an exact sequence

$$0 \longrightarrow F \otimes L_1^{-1} \longrightarrow F \longrightarrow F_D \longrightarrow 0$$

(if $D = \mathrm{div}_V(s)$ with $s \in H^0(L_1)$, then the condition $D \cap \mathrm{Ass}(F) = \varnothing$ means that s is a nonzerodivisor in the stalk F_x of F at every point $x \in V$). Tensoring with $L_1^{n_1} \otimes \cdots \otimes L_t^{n_t}$, we get an exact sequence

$$0 \longrightarrow F \otimes L_1^{n_1-1} \otimes L_2^{n_2} \otimes \cdots \otimes L_t^{n_t} \longrightarrow F \otimes L_1^{n_1} \otimes \cdots \otimes L_t^{n_t}$$
$$\longrightarrow F_D \otimes L_1^{n_1} \otimes \cdots L_t^{n_t} \longrightarrow 0,$$

and from this we get

$$\chi(F_D \otimes L_1^{n_1} \otimes \cdots \otimes L_t^{n_t})$$
$$= \chi(F \otimes L_1^{n_1} \otimes \cdots \otimes L_t^{n_t}) - \chi(F \otimes L_1^{n_1-1} \otimes \cdots \otimes L_t^{n_t})$$
$$= [(L_1 \cdots L_t \cdot F) \cdot n_1 \cdots n_t + \cdots]$$
$$\quad - [(L_1 \cdots L_t \cdot F) \cdot (n_1 - 1)n_2 \cdots n_t + \cdots]$$
$$= (L_1 \cdots L_t \cdot F)n_2 \cdots n_t + \cdots .$$

But taking $n_1 = 0$ in the polynomial $\chi(F_D \otimes L_1^{n_1} \otimes \cdots L_t^{n_t})$ we see that

$$\chi(F_D \otimes L_1^{n_1} \otimes \cdots \otimes L_t^{n_t}) = (L_2 \cdots L_t \cdot F_D) \cdot n_2 \cdots n_t + \cdots .$$

The result follows by considering the coefficient of $n_2 \cdots n_t$ in these two identities. $\square$

Now assume that $X = \mathrm{Supp}(F)$ is contained in a closed subscheme W of V, so that the closed subscheme X defined by the ideal sheaf $\mathrm{Ann}(F)$ is also a closed subscheme of W. Then F can also be considered a coherent $\mathcal{O}_W$-module, and we have

$$F \otimes L_1^{n_1} \otimes \cdots \otimes L_t^{n_t} = (F \otimes \mathcal{O}_W) \otimes L_1^{n_1} \otimes \cdots \otimes L_t^{n_t} =$$
$$= F \otimes L_{1,W}^{n_1} \otimes \cdots \otimes L_{t,W}^{n_t},$$

where by $L_{i,W}$ we denote the sheaf $L_i \otimes \mathcal{O}_W$. Thus we obtain:

Lemma 1.10. *In the situation described earlier, we have*

$$(L_1 \cdots L_t \cdot F)_V = (L_{1,W} \cdots L_{t,W} \cdot F)_W.$$

In particular,

$$(L_1 \cdots L_t \cdot W)_V = (L_{1,W} \cdots L_{t,W})_W.$$

$\square$

Now let $V_1, \ldots, V_n$ be the irreducible components of V. Let $V_i^0 = V_i - \bigcup_{j \neq i} V_j = V - \bigcup_{j \neq i} V_j$, $i = 1, \ldots, n$. Since V_i^0 is open in V and dense (open) in V_i, it follows from [EGA I, (9.5.10)] that V_i has a well-determined scheme structure. Thus we get the following.

Corollary 1.11. *Assume that $V = \operatorname{Supp}(F)$, and let $V_1, \ldots, V_n$ be the irreducible components of V, with the subscheme structures described earlier. If we write $F_i = F \otimes \mathcal{O}_{V_i}$, we have:*

$$(L_1 \cdots L_t \cdot F) = (L_1 \cdots L_t \cdot F_1) + \cdots + (L_1 \cdots L_t \cdot F_n).$$

Proof. In the exact sequence

$$0 \longrightarrow K \longrightarrow F \overset{\alpha}{\longrightarrow} \bigoplus_{i=1}^{n} F_i \longrightarrow C \longrightarrow 0$$

we have $\operatorname{Supp}(K)$, $\operatorname{Supp}(C) \subset \bigcup_{i \neq j} V_i \cap V_j$, and therefore $\dim \operatorname{Supp}(K)$, $\dim \operatorname{Supp}(C) < t$. Taking into account (1.5) and (1.8) we have successively:

$$(L_1 \cdots L_t \cdot F) = (L_1 \cdots L_t \cdot K) + (L_1 \cdots L_t \cdot \operatorname{Im}(\alpha))$$

$$= (L_1 \cdots L_t \cdot \bigoplus_{i=1}^{n} F_i) - (L_1 \cdots L_t \cdot C) = \sum_{i=1}^{n}(L_1 \cdots L_t \cdot F_i).$$

$\square$

Corollary 1.12. *Assume that V is irreducible and $\dim(V) \leq t$. Let $x \in V$ be the generic point of V and let $d = \operatorname{length}_{\mathcal{O}_x}(F_x)$ be the length of the $\mathcal{O}_x$-module F_x. Then*

$$(L_1 \cdots L_t \cdot F) = d(L_1 \cdots L_t \cdot V_{\mathrm{red}}).$$

Proof. Since $\mathcal{O}_x$ is an Artinian ring and F is a coherent $\mathcal{O}_X$-module, the stalk F_x is always an $\mathcal{O}_x$-module of finite length. If $\dim \operatorname{Supp}(F) < t$, the corollary is obvious. The corollary is also obvious if $F = \mathcal{O}_{V_{\mathrm{red}}}$. Let $\mathcal{L}'$ denote the class of all coherent $\mathcal{O}_V$-modules for which Corollary 1.12 is true. As $(L_1 \cdots L_t \cdot F)$ and $\operatorname{length}_{\mathcal{O}_x}(F_x)$ are both additive functions in F, $\mathcal{L}'$ is an exact subclass of the category $\mathcal{L}$ of all coherent $\mathcal{O}_V$-modules. The corollary now follows from the Dévissage lemma ([EGA III, (3.1.2)]). $\square$

Corollary 1.13. *Let W be a closed subscheme of V, of dimension t, and let $D \in |L_{1,W}|$, where $L_{1,W} = L_1 \otimes \mathcal{O}_W$. Then*

$$(L_1 \cdots L_t \cdot W) = (L_2 \cdots L_t \cdot D).$$

Proof. We have:

$$
\begin{aligned}
(L_1 \cdots L_t \cdot W) &= (L_{1,W} \cdots L_{t,W}) & \text{by (1.10)} \\
&= (L_{2,W} \cdots L_{t,W} \cdot D) & \text{by (1.9)} \\
&= (L_2 \cdots L_t \cdot D) & \text{by (1.10).}
\end{aligned}
$$

$\square$

1.14. Let $f : V' \longrightarrow V$ be a morphism between two complete and irreducible algebraic $\Bbbk$-schemes V' and V with generic points x' and x, respectively. We have:

(a) $f(x') = x \iff \dim f(V') = \dim(V)$;

(b) if $f(x') = x$, then [every $\mathcal{O}_{x'}$-module M of finite length is an $\mathcal{O}_x$-module of finite length] $\iff$ [$\dim(V') = \dim(V)$], and in that case we have the formula:

$$
\text{length}_{\mathcal{O}_x}(M) = [\Bbbk(x') : \Bbbk(x)] \cdot \text{length}_{\mathcal{O}_{x'}}(M).
$$

Definition 1.15. Let $f : V' \longrightarrow V$ be a morphism between two complete and irreducible algebraic $\Bbbk$-schemes V' and V with generic points x' and x, respectively. The degree of f is by definition the following rational number:

$$
\deg(f) = \begin{cases} \dfrac{\text{length}_{\mathcal{O}_x}(\mathcal{O}_{x'})}{\text{length}_{\mathcal{O}_x}(\mathcal{O}_x)} & \text{if } \dim(V') = \dim(V) = \dim f(V'), \\ 0 & \text{otherwise.} \end{cases}
$$

Examples 1.16. (a) Let $f : V_{\text{red}} \hookrightarrow V$ be the canonical inclusion morphism, and let $n = \text{length}_{\mathcal{O}_x}(\mathcal{O}_x)$. Then $\deg(f) = 1/n$.

(b) Let $f : V' \longrightarrow V$ be a birational morphism, that is, a morphism f such that $f(x') = x$ and the local homomorphism $\mathcal{O}_x \longrightarrow \mathcal{O}_{x'}$ is an isomorphism. Then $\deg(f) = 1$.

(c) Let $f : V' \longrightarrow V$ be a morphism with $\deg(f) > 0$. If V' and V are integral, then $\deg(f) = [\Bbbk(x') : \Bbbk(x)]$.

Lemma 1.17. *Let* $f : V' \longrightarrow V$, $g : V'' \longrightarrow V'$ *and* $h = f \circ g$, *where* V'', V', *and* V *are complete and irreducible algebraic* $\Bbbk$-*schemes. Then* $\deg(h) = \deg(f) \cdot \deg(g)$, *except when* $\dim(V') > \dim(V'') = \dim(V) = \dim h(V'')$.

Proof. We may assume that $\deg(h) > 0$. Then $\dim(V'') = \dim(V) = \dim h(V'') \leq \dim g(V''), \dim f(V') \leq \dim(V')$. Thus $\deg(f), \deg(g) > 0$, unless $\dim(V') > \dim(V'') = \dim(V) = \dim h(V'')$. Assuming therefore that $\deg(f) > 0, \deg(g) > 0$, we have $\deg(h) > 0$ and $\text{length}_{\mathcal{O}_x}(\mathcal{O}_{x''}) = [\Bbbk(x') : \Bbbk(x)] \cdot \text{length}_{\mathcal{O}_{x'}}(\mathcal{O}_{x''}) = [\Bbbk(x') : \Bbbk(x)] \cdot \text{length}_{\mathcal{O}_{x'}}(\mathcal{O}_{x'}) \cdot \deg(g) = \text{length}_{\mathcal{O}_x}(\mathcal{O}_{x'}) \cdot \deg(g)$ (where x'', x', and x are the generic points of V'', V', and V, respectively). $\square$

Lemma 1.18. *Let $f : V' \longrightarrow V$ be a morphism of complete and irreducible algebraic schemes over $\Bbbk$, and assume that $t \geq \dim(V'), \dim(V)$. Let $L_1, \ldots, L_t$ be t invertible $\mathcal{O}_V$-modules and $L'_i = f^*(L_i)$, $i = 1, \ldots, t$. Then we have:*

$$(L'_1 \cdots L'_t)_{V'} = \deg(f) \cdot (L_1 \cdots L_t)_V.$$

Proof. Put $L = L_1^{n_1} \otimes \cdots \otimes L_t^{n_t}$ and $L' = L_1'^{n_1} \otimes \cdots \otimes L_t'^{n_t}$. We have $L' = f^*(L)$ and $R^q f_*(L') = R^q f_*(\mathcal{O}_{V'} \otimes L') = R^q f_*(\mathcal{O}_{V'}) \otimes L$ (by the projection formula). Looking at the Leray spectral sequence

$$E_2^{pq} = H^p(V, R^q f_*(\mathcal{O}_{V'}) \otimes L) \implies H^{p+q}(L')$$

and keeping in mind that the term E_{n+1} is obtained by taking the homology of E_n, we get

$$\chi(L') = \sum_{q=0}^{\infty} (-1)^q \chi(R^q f_*(\mathcal{O}_{V'}) \otimes L). \qquad (*)$$

Step 1: $\deg(f) = 0$.

Then $\dim f(V') < t$. Hence $\dim \operatorname{Supp} R^q f_*(\mathcal{O}_{V'}) < t$ for every q, so that $\chi(R^q f_*(\mathcal{O}_{V'}) \otimes L_1^{n_1} \otimes \cdots \otimes L_t^{n_t})$ is a numerical polynomial of degree $< t$; therefore $\chi(L_1'^{n_1} \otimes \cdots \otimes L_t'^{n_t})$ also has degree $< t$. Therefore we see that $(L'_1 \cdots L'_t) = 0$.

Step 2: f *is a birational morphism.*

Then f induces a biregular isomorphism from an open subscheme of V' onto an open subscheme of V, and therefore $\dim \operatorname{Supp} R^q f_*(\mathcal{O}_{V'}) < t$ for every $q > 0$. Therefore the homogeneous part of degree t of the polynomial $\chi(L_1'^{n_1} \otimes \cdots \otimes L_t'^{n_t})$ coincides with the homogeneous part of degree t of the polynomial $\chi(f_*(\mathcal{O}_{V'}) \otimes L_1^{n_1} \otimes \cdots \otimes L_t^{n_t})$. On the other hand we have the exact sequence

$$0 \longrightarrow K \longrightarrow \mathcal{O}_V \xrightarrow{\alpha} f_*(\mathcal{O}_{V'}) \longrightarrow C \longrightarrow 0$$

with $\dim \operatorname{Supp}(K), \dim \operatorname{Supp}(C) < t$. From (1.5) and (1.8) we see immediately that the homogeneous part of degree t of the polynomial $\chi(L_1^{n_1} \otimes \cdots \otimes L_t^{n_t})$ coincides with the homogeneous part of degree t of the polynomial $\chi(f_*(\mathcal{O}_{V'}) \otimes L_1^{n_1} \otimes \cdots \otimes L_t^{n_t})$. Therefore we get

$$(L'_1 \cdots L'_t)_{V'} = (L_1 \cdots L_t)_V.$$

Step 3: f *is a finite morphism,* $\deg(f) > 0$, *and V' and V are reduced.*

Then f is an affine morphism, so that $R^q f_*(\mathcal{O}_{V'}) = 0$ for all $q > 0$. Therefore $(L'_1 \cdots L'_t)_{V'} = (L_1 \cdots L_t \cdot f_*(\mathcal{O}_{V'}))_V$.

Applying (1.12) to the last term , we get $d = \operatorname{length}_{\mathcal{O}_x} f_*(\mathcal{O}_{V'})_x = [\Bbbk(x') : \Bbbk(x)] = \deg(f)$, so the conclusion follows in this case as well.

Step 4: $V' = V_{\text{red}}$ and $f : V_{\text{red}} \hookrightarrow V$ is the canonical inclusion.
Then the conclusion follows from (1.12).
Step 5: the general case, with $\deg(f) > 0$.
Consider the following decomposition of the morphism f:

$$
\begin{array}{ccccc}
Y & \xrightarrow{\ \text{birat}\ } & V'_{\text{red}} & \hookrightarrow & V' \\
\Big\downarrow{\scriptstyle\text{birat}} & & & & \Big\downarrow{\scriptstyle f} \\
X & \xrightarrow[\text{finite}]{} & V_{\text{red}} & \hookrightarrow & V
\end{array}
$$

where X is the normalization of V_{red} in the field of rational functions of V'_{red} and Y is the "join" of X and V'_{red}. Step 5 follows from this decomposition and Steps 2–4. □

Lemma 1.19. *The symbol $(L_1 \cdots L_t \cdot F)_V$ is uniquely determined by the properties in Lemmas 1.5, 1.6, 1.8, 1.9, 1.10, and 1.18.*

Proof. Let $\langle L_1 \cdots L_t \cdot F \rangle_V$ be another symbol satisfying Lemmas 1.5, 1.6, 1.8, 1.9, 1.10, and 1.18. Then this symbol must also satisfy Corollaries 1.11, 1.12, and 1.13, which follow formally from those lemmas. We will show that

$$\langle L_1 \cdots L_t \cdot F \rangle_V = (L_1 \cdots L_t \cdot F)_V.$$

By (1.10) we may assume that $V = \text{Supp}(F)$, by (1.11) that V is irreducible, by (1.12) that V is integral and $F = \mathcal{O}_V$, and by Chow's lemma [EGA II (5.6.1)] and (1.18) that V is projective. We will show that $\langle L_1 \cdots L_t \cdot W \rangle_V = (L_1 \cdots L_t \cdot W)_V$, with W an arbitrary closed subscheme of dimension t of V. When $t = 0$ everything is clear from (1.5). Therefore we will assume that $t > 0$. Then L_1 can be written in the form $L_1 = \mathcal{O}_V(H - H')$ with H and H' two very ample Cartier divisors on V. Let $M = \mathcal{O}_V(H)$ and $N = \mathcal{O}_V(H')$. Then

$$
\begin{aligned}
&(L_1 \cdots L_t \cdot W)_V \\
&= (M \cdot L_2 \cdots L_t \cdot W)_V - (N \cdot L_2 \cdots L_t \cdot W)_V && \text{by (1.6),} \\
&= (L_2 \cdots L_t \cdot W \cap H)_V - (L_2 \cdots L_t \cdot W \cap H')_V && \text{by (1.9),} \\
&= \langle L_2 \cdots L_t \cdot W \cap H \rangle_V - \langle L_2 \cdots L_t \cdot W \cap H' \rangle_V && \text{inductive hypothesis,} \\
&= \langle L_1 \cdots L_t \cdot W \rangle_V && \text{by (1.9) and (1.6).}
\end{aligned}
$$

□

Corollary 1.20. *If V is a nonsingular projective surface, then $(L_1 \cdot L_2)$ coincides with the classical intersection number.*

Lemma 1.21. *If V is a complete algebraic scheme over $\Bbbk$ and W is a closed subscheme of dimension t of V, then for every invertible $\mathcal{O}_V$-module L the number $(L^{\cdot t} \cdot W)$ coincides with $t!\,a$, where a is the coefficient of n^t in $\chi(L^{\otimes n} \otimes \mathcal{O}_W)$.*

Proof. Replacing V with W and L with $L \otimes \mathcal{O}_W$, we may assume by 1.10 that $W = V$. Let $n = n_1 + \cdots + n_t$. Then $\chi(L^{n_1} \otimes \cdots \otimes L^{n_t}) = \chi(L^n)$, so that $an^t + a_{t-1}n^{t-1} + \cdots = a(n_1 + \cdots + n_t)^t + \cdots$. Since the coefficient of $n_1 \ldots n_t$ in $(n_1 + \cdots + n_t)^t$ is equal to $t!$, Lemma 1.21 follows directly from the definition of the intersection number. $\qquad\square$

Now we are ready to prove the following important criterion of ampleness.

Theorem 1.22 (The Nakai–Moishezon Criterion). *Let V be a complete algebraic scheme over $\Bbbk$ and let L be an invertible $\mathcal{O}_V$-module. Then L is ample if and only if for every integral closed subscheme $W \subseteq V$ of dimension $t > 0$ we have $(L^{\cdot t} \cdot W) > 0$.*

Proof. First a few observations:

(a) Let $V_1, \ldots, V_n$ be the irreducible components of V, with the closed subscheme structures described before (1.11). Then, by [Har3, p. 24], L is ample if and only if $L \otimes \mathcal{O}_{V_i}$ is ample for each $i = 1, \ldots, n$.

(b) Let V be irreducible. Then L is ample if and only if $L \otimes \mathcal{O}_{V_{\mathrm{red}}}$ is ample (cf. [EGA II, (4.5.14)]).

(c) L satisfies the numerical conditions in the theorem if and only if for each integral component $V_{i,\mathrm{red}}$ of V, $L \otimes \mathcal{O}_{V_{i,\mathrm{red}}}$ satisfies the same numerical conditions on $V_{i,\mathrm{red}}$.

From these three observations we see that, without loss of generality, we may also assume that V is integral.

Assume first that L is ample. Then there is an $n > 0$ such that L^n is very ample. Therefore, since V is integral, L is of the form $\mathcal{O}_V(D)$ with D a very ample Cartier divisor on V. In other words, there is an embedding $V \hookrightarrow \mathbb{P}^m$ and a hyperplane H in $\mathbb{P}^m$ such that $D = H \cap V$ (in the sense of inverse images of Cartier divisors). Then let W be an integral closed subscheme of dimension $t > 0$ of V. We have

$$(L^{\cdot t} \cdot W) = 1/n^t \cdot (\mathcal{O}_V(D)^{\cdot t} \cdot W)_V = 1/n^t \cdot (\underbrace{D \cdots D}_{t \text{ times}} \cdot W)_V$$

$$= 1/n^t \cdot (\underbrace{H \cdots H}_{t \text{ times}} \cdot W)_{\mathbb{P}^m} = 1/n^t \cdot \deg(W) > 0.$$

Conversely, assume that for every integral closed subscheme W of dimension $t > 0$ of V we have $(L^{\cdot t} \cdot W) > 0$. We must show that L is ample. We will argue by induction on $d = \dim(V)$. If $d = 0$ the theorem is trivial. If V is an integral curve, we know that $\deg(L) = (L)_V > 0$. Let $f : V' \longrightarrow V$ be the normalization morphism (which is a birational morphism). Then V' is a nonsingular projective curve, and by 1.18 we have deg $f^*(L) = \deg(L) > 0$. By the Riemann–Roch theorem for the curve V' we have that $(f^*(L))^n$ is very ample if $n > 2p_a(V') + 1$, whence $f^*(L)$ is ample. Since f is a finite morphism, we see that L is also ample (this follows, for example, from the cohomological criterion of ampleness, [EGA III, (2.3.1)], which may also be used to prove observation (a)).

Now assume that $d \geq 2$, V is integral, and $L|_W$ is ample for every closed subscheme $W \subsetneq V$ (by the inductive hypothesis, for $\dim(W) < \dim(V)$).

Step 1. $\chi(L^n) \to \infty$ *for* $n \to \infty$.

Indeed, $(L^{\cdot d}) = d!\, a > 0$ by (1.21), where $\chi(L^n) = a \cdot n^d + \cdots$. Therefore $a > 0$, and the assertion is clear.

Step 2. We may assume that $L = \mathcal{O}_V(D)$ *with* D *an effective divisor.*

To show this, it will suffice (replacing L with L^n) to prove that $H^0(L^n) \neq 0$ for $n \gg 0$. As V is integral, we may regard L as a subsheaf of the constant sheaf $\Bbbk(V)$ of all rational functions on V; put $I = L^{-1} \cap \mathcal{O}_V$ and $J = I \cdot L = I \otimes L \subseteq \mathcal{O}_V$. Let X and Y be the closed subschemes of V defined, respectively, by the ideals I and J. Since I and J are nonzero ideals, X and Y are proper closed subschemes of V, so that $L_X = L \otimes \mathcal{O}_X$ and $L_Y = L \otimes \mathcal{O}_Y$ are ample on X and Y, respectively, by the inductive hypothesis. By [EGA III, (2.3.1)] we have $H^q(L_X^n) = H^q(L_Y^n) = 0$ for every $q > 0$ and $n \gg 0$. But we infer from the exact sequences

$$0 \longrightarrow I \otimes L^n \longrightarrow L^n \longrightarrow L_X^n \longrightarrow 0$$

$$0 \longrightarrow J \otimes L^{n-1} \longrightarrow L^{n-1} \longrightarrow L_Y^{n-1} \longrightarrow 0$$

that $H^q(L^n) = H^q(L^{n-1})$ for every $q \geq 2$ and $n \gg 0$. From this we see that, if $n \gg 0$, then $\chi(L^n) = \dim H^0(L^n) - \dim H^1(L^n) + \text{const}$. The conclusion of Step 2 follows from Step 1.

Step 3. L^n *is generated by its global sections for* $n \gg 0$.

Indeed, by Step 2 we may assume that $L = \mathcal{O}_V(D)$ for some effective Cartier divisor D. From the exact sequence

$$0 \longrightarrow L^{n-1} \longrightarrow L^n \longrightarrow L_D^n = L^n \otimes \mathcal{O}_D \longrightarrow 0$$

and $H^1(L_D^n) = 0$ for $n \gg 0$ (L_D is ample on D by the inductive hypothesis), we get the exact sequence of cohomology spaces (for $n \gg 0$):

$$H^0(L^n) \longrightarrow H^0(L_D^n) \longrightarrow H^1(L^{n-1}) \longrightarrow H^1(L^n) \longrightarrow 0. \qquad (*)$$

In particular, for $n \gg 0$ we have $\dim H^1(L^n) \leq \dim H^1(L^{n-1})$; since $H^1(L^{n-1})$ is finite-dimensional, we see that for $n \gg 0$ the linear maps $H^1(L^{n-1}) \longrightarrow H^1(L^n)$ are isomorphisms. Going back to the exact sequence $(*)$, we see that for $n \gg 0$ the restriction map $H^0(L^n) \longrightarrow H^0(L_D^n)$ is surjective. Since L_D is ample, L_D^n is generated by its global sections for $n \gg 0$, and therefore we can find p global sections $s_1, \ldots, s_p \in H^0(L_D^n)$ such that for every $x \in D$ at least one of the s_i does not vanish at x. Lifting these sections to $s_1', \ldots, s_p' \in H^0(L^n)$ and adding a section $s_0' \in H^0(L^n)$ such that $nD = \mathrm{div}_V(s_0')$, we get $p+1$ global sections of L^n such that for every $x \in V$ at least one of these sections does not vanish at x. This completes the proof of Step 3.

Step 4. Conclusion.

By Step 3 there exists a morphism $f : V \longrightarrow \mathbb{P}^m$ with $f^*(\mathcal{O}_{\mathbb{P}^m}(1)) \cong L^n$ for some $n > 0$. Then f is a finite morphism; otherwise there would exist an integral curve $C \subset V$ with $f(C)$ a point. But then $(L \cdot C)$ would be equal to zero, contradicting the numerical hypothesis on L. As f is a finite morphism and $\mathcal{O}_{\mathbb{P}^m}(1)$ is ample, we get, by a criterion we have already used, that L^n is ample, so that L itself is ample.   □

Remark 1.23. To prove that L is ample when V is an integral curve and $\deg(L) > 0$, we used the Riemann–Roch theorem for nonsingular curves. We notice, however, that we could have avoided using that theorem, because Steps 1–4 prove the same thing directly.

Corollary 1.24. *If V is a complete surface over $\Bbbk$ and L is an invertible $\mathcal{O}_V$-module, then L is ample if and only if $(L^2) > 0$ and $(L \cdot C) > 0$ for every integral curve $C \subset V$. If in addition $H^0(L) \neq 0$, then the condition $(L^2) > 0$ is not needed.*

Proof. Analyzing the proof of Theorem 1.22, we see that we used the condition $(L^2) > 0$ only to reduce the proof to the case $H^0(L) \neq 0$.   □

Theorem 1.25. *Let V be a projective surface over $\Bbbk$ and let L be an invertible $\mathcal{O}_V$-module such that $(L \cdot C) \geq 0$ for every integral curve $C \subset V$. Then $(L^2) \geq 0$.*

Proof. Let a, b be natural numbers and $L' = L^a \otimes H^b$, where H is a hyperplane section on V. Then

$$(L'^2) = a^2(L^2) + 2ab\,(L \cdot H) + b^2(H^2).$$

Let $P(t) = m_2 t^2 + 2m_1 t + m_0$, with $m_0 = (L^2), m_1 = (L \cdot H)$, and $m_2 = (H^2)$. Since $m_1 \geq 0$ and $m_2 > 0$ by hypothesis (H is an effective, very ample Cartier divisor on V), we have that $P(t)$ is a strictly increasing function on $[0, \infty)$. By way of contradiction, if $(L^2) = m_0 < 0$, then P would have exactly one strictly positive real root t_0. If $b/a > t_0$, then $0 < a^2 P(b/a) = (L'^2)$. Since on the other hand $(L' \cdot C) = a \cdot (L \cdot C) + b \cdot (H \cdot C) > 0$ for every integral curve $C \subset V$, we see (using (1.24)) that L' is ample. So L'^n is very ample for $n \gg 0$, and therefore

$$(L' \cdot L) = 1/n \cdot (L'^n \cdot L) \geq 0.$$

Let $Q(t) = m_1 t + m_0$. We have $a \cdot Q(b/a) = am_0 + bm_1 = a(L \cdot L) + b(L \cdot H) = (L' \cdot L) \geq 0$. Since $Q(t)$ is continuous and $Q(b/a) \geq 0$ whenever $b/a > t_0$, we get $Q(t_0) \geq 0$. But then we get $P(t_0) \geq P(t_0) - Q(t_0) = m_2 t_0^2 + m_1 t_0 \geq m_2 t_0^2 > 0$, which is a contradiction.   □

Using the intersection theory developed earlier one could prove with the same arguments the following generalization of (1.25).

Theorem 1.26 (Kleiman [Kle]). *Let V be a complete $\Bbbk$-scheme of dimension ≥ 2, and let L be an invertible $\mathcal{O}_V$-module. Then $(L^t \cdot W) \geq 0$ for every integral closed subscheme W of dimension $t > 0$, if and only if $(L \cdot C) \geq 0$ for every integral curve $C \subset V$.*

Remark 1.27. As shown by examples given by Mumford and Ramanujam, the condition "$(L^t \cdot W) > 0$ for every integral closed subscheme $W \subseteq V$ of dimension $t > 0$" does not follow in general from the condition "$(L \cdot C) > 0$ for every integral curve $C \subset V$." More precisely:

(a) Mumford constructed an example of a nonsingular projective surface V over the complex field and a divisor D such that $(D \cdot C) > 0$ for every integral curve $C \subset V$, yet $(D^2) = 0$. See [Har3, p. 56].

(b) Ramanujam constructed an example of a nonsingular projective variety of dimension three and an *effective* divisor D such that $(D \cdot C) > 0$ for every integral curve C on V, without D being ample. See [Har3, p. 57].

As an application of the Nakai–Moishezon criterion for surfaces, we prove the following theorem.

Theorem 1.28 (Zariski–Goodman). *Let X be a nonsingular complete surface and let $U \subset X$ be a nonempty affine open subset. Then $F = X \setminus U$ is a connected algebraic set of pure codimension one in X, supporting an ample effective divisor. In particular, X is projective.*

Proof. Let $F_1, \ldots, F_n$ be the irreducible (and reduced) components of F.

Step 1. $\dim(F_i) = 1$ *for every* $i = 1 \ldots, n$.

Let $x_i \in F_i - \bigcup_{j \neq i} F_j$ be a closed point, $i = 1, \ldots, n$, and let V_i be an affine open neighborhood of x_i such that $V_i \cap \left(\bigcup_{j \neq i} F_j \right) = \varnothing$. Then $V_i \setminus V_i \cap U = V_i \cap F_i$, and $V_i \cap U$ is an affine open subset of X. If F_i is not a curve, then F_i must be a point, namely, the point x_i. From Serre's criterion for normality we see easily that the restriction homomorphism $H^0(V_i, \mathcal{O}_X) \longrightarrow H^0(V_i \setminus \{x_i\}, \mathcal{O}_X)$ is an isomorphism; this contradicts the fact that V_i and $V_i \cap U = V_i \setminus \{x_i\}$ are affine varieties and the inclusion $V_i \cap U \hookrightarrow V_i$ is not an isomorphism.

Step 2. F *is connected.*

Let Y be a projective closure of the affine open set U. We get a birational map $f : X \longrightarrow Y$ such that $f|_U$ is an isomorphism. Using the theorem on elimination of indeterminacies [Sha1, p. 296], we see that after a finite number of blow-ups of points outside U we get a birational morphism $g : X' \longrightarrow X$ such that $f \circ g$ is a morphism and $g^{-1}(U) = U'$ is isomorphic via g to U. It suffices to show that $X' \setminus U'$ is connected; thus, without loss of generality, we may assume that f is a birational morphism. We may further assume that Y is normal, replacing Y if necessary with its normalization via the Stein factorization of the morphism f. Since $Y \setminus f(U)$ obviously supports an ample divisor (namely, the intersection of Y with the hyperplane at infinity), $Y \setminus f(U)$ is connected by the Enriques–Severi–Zariski–Serre theorem [Ser1, § 76, Theorem 4]. Since Zariski's connectedness theorem [EGA III, (4.3.1)] shows that f has connected fibers, we see that $X \setminus U = f^{-1}(Y \setminus f(U))$ is connected as well.

Step 3. There is a divisor $D' = \sum_{i=1}^{n} m_i F_i$, with $m_i \geq 0$, such that $(D' \cdot F_i) \geq 0$ for every i and $(D' \cdot F_i) > 0$ for at least one i.

Indeed, let $f \in \Gamma(U, \mathcal{O}_X)$ be a nonconstant function; adding a suitable constant to f if necessary, we may assume that for every i there is a point $x_i \in F_i$ such that $f(x_i) \neq 0$. Let $D = \mathrm{div}(f)_0$ and $D' = \mathrm{div}(f)_\infty$, the divisor of zeroes and, respectively, the divisor of poles of f. We have $D - D' = \mathrm{div}(f)$, and $\mathrm{Supp}(D') \subseteq F$, so that $D' = \sum_{i=1}^{n} m_i F_i$ with $m_i \geq 0$ for every i and $m_i > 0$ for at least one i. As $(D \cdot F_i) \geq 0$ for every i and $(D \cdot F_i) > 0$ for at least one i (otherwise U would contain complete curves), Step 3 follows from $D' \sim D$.

Step 4. Using Steps 2 and 3, we can reorder the components F_i (allowing repetitions) so that $(D' \cdot F_1) > 0$ and $(F_i \cdot F_{i+1}) > 0$ for all i. Let $F_1, \dots, F_m$ be the components of F in this new order (with repetitions allowed).

Step 5. We prove by induction on $r = 1, \dots, m+1$ that there exists a divisor $D_r = \sum_{i=1}^{m} n_i F_i$, $n_i \geq 0$, such that

(a) $(D_r \cdot F_i) \geq 0$ for every $i = 1 \dots, m$;

(b) $(D_r \cdot F_i) > 0$ for every $i = 1, \dots, \min(r, m)$;

(c) $F_1, \dots, F_{r-1} \subseteq \mathrm{Supp}(D_r)$.

Indeed, for $r = 1$ we can take $D_1 = D'$ (constructed in Step 3). Now assume that $D_1, \dots, D_r$ have already been constructed, $r \leq m$. By (b) we have $(D_r \cdot F_r) > 0$. Therefore we can find an integer $e_r > 0$ such that $e_r(D_r \cdot F_r) + (F_r^2) > 0$; then put $D_{r+1} = e_r D_r + F_r$. It is clear that $F_1, \dots, F_r$ have strictly positive coefficients in D_{r+1} and that $\mathrm{Supp}(D_{r+1}) = \mathrm{Supp}(D_r) \cup F_r$. We have $(D_{r+1} \cdot F_i) = e_r(D_r \cdot F_i) + (F_r \cdot F_i)$, and therefore $(D_{r+1} \cdot F_r) > 0$ by the choice of e_r. By induction and Step 4 we see immediately that $(D_{r+1} \cdot F_i) \geq 0$ for every i. It only remains to check (b) for D_{r+1}. We distinguish two cases:

(1) $r < m$. We have $\min(r + 1, m) = r + 1$. Then $(D_{r+1} \cdot F_i) > 0$ by induction if $i < r$ and $F_i \neq F_r$, by the choice of e_r if $F_i = F_r$, and by induction and the ordering in Step 4 if $i = r + 1$.

(2) $r = m$. Then we have $(D_{r+1} \cdot F_i) > 0$ for every $i = 1, \ldots, m$: by induction if $F_i \neq F_m$ and by the choice of e_m if $F_i = F_m$.

Step 6. (Conclusion) D_{m+1} *is ample.* We use the Nakai–Moishezon criterion. Let $C \subset X$ be any integral curve. Since U is affine, $C \not\subset U$. If $C \neq F_i$ for all i, we have $(C \cdot F_i) \geq 0$ for every i, with strict inequality for at least one i. By Step 5 (c) we therefore have $(D_{m+1} \cdot C) > 0$. On the other hand, if $C = F_i$ for some i then $(D_{m+1} \cdot C) > 0$ by Step 5 (b). Finally, since D_{m+1} is effective, we may apply the last part of Corollary 1.24 to conclude that D_{m+1} is ample.   □

Remark 1.29. The analogue of Theorem 1.28 in dimension ≥ 3 is false. Indeed, there are elementary examples (due to Hironaka) of algebraic varieties of dimension three that are complete and nonsingular but not projective (see [Sha1, p. 399]).

EXERCISES

Exercise 1.1. Let X be a nonsingular projective surface and let K be a canonical divisor on X. Compute (K^2) in the following cases:

(i) X is a surface of degree d in $\mathbb{P}^3$.

(ii) $X = C \times C'$, with C and C' nonsingular curves of genus g and g', respectively.

Exercise 1.2. Let C be a nonsingular projective curve of genus g. Show that $(\Delta^2) = 2 - 2g$, where Δ is the diagonal of the surface $X = C \times C$. Generalize this to the case when $\Gamma_f \subset C_1 \times C_2$ is the graph of a surjective morphism $f : C_1 \to C_2$ of degree d, by showing that $(\Gamma_f^2) = d(2 - 2g_2)$, where g_2 is the genus of C_2.

Exercise 1.3. Let C_1 and C_2 be two nonsingular projective curves, and let $x_1 \in C_1$ and $x_2 \in C_2$ be two points. Use the Nakai–Moishezon criterion to show that the divisor $x_1 \times C_2 + C_1 \times x_2$ is ample on the surface $X = C_1 \times C_2$.

Exercise 1.4. Let C be a curve on a nonsingular projective surface X and let $x \in C$. Let $f : X' \to X$ be the quadratic transformation (or blow-up) of X of center x. Prove that

$$f^*(C) = C' + (E \cdot C')E,$$

where $E = f^{-1}(x)$ and C' is the proper transform of C under f.

Show that x is a nonsingular point of C if and only if $(E \cdot C') = 1$. More generally, show that $(E \cdot C') = k$, where k is the multiplicity of C at x (defined as the unique integer $k \geq 1$ such that $\xi \in \mathfrak{m}_x^k \setminus \mathfrak{m}_x^{k+1}$, where ξ is a local equation of C at x and $\mathfrak{m}_x$ is the maximal ideal of the local ring $\mathcal{O}_{X,x}$).

Exercise 1.5. Prove that the self-intersection number of a curve C on a nonsingular surface X of even degree in $\mathbb{P}^3$ is always even. If the degree of X is 4 and $(C^2) < 0$, show that C is rational and $(C^2) = -2$.

Exercise 1.6. Assume that the nonsingular surface X of degree d in $\mathbb{P}^3$ contains a straight line $C \subset \mathbb{P}^3$. Show that $(C^2) = 2 - d$.

Exercise 1.7. Let $H_X = X \cap H$ be a plane section of a nonsingular surface X of degree h in $\mathbb{P}^3$. Assume that $H_X = C + D$, with C and D curves of degree c and d, respectively. Compute the following intersection numbers on X: (C^2), (D^2), and $(C \cdot D)$. Show that $(C \cdot D)$ (computed on X) is the same as $(C \cdot D)_H$ (computed on $H = \mathbb{P}^2$).

Exercise 1.8. Let Y be an irreducible curve on a nonsingular projective surface X, with $(Y^2) > 0$. Show that the complete linear system $|nY|$ is base-point free for n sufficiently large, and the corresponding morphism $\phi_{|nY|} : X \to \mathbb{P}(H^0(X, \mathcal{O}_X(nY))^{\check{}})$ is birational onto its image $S_n = \phi_{|nY|}(X)$. Moreover, show that $\phi_{|nY|}$ is an isomorphism between an open neighborhood of Y in X and an open subset of S_n (in particular, Y can be embedded also in S_n), and Y is an ample Cartier divisor on the surface S_n. Prove also that S_n is normal for $n \gg 0$. [*Hint:* Adapt part of the proof of Theorem 1.22.]

Exercise 1.9. Let L be an ample invertible $\mathcal{O}_X$-module on a nonsingular projective surface X, and let L' be another invertible $\mathcal{O}_X$-module that is numerically equivalent to L. Show that L' is also ample. Show by example that if L is very ample, L' need not be very ample.

Bibliographic References. The main bibliographic source for this chapter is [Kle]. Theorem 1.28 was first proved by Zariski; the proof given here belongs to Goodman (see [Har1, pp. 69–71]).

2

The Hodge Index Theorem and the Structure of the Intersection Matrix of a Fiber

Throughout this chapter X will denote a nonsingular projective surface defined over an algebraically closed field $\Bbbk$ of arbitrary characteristic, and K will denote a canonical divisor on X.

Lemma 2.1. *Let D_1 and D_2 be two divisors on the surface X, and let $E_2 \in |D_2|$. Then the map $E \mapsto E + E_2 : |D_1| \longrightarrow |D_1 + D_2|$ is injective, and therefore $\dim |D_1| \leq \dim |D_1 + D_2|$.*

The proof is obvious.

Proposition 2.2. *If D is a divisor on X with $(D^2) > 0$ and H is a hyperplane section on X, then exactly one of the following two statements is true:*

(i) $(D \cdot H) > 0$ *and* $\dim |nD| \longrightarrow \infty$ *as* $n \longrightarrow \infty$, *or*

(ii) $(D \cdot H) < 0$ *and* $\dim |nD| \longrightarrow \infty$ *as* $n \longrightarrow -\infty$.

Proof. From the Riemann–Roch theorem we have:

$$\dim |nD| + \dim |K - nD| \geq \frac{1}{2}n^2(D^2) - \frac{1}{2}n(D \cdot K) + \chi(\mathcal{O}_X) - 2,$$

and therefore $\dim |nD| + \dim |K - nD| \longrightarrow \infty$ as $n \longrightarrow \pm\infty$ (for $(D^2) > 0$).
  Claim 1: We cannot have both $\dim |nD| \longrightarrow \infty$ and $\dim |K - nD| \longrightarrow \infty$ as $n \longrightarrow \infty$ (or as $n \longrightarrow -\infty$).
  Indeed, otherwise there would exist an $E \in |nD|$ for $n \gg 0$ (resp. $n \ll 0$), and therefore, using (2.1), $\dim |K - nD| \leq \dim |K|$, which is a contradiction.

Claim 2: dim $|nD|$ *cannot go to* ∞ *for both* $n \longrightarrow \infty$ *and* $n \longrightarrow -\infty$.

Indeed, if $|nD| \neq \varnothing$ for some $n > 0$, then $(D \cdot H) > 0$, because H is ample. Then $|nD| = \varnothing$ for all $n < 0$ (otherwise we would have $(D \cdot H) < 0$, which is a contradiction).

Claim 3: dim $|K - nD|$ *cannot go to* ∞ *for both* $n \longrightarrow \infty$ *and* $n \longrightarrow -\infty$.

Indeed, otherwise for $n \gg 0$ there would exist an $E \in |K - nD|$, and then by (2.1) we would have dim $|2K| = $ dim $|(K + nD) + (K - nD)| \geq$ dim $|K + nD| \longrightarrow \infty$, which is a contradiction.

The conclusion follows immediately from these three claims. $\square$

Corollary 2.3. *If D is a divisor on X and H is a hyperplane section on X such that $(D \cdot H) = 0$, then $(D^2) \leq 0$, and $(D^2) = 0$ if and only if $D \equiv 0$.*

Proof. Only the last statement requires proof. If D would not be numerically equivalent to zero, then there would exist a divisor E on X such that $(D \cdot E) \neq 0$. Taking $E' = (H^2)E - (E \cdot H)H$, we have:

$$(D \cdot E') = (H^2) \cdot (D \cdot E) - (E \cdot H) \cdot (H \cdot D) = (H^2) \cdot (D \cdot E) \neq 0,$$
$$(H \cdot E') = (H^2) \cdot (H \cdot E) - (H \cdot E) \cdot (H^2) = 0.$$

Replacing E with E' we may therefore also assume that $(H \cdot E) = 0$. Now put $D' = nD + E$. We have $(D' \cdot H) = 0$ and $(D'^2) = 2n \cdot (D \cdot E) + (E^2)$. Taking $n \gg 0$ if $(D \cdot E) > 0$, resp. $n \ll 0$ if $(D \cdot E) < 0$, we get $(D'^2) > 0$ and $(D' \cdot H) = 0$, contradicting Proposition 2.2. $\square$

Corollary 2.4 (The Hodge Index Theorem). *If E is a divisor on X such that $(E^2) > 0$, then for every divisor D on X such that $(E \cdot D) = 0$ we have $(D^2) \leq 0$, and furthermore $(D^2) = 0$ if and only if $D \equiv 0$.*

Proof. Let $M = \mathrm{Num}(X) \otimes_{\mathbb{Z}} \mathbb{R}$; then, by the Néron–Severi theorem, M is a finite-dimensional vector space over $\mathbb{R}$ (of dimension $\rho = \mathrm{rank}\,\mathrm{NS}(X)$). Moreover, the intersection number defines a nondegenerate bilinear symmetric form on M. If h is the class in M of a hyperplane section of X, and if we complete h to a basis of M over $\mathbb{R}$, say, $h_1 = h, h_2, \ldots, h_\rho$, such that $(h \cdot h_i) = 0$ for every $i \geq 2$, then Corollary 2.3 shows that $(\cdot)$ has signature $(1, \rho - 1)$ in this basis. Corollary 2.4 follows from Sylvester's well-known theorem on the invariance of the signature of a symmetric bilinear form on M. $\square$

Proposition 2.5. *Let $x \cdot y$ be a symmetric bilinear form on a $\mathbb{Q}$-vector space M, and let $\{e_i\}_{i \in I}$ be a finite family of elements of M that generate M and such that $e_i \cdot e_j \geq 0$ for all $i \neq j$. Assume that there exists $z \in M, z = \sum_j a_j e_j, a_j > 0$ for all j, such that $z \cdot e_i = 0$ for all i. Then $x \cdot x \leq 0$ for every $x \in M$, and $\{x \in M \mid x \cdot x = 0\}$ is a linear subspace of M. If the e_i are linearly independent over $\mathbb{Q}$, then the dimension of this subspace is equal to the number of connected components of the graph whose vertices* are the elements of I and whose arrows $\{i, j\}$ correspond to pairs (i, j) with $e_i \cdot e_j > 0$.

Proof. Replacing each e_j with $a_j e_j$, we may assume that $z = \sum_j e_j$. Every $x \in M$ can be written in the form $x = \sum_i c_i e_i, c_i \in \mathbb{Q}$. Then

$$
\begin{aligned}
x \cdot x &= \sum_i c_i^2 (e_i \cdot e_i) + 2 \sum_{i<j} c_i c_j (e_i \cdot e_j) \\
&\leq \sum_i c_i^2 (e_i \cdot e_i) + \sum_{i<j} (c_i^2 + c_j^2)(e_i \cdot e_j) \\
&= \sum_{i,j} c_i^2 (e_i \cdot e_j) = \sum_i c_i^2 (z \cdot e_i) = 0.
\end{aligned}
$$

For the other two statements in the corollary, we may assume that the e_i are linearly independent. Then the c_i are uniquely determined by x, and going through the computation we see that $x \cdot x = 0$ if and only if $c_i = c_j$ whenever $e_i \cdot e_j > 0$; this clearly implies both statements about $\{x \in M \mid x \cdot x = 0\}$. $\qquad \square$

Corollary 2.6. *Let $f : X \longrightarrow B$ be a morphism from a surface X to a nonsingular curve B, and let $D = \sum_i n_i E_i$ be a fiber of f, with E_i distinct integral curves. (Thus $n_i > 0$ for all i.) Then for every divisor of the form $D' = \sum_i n_i' E_i$, with $n_i' \in \mathbb{Z}$, we have $(D'^2) \leq 0$. If f also has connected fibers then $(D'^2) = 0$ if and only if there exists $a \in \mathbb{Q}$ such that $D' = aD$.*

Proof. This follows directly from (2.5), taking $M = \oplus_i \mathbb{Q} E_i$, the intersection number as the symmetric bilinear form, and $z = D$, since it is clear that $(D \cdot E_i) = 0$ for every i. $\qquad \square$

Corollary 2.7. *Let $f : X \longrightarrow Y$ be a birational morphism from a nonsingular projective surface X to a normal surface Y. Let y be a point on Y, and let $f^{-1}(y) = \sum_{i=1}^n r_i E_i$ be the corresponding fiber, where we assume that the E_i are pairwise distinct integral curves. Then the intersection matrix $\|(E_i \cdot E_j)\|$ is negative definite.*

Proof. For each i there exists an integral curve D_i on X, distinct from all E_js, such that $(D_i \cdot E_i) > 0$ (that is, such that $D_i \cap E_i \neq \varnothing$). Since f is a proper morphism, $f(\bigcup_{i=1}^n D_i)$ is a closed subset of Y containing y. Let g be a nonzero rational function on Y, defined on an open neighborhood U of y, which vanishes identically on $f(\bigcup_{i=1}^n D_i) \cap U$. Then $\operatorname{div}_X(f^*(g)) = \sum_{i=1}^n m_i E_i + D$, with $m_i > 0$ for all i, $\bigcup_{i=1}^n D_i \subseteq \operatorname{Supp}(D)$, and the E_i are not components of $\operatorname{Supp}(D)$. Since $D|_{f^{-1}(U)}$ is effective and $E_i \subset f^{-1}(U)$

for every i, we have $(E_i \cdot D) > 0$ for every i. Also:

$$\left(E_j \cdot \sum_{i=1}^{n} m_i E_i + D\right) = (E_j \cdot \operatorname{div}_X(f^*(g))) = 0 \quad \text{for every } j,$$

$$\left(D \cdot \sum_{i=1}^{n} m_i E_i + D\right) = (D \cdot \operatorname{div}_X(f^*(g))) = 0.$$

Taking for M in (2.5) the $\mathbb{Q}$-vector space spanned by the basis $e_1 = E_1, \ldots, e_n = E_n$, $e_{n+1} = D$, $a_1 = m_1, \ldots, a_n = e_n, a_{n+1} = 1$, the intersection number as the symmetric bilinear form, and $z = \sum_{i=1}^{n} m_i E_i + D$, we see that $(D'^2) \leq 0$ for every $D' = \sum c_i E_i$, and moreover $(D'^2) = 0$ if and only if there exists a rational number a such that $D' = a \cdot \left(\sum_{i=1}^{n} m_i E_i + D\right)$, that is, if and only if $D' = 0$. $\square$

EXERCISES

Exercise 2.1. Prove the following variant of the Hodge index theorem: For every two divisors D_1 and D_2 on a nonsingular projective surface X with $(D_1^2) > 0$, the following inequality holds:

$$(D_1^2)(D_2^2) \leq (D_1 \cdot D_2)^2.$$

[*Hint:* Find two integers m and n such that $(D_1 \cdot mD_1 + nD_2) = 0$, and then apply Corollary 2.4.]

Exercise 2.2. Let C_1 and C_2 be two nonsingular projective curves, and let D be a divisor on the surface $X = C_1 \times C_2$. Use the Hodge index theorem to prove the following inequality of Castelnuovo–Severi–Weil (cf. [Gro1]):

$$(D^2) \leq 2(C_1 \times \{x_2\} \cdot D)(\{x_1\} \times C_2 \cdot D),$$

where $x_1 \in C_1$ and $x_2 \in C_2$ are points. [*Hint:* Find integers m and n such that the divisor $D' = D + m(C_1 \times \{x_2\}) + n(\{x_1\} \times C_2)$ satisfies $(D' \cdot C_1 \times \{x_2\}) = (D' \cdot \{x_1\} \times C_2) = 0$, and use Exercise 2.1.]

Exercise 2.3. Let C be a smooth projective curve of genus g, and let $f : C \to C$ be a surjective morphism of degree d. If $\Gamma_f \subset C \times C$ is the graph of f and Δ is the diagonal of $C \times C$, show that

$$|(\Gamma_f \cdot \Delta) - d - 1| \leq 2g\sqrt{d}.$$

[*Hint:* Put $D = m\Delta + n\Gamma_f$, write $D^2 - 2(C \times \{c\} \cdot D)(\{c\} \times C \cdot D)$ (with $c \in C$) as a quadratic form in m and n, and use the fact that this quadratic form is negative definite by Exercise 2.2.]

Exercise 2.4. *Weil's proof of the analogue of the Riemann hypothesis for curves* (cf. [Wei*, Gro1]). Let C be a smooth projective curve of genus

g defined over the finite field $\Bbbk$ with $q = p^n$ elements ($p \geq 2$ prime), and denote by N the number of $\Bbbk$-rational points of C. Prove that $N = 1-a+q$, with $|a| \leq 2g\sqrt{q}$. [*Hint:* In Exercise 2.3 observe that $(\Gamma_f \cdot \Delta)$ is just the number of fixed points of f, provided that $f \neq \mathrm{id}_C$. Then apply Exercise 2.3 to the nth power $f = F^n$ of the Frobenius morphism $F : C \to C$.]

Bibliographic References. The first algebraic (and very elementary) proof of the Hodge index theorem was given by Grothendieck [Gro1]. The proof given here is essentially that of Grothendieck (cf. [Mum1] or [BH]). Proposition 2.5 is in [Sha2], with a more complicated proof. The proof given here is taken from [BH]. Corollary 2.7 was proved by Mumford in [Mum5], but this property was known much earlier to P. Du Val, cf. [DuV].

3

Criteria of Contractability and Rational Singularities

In what follows, *surface* will mean nonsingular projective surface. If other types of surfaces are to be considered, we will specify that explicitly; for example, we will say normal surfaces, affine surfaces, etc.

If X is a surface and $E = E_1 \cup E_2 \cup \cdots \cup E_n$ is a connected curve on X with pairwise distinct integral components $E_1, \ldots, E_n$, then (2.7) shows that a necessary condition for E to be contractable to a point is that the intersection matrix $\|(E_i \cdot E_j)\|_{i,j}$ be negative definite. In general this condition is not sufficient, as illustrated by Example 3.1 below (due to Nagata). However, if we place ourselves in the larger category of two-dimensional algebraic spaces, then this condition is also sufficient, by a result of Grauert [Gra*, Art3]. In other words, if the intersection matrix $\|(E_i \cdot E_j)\|_{i,j}$ is negative definite, then there exists a birational morphism $f : X \to Y$ with Y a normal algebraic space (of dimension two) such that $f(E)$ is a point on Y, and the restriction of f to $X \setminus E$ is an isomorphism onto $Y \setminus f(E)$.

Example 3.1. (Nagata). Let $\Bbbk = \mathbb{C}$, the field of complex numbers; let E_0 be a nonsingular curve of degree three in $X_0 = \mathbb{P}^2$; and let $0 \in E_0$ be an inflection point (that is, a point with the property that the tangent line to E_0 at 0 does not intersect E_0 at any other point), for example, the curve given by the equation

$$x_1^2 x_2 - x_0^3 + x_0 x_2^2 = 0$$

and the point $0 = (0, 1, 0)$. It is well known that E_0 admits a unique structure of algebraic group such that 0 is the neutral element (see, for example, [Sha1, p. 215]). Furthermore, from the geometric characterization of this group structure we see that if F_0 is another curve in $X_0 = \mathbb{P}^2$ such that $E_0 \cap F_0 = \{y_1, \ldots, y_m\}$, with each point y_i counted $(E_0 \cdot F_0)_{y_i}$ times, then $y_1 + \cdots + y_m = 0$. As an abstract group, E_0 is isomorphic to the additive group $\mathbb{R}/\mathbb{Z} \times \mathbb{R}/\mathbb{Z}$; therefore there exists a point $x_0 \in E_0$ of infinite order (the fact that we are considering the complex field is essential here; for example, if $\Bbbk$ is instead the algebraic closure of a finite field, then every point of E_0 has finite order!).

Let X_1 be the surface obtained by blowing up $X_0 = \mathbb{P}^2$ at x_0, and let E_1 be the proper transform of the curve E_0. As $(E_0^2) = 9$, we have $(E_1^2) = 8$. Let $x_1 \in E_1$ be the unique point of E_1 above x_0, and let X_2 be the surface obtained by blowing up X_1 at x_1. Let E_2 be the proper transform of E_1, and let $x_2 \in E_2$ be the unique point of E_2 above x_1. Then $(E_2^2) = 7$. Let X_3 be the blow-up of X_2 at x_2; etc. Eventually we find a surface $X = X_{10}$ and an elliptic curve $E = E_{10}$ on X_{10} with the property that $(E_{10}^2) = -1$.

Claim: The curve E on X is not contractable to a point.

By way of contradiction, if there exists a morphism $f : X \to Y$ with the properties: $f(E)$ is a normal point y on Y and the restriction of f to $X \setminus E$ is an isomorphism onto $Y \setminus \{y\}$, then there would obviously exist a curve D' on Y that would not contain y. Hence there would exist a curve D'' on X, disjoint from $E = E_{10}$. Then let D be the projection of D'' on $X_0 = \mathbb{P}^2$. Since D'' is disjoint from E, D cannot intersect E_0 at any point other than x_0. Now put $m = (D \cdot E_0) = (D \cdot E_0)_{x_0}$. From what we said earlier we see that $m \cdot x_0 = 0$, which contradicts the fact that x_0 has infinite order on the elliptic curve E_0. □

3.2. In what follows we would like to give sufficient criteria for contractability in the category of algebraic surfaces; these criteria will play an essential role later in this book.

Thus let X be a surface (over an algebraically closed field of arbitrary characteristic), and let $E = E_1 \cup \cdots \cup E_n$ be a connected curve on X, with pairwise distinct integral components $E_1, \ldots, E_n$. Let $Z = \sum_{i=1}^{n} r_i E_i$, $r_i \geq 0$, be an effective divisor with support contained in E.

Lemma 3.3. *In the situation described in (3.2), the following conditions are equivalent:*

(a) $H^1(\mathcal{O}_Z) = 0$;

(b) *for every divisor Z' with $0 < Z' \leq Z$ we have $p_a(Z') \leq 0$.*

Proof. The arithmetic genus of Z' is given by the formula

$$p_a(Z') = \tfrac{1}{2}((Z' + K) \cdot Z') + 1 = 1 - \chi(\mathcal{O}_{Z'}),$$

where K is a canonical divisor on X.

(a) $\implies$ (b): Since the dimension of Z is one, the surjective homomorphism $\mathcal{O}_Z \to \mathcal{O}_{Z'}$ induces a surjective homomorphism $H^1(\mathcal{O}_Z) \to H^1(\mathcal{O}_{Z'})$, so that $H^1(\mathcal{O}_{Z'}) = 0$.

Now $p_a(Z') = 1 - \dim H^0(\mathcal{O}_{Z'}) + \dim H^1(\mathcal{O}_{Z'}) = 1 - \dim H^0(\mathcal{O}_{Z'}) \leq 0$, as stated.

(b) $\implies$ (a): From the hypotheses of (b) we see in particular that $p_a(E_i) = 0$ for each i such that E_i is in the support of Z, and therefore each such E_i is a smooth rational curve. We prove the implication (b) $\implies$ (a) by induction on $\sum_{i=1}^{n} r_i$, noting that we have already proved the claim when $\sum_{i=1}^{n} r_i = 1$. Assume therefore that $\sum_{i=1}^{n} r_i > 1$. We may assume that $r_1 > 0$. Since the implication is valid for $Z_1 = Z - E_1$, we have $H^1(\mathcal{O}_{Z_1}) = 0$. Consider the surjective homomorphism $\mathcal{O}_Z \to \mathcal{O}_{Z_1}$. Its kernel is the ideal $I_{Z,Z_1} = \mathcal{O}_X(-Z_1)/\mathcal{O}_X(-Z) = \mathcal{O}_X(-Z_1) \otimes \mathcal{O}_{E_1}$; we have $\deg_{E_1}(I_{Z_1,Z}) = -(Z_1 \cdot E_1)$. Taking cohomology, we get the exact sequence

$$H^1(I_{Z_1,Z}) \longrightarrow H^1(\mathcal{O}_Z) \longrightarrow H^1(\mathcal{O}_{Z_1}) \longrightarrow 0.$$

If we can show that there exists a component E_1 of Z such that $-(Z_1 \cdot E_1) \geq -1$, then we have $H^1(I_{Z_1,Z}) = 0$ (because E_1 is isomorphic to the projective line), and then $H^1(\mathcal{O}_Z) = 0$, as required. The inequality $-(Z_1 \cdot E_1) \geq -1$ is equivalent to $(Z \cdot E_1) \leq 1 + (E_1^2)$. By way of contradiction, suppose that $(Z \cdot E_i) \geq 2 + (E_i^2)$ for all i. Since $p_a(E_i) = 0$, we have $2 + (E_i^2) = -(K \cdot E_i)$; so we can write $(Z \cdot E_i) \geq -(K \cdot E_i)$, or $(Z + K \cdot E_i) \geq 0$, for all i. But then

$$p_a(Z) = \tfrac{1}{2}(Z + K \cdot Z) + 1$$

$$= \tfrac{1}{2}(Z + K \cdot \sum_{i=1}^{n} r_i E_i) + 1$$

$$= \tfrac{1}{2} \sum_{i=1}^{n} r_i(Z + K \cdot E_i) + 1$$

$$\geq 1,$$

contradicting the hypothesis. $\square$

Lemma 3.4. *With the same hypotheses and notation as in Lemma 3.3, if the support of Z contains all the curves E_i, then conditions (a) and (b) are also equivalent to the following two conditions:*

(c) *the canonical homomorphism $d : H^1(\mathcal{O}_Z^*) \to \mathbb{Z}^n$, which associates the n-tuple $(\deg_{E_1}(L|E_1), \ldots, \deg_{E_n}(L|E_n))$ to every invertible $\mathcal{O}_X$-module L, is an isomorphism;*

(c') *the homomorphism d described earlier has finite kernel.*

Remark 3.5. In the following chapters we will make extensive use of the theory of the Picard scheme of curves and surfaces, so we note that, in the context of that theory, Lemma 3.4 can be proved immediately as follows. Since Z is a curve, $\underline{\mathrm{Pic}}^0(Z)$, the connected component of the origin of the Picard scheme $\underline{\mathrm{Pic}}(Z)$, is reduced; as $\underline{\mathrm{Pic}}^0(Z)$ is a commutative group scheme, it is a commutative algebraic group. Moreover, the tangent space to $\underline{\mathrm{Pic}}^0(Z)$ at the origin is canonically identified to $H^1(\mathcal{O}_Z)$, and therefore condition (a) in (3.3) means exactly that $\underline{\mathrm{Pic}}^0(Z) = 0$, which is (c$'$).

However, we will also prove this key lemma directly, following [Art1].

Proof of Lemma 3.4. Let $Z = \sum_{i=1}^n r_i E_i$, with $r_i \geq 1$ for all i. Let $E = Z_{\mathrm{red}} = \sum_{i=1}^n E_i$. We have an exact sequence

$$0 \longrightarrow I \longrightarrow \mathcal{O}_Z \longrightarrow \mathcal{O}_E \longrightarrow 0,$$

where the ideal I is nilpotent. We get an exact sequence of sheaves of multiplicative groups:

$$1 \longrightarrow J \longrightarrow \mathcal{O}_Z^* \longrightarrow \mathcal{O}_E^* \longrightarrow 1.$$

Since Z is a curve and the homomorphisms $H^0(\mathcal{O}_Z) \to H^0(\mathcal{O}_E)$ and $H^0(\mathcal{O}_Z^*) \to H^0(\mathcal{O}_E^*)$ are surjective (E being connected and reduced, we have $H^0(\mathcal{O}_E) = \Bbbk$ and $H^0(\mathcal{O}_E^*) = \Bbbk^*$), we get exact sequences:

$$0 \to H^1(I) \to H^1(\mathcal{O}_Z) \to H^1(\mathcal{O}_E) \to 0,$$
$$1 \to H^1(J) \to H^1(\mathcal{O}_Z^*) \to H^1(\mathcal{O}_E^*) \to 1.$$

If $I^2 = 0$, then I is isomorphic to J by the map $x \mapsto 1 + x$, so that $H^1(I) \cong H^1(J)$. Even if I^2 is nonzero, there exists an integer $t > 0$ such that $I^t = 0$ (because $Z_{\mathrm{red}} = E$). Let Z_r be the closed subscheme of Z with ideal I^r, $1 \leq r \leq t$. Assume that $r < t$. Then we have a commutative diagram with exact rows and columns:

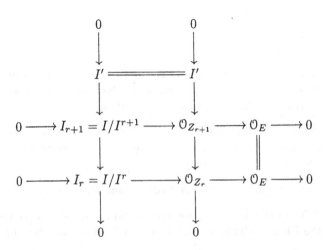

with $I'^2 = 0$. Similarly, we have another commutative diagram with exact rows and columns:

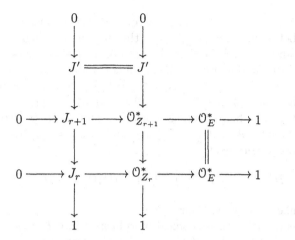

Since $I'^2 = 0$, we have an isomorphism $\epsilon : I' \xrightarrow{\cong} J'$, $\epsilon : x \mapsto 1 + x$, and therefore we have a diagram with exact rows:

$$
\begin{array}{ccccccccc}
H^0(I_r) & \xrightarrow{\delta} & H^1(I') & \longrightarrow & H^1(I_{r+1}) & \longrightarrow & H^1(I_r) & \longrightarrow & 0 \\
\epsilon' \downarrow & & \epsilon \downarrow \cong & & & & & & \\
H^0(J_r) & \xrightarrow{\delta'} & H^1(J') & \longrightarrow & H^1(J_{r+1}) & \longrightarrow & H^1(J_r) & \longrightarrow & 0
\end{array}
$$

where ϵ' is the bijective map $u \mapsto 1 + u$, which is *not* a group homomorphism if $I_r^2 \neq 0$. Still, if we can show that the square in this diagram is commutative, then the homomorphism ϵ induces a group isomorphism $\bar{\epsilon} : \mathrm{Coker}(\delta) \xrightarrow{\cong} \mathrm{Coker}(\delta')$, and then (by induction on r) we would get the following.

(3.4.1) If $I^t = 0$, there exist two towers of subgroups:

$$
H^1(I) = I_0' \supseteq I_1' \supseteq \cdots \supseteq I_t' = 0 \quad \text{and}
$$
$$
H^1(J) = J_0' \supseteq J_1' \supseteq \cdots \supseteq J_t' = 0,
$$

with $I_r'/I_{r+1}' \cong J_r'/J_{r+1}'$, $r = 0, 1, \ldots, t-1$.

To check the commutativity of the square in the preceding diagram, let $u \in H^0(I_r)$ be an arbitrary element and let $\{U_i\}$ be a finite open cover of E, with sections $u_i \in \Gamma(U_i, I_{r+1})$ such that $u_i|_{Z_r} = u$. Then the $u_i - u_j$ form a cocycle that represents $\delta(u)$, and therefore $\epsilon\delta(u)$ is the class of the cocycle $\{1 + u_i - u_j\}_{i,j}$. On the other hand, $\epsilon'(u) = 1 + u$ can be lifted over the U_i to the sections $1 + u_i \in \Gamma(U_i, J_{r+1})$, so that $\delta'\epsilon'(u)$ is represented by the cocycle $\{(1 + u_i)/(1 + u_j)\}_{i,j}$. However, since $u_i - u_j \in \Gamma(U_i \cap U_j, I')$ and $I' \cdot I_{r+1} = 0$, we have $(1 + u_j)(1 + u_i - u_j) = 1 + u_i$, or equivalently, $(1 + u_i)/(1 + u_j) = 1 + u_i - u_j$.

By (3.4.1), there only remains to consider the special case $Z = E = \sum_{i=1}^n E_i$. Let then $\bar{E} = \coprod_{i=1}^n E_i$ be the disjoint topological sum of the curves E_i, and let $\pi : \bar{E} \to E$ be the canonical surjective morphism. π is a finite morphism, and an isomorphism outside a finite number of points. We have $\bar{E} = \operatorname{Spec} \prod_{i=1}^n \mathcal{O}_{E_i}$, because $\pi_* \mathcal{O}_{\bar{E}} = \prod_{i=1}^n \mathcal{O}_{E_i}$. As E is reduced, the canonical homomorphisms $\mathcal{O}_E \to \prod_i \mathcal{O}_{E_i}$ and $\mathcal{O}_E^* \to (\prod_i \mathcal{O}_{E_i})^* = \prod_i \mathcal{O}_{E_i}^*$ are injective, and their cokernels are coherent and supported at a finite number of points. As $\pi_* \mathcal{O}_{\bar{E}} = \prod_i \mathcal{O}_{E_i}$ and $\pi_* \mathcal{O}_{\bar{E}}^* = \prod_i \mathcal{O}_{E_i}^*$, we get the surjective homomorphisms

$$H^1(\mathcal{O}_E) \to H^1(\mathcal{O}_{\bar{E}}) \quad \text{and} \quad H^1(\mathcal{O}_E^*) \to H^1(\mathcal{O}_{\bar{E}}^*);$$

we would like to compare their kernels.

To this end, factor the canonical morphism π into $\bar{E} \to E' \to E$, where $\mathcal{O}_{E'}$ is the $\mathcal{O}_E$-subalgebra of $\prod_i \mathcal{O}_{E_i}$ whose sections are the n-tuples $f = (f_1, \dots, f_n) \in \Gamma(U, \prod_i \mathcal{O}_{E_i})$, with U open in E, such that $f_i(P) = f_j(P)$ if $P \in E_i \cap E_j \cap U$. From the definition of E' we get an exact sequence

$$0 \longrightarrow \mathcal{O}_{E'} \longrightarrow \prod_i \mathcal{O}_{E_i} \longrightarrow F \longrightarrow 0, \qquad (3.4.2)$$

where F is defined as follows: for every $P \in E$, F_P is the direct sum of $n_P - 1$ linear spaces equal to $\Bbbk$, where n_P is the number of components E_i passing through P (so that $F_P = 0$ if $n_P = 1$). As $H^1(F) = 0$, (3.4.2) shows that the kernel of the surjective homomorphism $H^1(\mathcal{O}_{E'}) \to H^1(\mathcal{O}_{\bar{E}})$ is a $\Bbbk$-vector space of dimension $N = \sum_P (n_P - 1) - n + 1$.

Analogously, considering the exact sequence

$$0 \longrightarrow \mathcal{O}_{E'}^* \longrightarrow \prod_i \mathcal{O}_{E_i}^* \longrightarrow G \longrightarrow 0 \qquad (3.4.3)$$

we see that the homomorphism $H^1(\mathcal{O}_{E'}^*) \to H^1(\mathcal{O}_{\bar{E}}^*)$ is surjective, and its kernel is isomorphic to the direct sum of N copies of the multiplicative group $\Bbbk^*$.

This leads us to consider the injective homomorphisms $\lambda : \mathcal{O}_E \to \mathcal{O}_{E'}$ and $\lambda^* : \mathcal{O}_E^* \to \mathcal{O}_{E'}^*$, whose cokernels are again supported at a finite number of points. (Note that the morphism $E' \to E$ is bijective.) As $H^0(\mathcal{O}_E) = H^0(\mathcal{O}_{E'}) = \Bbbk$, we infer that $\operatorname{Ker}\left(H^1(\mathcal{O}_E) \xrightarrow{\bar{\lambda}} H^1(\mathcal{O}_{E'}) \right)$ is the direct sum of the additive groups $\mathcal{O}_{E',P}/\mathcal{O}_{E,P}$ for $P \in E$. Similarly, $\operatorname{Ker}\left(H^1(\mathcal{O}_E^*) \xrightarrow{\bar{\lambda}^*} H^1(\mathcal{O}_{E'}^*) \right)$ is the direct sum of the multiplicative groups $\mathcal{O}_{E',P}^*/\mathcal{O}_{E,P}^*$ for $P \in E$. Moreover, the maps $\bar{\lambda}$ and $\bar{\lambda}^*$ are surjective.

Considering (3.4.1) and the conclusions obtained from the exact sequences (3.4.2) and (3.4.3), the proof of Lemma 3.4 can be easily completed if we prove the following.

(3.4.4) For each P, there exists an integer $p \geq 0$ and a tower of subrings $\mathcal{O}_{E',P} = A_0 \supseteq A_1 \supseteq \cdots \supseteq A_p = \mathcal{O}_{E,P}$ such that we have group isomorphisms $A_r/A_{r+1} \cong A_r^*/A_{r+1}^*$, $r = 0, 1, \ldots, p-1$.

Proof of (3.4.4). If $n_P = 1$, take $p = 0$. Otherwise let $A' = \mathcal{O}_{E',P}$ and $A = \mathcal{O}_{E,P}$, with maximal ideals $\mathfrak{m}'$ and respectively $\mathfrak{m}$. Since $\dim_\Bbbk(A'/A) < \infty$, the conductor of A' in A is an $\mathfrak{m}$-primary ideal (see [Ser3, Chapter IV]). Therefore there exists a $p \geq 0$ such that $\mathfrak{m}'^{p+1} \subseteq \mathfrak{m}$. Put $A_r = A + \mathfrak{m}'^{r+1}$, $r = 0, 1, \ldots, p$. Then A_r is a local ring with maximal ideal $\mathfrak{m}_r$, $A_p = A$, $A_0 = A'$, and $\mathfrak{m}_r^2 \subseteq \mathfrak{m}_{r+1}, r = 0, 1, \ldots, p-1$. It is also clear that we have isomorphisms $\mathfrak{m}_r/\mathfrak{m}_{r+1} \cong A_r/A_{r+1}$, $r = 0, 1, \ldots, p-1$. Let $\mathfrak{m}_r^* = 1 + \mathfrak{m}_r^2$. We also have $\mathfrak{m}_r^*/\mathfrak{m}_{r+1}^* \cong A_r^*/A_{r+1}^*$, $r = 0, 1, \ldots, p-1$. Define a map $\phi : \mathfrak{m}_r \to \mathfrak{m}_r^*/\mathfrak{m}_{r+1}^*$ by $\phi(x) = 1 + x \mod \mathfrak{m}_{r+1}^*$. Note that

$$(1+x)(1+y)/(1+x+y) = 1 + (xy/(1+x+y)) \quad \text{and}$$
$$xy/(1+x+y) \in \mathfrak{m}_r^2 \subseteq \mathfrak{m}_{r+1},$$

and therefore ϕ is a group homomorphism. It is clear that ϕ is surjective and $\mathrm{Ker}(\phi) = \mathfrak{m}_{r+1}$, whence the isomorphism $\mathfrak{m}_r/\mathfrak{m}_{r+1} \cong \mathfrak{m}_r^*/\mathfrak{m}_{r+1}^*$; this concludes the proof of (3.4.4).

From the preceding discussion we see that the surjective homomorphism $H^1(\mathcal{O}_Z) \to \prod_{i=1}^n H^1(\mathcal{O}_{E_i})$ is an isomorphism if and only if the surjective homomorphism $H^1(\mathcal{O}_Z^*) \to \prod_{i=1}^n H^1(\mathcal{O}_{E_i}^*)$ is an isomorphism. Combining this conclusion with the observation that $H^1(\mathcal{O}_{E_i}) = 0$ if and only if the map $\deg : H^1(\mathcal{O}_{E_i}) \to \mathbb{Z}$ is an isomorphism, that is, if and only if E_i is isomorphic to the projective line, we see that the proof of Lemma 3.4 is complete. ☐

Corollary 3.6. *With the hypotheses of (3.3), assume also that $H^1(\mathcal{O}_Z) = 0$. Then every invertible $\mathcal{O}_Z$-module L such that $d_i = \deg_{E_i}(L) \geq 0, \forall i$, is generated by its global sections and has $H^1(L) = 0$.*

Proof. Let $x_i, y_i \in E_i$ be two distinct points on the nonsingular rational curve E_i, which are not on $\bigcup_{j \neq i} E_j$, $i = 1, \ldots, n$. Let f_i (resp. g_i) be a regular local parameter on E at the point x_i (resp. y_i), that is, a generator of the maximal ideal of the local ring $\mathcal{O}_{E,x_i}$ (resp. $\mathcal{O}_{E,y_i}$). Let $\bar{f}_i \in \mathcal{O}_{Z,x_i}$ (resp. $\bar{g}_i \in \mathcal{O}_{Z,y_i}$) be a preimage of f_i (resp. of g_i) via the surjective homomorphism (with nilpotent kernel) $\mathcal{O}_{Z,x_i} \to \mathcal{O}_{E,x_i}$ (resp. $\mathcal{O}_{Z,y_i} \to \mathcal{O}_{E,y_i}$). Let U_i (resp. V_i) be an open subset of $E_i \setminus \bigcup_{j \neq i} E_j$ that contains x_i (resp. y_i) and in which the function f_i (resp. g_i) is regular and has no zeros other than x_i (resp. y_i). It is then clear that

$$\{(U_1, \bar{f}_1^{d_1}), \ldots, (U_n, \bar{f}_n^{d_n}), (Z - \{x_1, \ldots, x_n\}, 1)\} \quad \text{and}$$
$$\{(V_1, \bar{g}_1^{d_1}), \ldots, (V_n, \bar{g}_n^{d_n}), (Z - \{y_1, \ldots, y_n\}, 1)\}$$

define two Cartier divisors D and D', respectively, whose restrictions to E are the divisors $\sum_i d_i x_i$ and $\sum_i d_i y_i$. Moreover, $\deg_{E_i} \mathcal{O}_Z(D) = d_i = \deg_{E_i} \mathcal{O}_Z(D')$. By Lemma 3.4, we have $L \cong \mathcal{O}_Z(D) \cong \mathcal{O}_Z(D')$. This implies immediately that L is generated by its global sections (namely, for each point $z \in Z$, at least one of the two divisors D and D' does not pass through z). Finally, if L is generated by its global sections, then there exists a surjection $\mathcal{O}_Z^{\oplus s} \to L \to 0$, which induces a surjective homomorphism $H^1(\mathcal{O}_Z^{\oplus s}) \to H^1(L) \to 0$, and the second statement of the corollary follows, for $H^1(\mathcal{O}_Z) = 0$ by hypothesis.   □

Corollary 3.7. *With the same hypotheses as in (3.6), if $Z = \sum_{i=1}^n r_i E_i$, $r_i \geq 1$, then*

$$\dim H^0(L) = \sum r_i d_i + \dim H^0(\mathcal{O}_Z).$$

Proof. By the definition of the intersection number $(L)_Z = \deg_Z(L)$, we have $\chi(L) = \deg_Z(L) + \chi(\mathcal{O}_Z)$, so by (3.6) we have

$$\dim H^0(L) = \deg_Z(L) + \dim H^0(\mathcal{O}_Z).$$

Finally, the fact that $\deg_Z(L) = \sum_{i=1}^n r_i d_i$ follows from (1.11) and (1.12).   □

Before proving M. Artin's "rational" contractability criterion, we need to prove the following lemma.

Lemma 3.8. *Let $f : X \to Y$ be a morphism from a surface X to a normal (and projective) surface Y, such that there exists a closed point $y \in Y$ with the following properties: f is an isomorphism of $X \backslash f^{-1}(y)$ onto $Y \backslash \{y\}$, and $f^{-1}(y)$ is a curve with integral components $E_1, \ldots, E_n$. (Such a morphism f will be called a* contraction *of X along the curve $E = E_1 + \cdots + E_n$.) Then the following conditions are equivalent:*

(a) $R^1 f_*(\mathcal{O}_X) = 0$,

(b) $\chi(\mathcal{O}_X) = \chi(\mathcal{O}_Y)$,

(c) $p_a(Z) = 0$ *for every positive divisor Z with support contained in E,*

(d) $H^1(\mathcal{O}_Z) = 0$ *for each Z as in condition (c), and*

(e) *for each $Z > 0$ with support equal to E, the canonical homomorphism d in Lemma 3.4 is an isomorphism.*

Proof. The equivalence of (c), (d), and (e) was proved in (3.3) and (3.4). Therefore it suffices to show that (a), (b), and (c) are equivalent. We prove their equivalence now.

The Leray spectral sequence associated with the morphism f induces an exact sequence (keeping in mind that $f_* \mathcal{O}_X = \mathcal{O}_Y$ and $R^q f_* \mathcal{O}_X = 0$ for $q \geq 2$):

$$0 \to H^1(\mathcal{O}_Y) \to H^1(\mathcal{O}_X) \to H^0(R^1 f_* \mathcal{O}_X) \to H^2(\mathcal{O}_Y) \to H^2(\mathcal{O}_X) \to 0,$$

and therefore

$$\chi(\mathcal{O}_Y) - \chi(\mathcal{O}_X) = \dim H^0(R^1 f_* \mathcal{O}_X) = \dim(R^1 f_* \mathcal{O}_X)_y. \qquad (3.8.1)$$

(Since f is an isomorphism outside y, $R^1 f_* \mathcal{O}_X$ is supported on the closed point y.) Then (3.8.1) proves the general inequality $\chi(\mathcal{O}_Y) \geq \chi(\mathcal{O}_X)$, as well as the equivalence of (a) and (b).

On the other hand, Zariski's holomorphic functions theorem, in the form given by Grothendieck [EGA III, (4.2.1)], shows that there exists a canonical isomorphism

$$(R^1 f_* \mathcal{O}_X)_y \cong \underset{Z_r}{\text{proj}\lim}\, H^1(\mathcal{O}_{Z_r}),$$

where $Z_r = \sum_{i=1}^n r_i E_i, r_i > 0$. Since $\dim(Z_r) = 1$, it follows that if $r' = (r'_1, \dots, r'_n) \geq r = (r_1, \dots, r_n)$, then the natural homomorphisms $H^1(\mathcal{O}_{Z_{r'}}) \to H^1(\mathcal{O}_{Z_r})$ in the projective system are surjective. Therefore, by (3) we get a canonical surjection $(R^1 f_* \mathcal{O}_X)_y \to H^1(\mathcal{O}_{Z_r})$ for every r. Therefore we have proved the equivalence of (a) and (c). □

Theorem 3.9 (M. Artin). *Let X be a (nonsingular projective) surface, and let E be a connected curve on X, with integral components $E_1, \dots, E_n$. Then the following conditions are equivalent:*

(1) *there exists a morphism $f : X \to Y$ such that Y is a normal projective surface, $f(E)$ is a point $y \in Y$, f is an isomorphism of $X \setminus E$ onto $Y \setminus \{y\}$, and $\chi(\mathcal{O}_X) = \chi(\mathcal{O}_Y)$; and*

(2) *the intersection matrix $\|(E_i \cdot E_j)\|_{i,j}$ is negative definite and $p_a(Z) \leq 0$ for every positive divisor Z with support contained in E.*

Furthermore, if (1) or (2) holds, and if $f_1 : X \to Y_1$ is another morphism that satisfies the same conditions as f, then there exists a unique isomorphism $u : Y \xrightarrow{\;\cong\;} Y_1$ such that $u \circ f = f_1$.

Proof. The existence of a unique isomorphism u as in the last part of the theorem follows from an easy, general property of normal surfaces—we leave the proof to the reader.

The implication (1) $\implies$ (2) follows from (3.8) and (2.7).

(2) $\implies$ (1). If we prove the existence of $f : X \to Y$ such that f is a contraction of X along E, then the fact that $\chi(\mathcal{O}_X) = \chi(\mathcal{O}_Y)$ is a consequence of (3.8).

Let G be the free Abelian group with basis $E_1, \ldots, E_n$. For each divisor $D \in \mathrm{Div}(X)$ we get a homomorphism $i_D \in \mathrm{Hom}_{\mathbb{Z}}(G, \mathbb{Z})$, defined by $i_D(Z) = (Z \cdot D)$ for every $Z \in G$. In particular, we get a group homomorphism $i : G \to \mathrm{Hom}_{\mathbb{Z}}(G, \mathbb{Z})$, $i : D \mapsto i_D$. This homomorphism is injective, because the matrix of i in the basis $E_1, \ldots, E_n$ of G and the dual basis $E_1^*, \ldots, E_n^*$ for $G^* = \mathrm{Hom}_{\mathbb{Z}}(G, \mathbb{Z})$ is exactly the intersection matrix $\|(E_i \cdot E_j)\|$, which by hypothesis is negative definite. Since G and G^* have the same rank, $\mathrm{Coker}(i)$ is a finite group. In particular, there exists an integer $m \geq 1$ such that, for every divisor $D \in \mathrm{Div}(X)$, we have $m \cdot i_D = i_{mD} = -i_Z$ for suitable $Z \in G$.

Now let H be a hyperplane section on X such that $H^q(\mathcal{O}_X(H)) = 0$ for all $q \geq 1$. Replacing H by a suitable multiple of H, we may also assume that $i_H = -i_Z$ for some $Z \in G$ (cf. earlier discussion). As $(H \cdot E_i) > 0$, $i = 1, \ldots, n$, we have $(Z \cdot E_i) < 0$ for all i. Therefore Z is an effective divisor, and in fact $Z \geq E = E_1 + \cdots + E_n$. Indeed, put $Z = Z_1 - Z_2$, with $Z_1, Z_2 \geq 0$ and without common components. Then $0 \geq (Z \cdot Z_2) = (Z_1 \cdot Z_2) - (Z_2^2)$. Since $(Z_1 \cdot Z_2) \geq 0$ and $-(Z_2^2) \geq 0$, we get $(Z_2^2) = 0$, and therefore $Z_2 = 0$, because the matrix $\|(E_i \cdot E_j)\|$ is negative definite. Thus $Z \geq 0$. Finally, since $(Z \cdot E_i) < 0$, Z must contain E_i as a component, so that $Z \geq E_i$.

Thus we have $i_{H+Z} = 0$, that is, $(H + Z \cdot E_i) = 0$ for all i. Put $L = \mathcal{O}_X(H + Z) \otimes \mathcal{O}_Z$. Then L is an invertible $\mathcal{O}_Z$-module such that $\deg_{E_i}(L|_{E_i}) = (L \cdot E_i) = 0$ for all i; therefore, by (3.3) and (3.4), we see that $L \cong \mathcal{O}_Z$ (this is the key point of the whole proof!). We get an exact sequence:

$$0 \longrightarrow \mathcal{O}_X(H) \longrightarrow \mathcal{O}_X(H + Z) \longrightarrow \mathcal{O}_Z \longrightarrow 0,$$

which, in view of $H^1(\mathcal{O}_X(H)) = 0$, shows that the canonical restriction homomorphism $H^0(\mathcal{O}_X(H + Z)) \to H^0(\mathcal{O}_Z)$ is surjective. Therefore there exists a global section $s \in H^0(\mathcal{O}_X(H + Z))$ whose restriction to Z is identically 1, that is—a divisor $D \in |H + Z|$ such that $D \cap E = \varnothing$. Since H is a hyperplane section, we have: (a) the complete linear system $|H + Z|$ does not have any fixed components or base points; (b) $|H + Z|$ separates points and tangent directions on $X \setminus E$. Moreover, from $(H + Z \cdot E_i) = 0$ for all i we infer that the morphism $f' : X \to \mathbb{P}(H^0(\mathcal{O}_X(H + Z))) = |H + Z|$ contracts the curve E to a point $f'(E) = y'$, and on the other hand f' is an isomorphism between $X \setminus E$ and $f'(X) \setminus \{y'\}$. Replacing $f'(X)$ if necessary with its normalization Y (which is a projective variety because $f'(X)$ is projective), and f' with the morphism $f : X \to Y$ given by the universal property of normalization, we get the contraction we need. $\square$

Corollary 3.10. *Let X be a surface and $E \subset X$ a connected curve that satisfies condition (2) in (3.9). Let D be a divisor on X such that $(D \cdot E_i) =$*

0 *for all* $i = 1, \cdots, n$. *Then there exists a divisor* D_1, *linearly equivalent to* D, *such that* $\mathrm{Supp}(D_1) \cap E = \varnothing$.

Proof. Using the same notation as in the proof of (3.9), let H be a hyperplane section such that $H^q(\mathcal{O}_X(H)) = H^q(\mathcal{O}_X(H + D)) = 0$ for all $q \geq 1$, and $i_H = -i_Z$ for an effective divisor $Z \geq E$ in G. From the hypotheses we also have that $i_{H+D} = -i_Z$. As in the proof of (3.9), we can prove the existence of two divisors $D' \in |H + Z|$ and $D'' \in |H + Z + D|$ such that $D' \cap E = D'' \cap E = \varnothing$. Taking $D_1 = D'' - D'$, we have $D_1 \sim D$ and $\mathrm{Supp}(D_1) \cap E = \varnothing$. □

3.11. We now make a small digression to review quickly the definition and main properties of the Grothendieck dualizing sheaf ω_Z associated to an algebraic $\Bbbk$-scheme Z of pure dimension $n \geq 1$. The reader can find details in [AK, Har2]. To define the dualizing sheaf, embed Z as a closed subscheme in a nonsingular variety P of dimension, say, m (with $m \geq n$), and then define

$$\omega_Z = \mathrm{Ext}^{m-n}_{\mathcal{O}_P}(\mathcal{O}_Z, \Omega^m_{P/\Bbbk}),$$

where $\Omega^m_{P/\Bbbk}$ is the sheaf of germs of regular differential forms of degree m on P. Here is a list of properties of ω_Z that will be used later:

(3.11.1) ω_Z is a coherent $\mathcal{O}_Z$-module that depends only on Z, that is, ω_Z is independent of the embedding $Z \subset P$;

(3.11.2) ω_Z coincides with $\Omega^n_{Z/\Bbbk}$ around the nonsingular points of Z;

(3.11.3) if Z is a Cohen–Macaulay projective variety, then for every locally free $\mathcal{O}_Z$-module F of finite rank there exists a nonsingular pairing

$$H^p(Z, F) \times H^{n-p}(Z, \check{F} \otimes \omega_Z) \to \Bbbk,$$

where $\check{F}$ is the dual of F, thus defining an algebraic duality between $H^p(Z, F)$ and $H^{n-p}(Z, \check{F} \otimes \omega_Z)$ (Grothendieck–Serre duality);

(3.11.4) if Z is Cohen–Macaulay, then the sheaf ω_Z is invertible in a neighborhood of a point $z \in Z$ if and only if the local ring $\mathcal{O}_{Z,z}$ is Gorenstein; and

(3.11.5) if D is an effective Cartier divisor on Z, then the adjunction formula holds:

$$\omega_D \cong \omega_Z \otimes \mathcal{O}_Z(D) \otimes \mathcal{O}_D.$$

In particular, if Z is a normal projective surface, then Z is automatically Cohen–Macaulay, and therefore ω_Z has all the properties listed above.

Lemma 3.12. *Let* A *be a local Noetherian ring of dimension* n, *quotient of a regular local ring* B *of dimension* $n + r$, *and let* I *be a proper ideal of* A *($I \neq A$), such that* $\mathrm{depth}_I(A) \geq 2$; *that is*, I *contains an* A-*regular sequence* f_1, f_2 *of length* 2. *If we put* $\Omega_A = \mathrm{Ext}^r_B(A, B)$, *then* $\mathrm{depth}_I(\Omega_A) \geq 2$.

Proof. Let $f_1, f_2 \in I$ be an A-regular sequence. We will show that f_1, f_2 is also an Ω_A-regular sequence. To see this, consider the exact sequence

$$0 \longrightarrow A \xrightarrow{f_1} A \longrightarrow A/f_1 A \longrightarrow 0,$$

where the map from A to A is multiplication by f_1. We get a long exact sequence of Ext:

$$\text{Ext}_X^r(A/f_1 A, B) \longrightarrow \Omega_A \xrightarrow{f_1} \Omega_A \longrightarrow \text{Ext}_B^{r+1}(A/f_1 A, B) = \Omega_{A/f_1 A}.$$

As $r < \dim(B) - \dim(A/f_1 A) = r + 1$, we have $\text{Ext}_B^r(A/f_1 A, B) = 0$ (see, for example, [AK, p. 50]). Therefore f_1 is a nonzerodivisor on Ω_A, and moreover $\Omega_A/f_1 \Omega_A$ is an A-submodule of $\Omega_{A/f_1 A}$. As f_2 is a nonzerodivisor on $A/f_1 A$, the same argument shows that f_2 is a nonzerodivisor on $\Omega_{A/f_1 A}$ and therefore on the submodule $\Omega_A/f_1 \Omega_A$ as well. $\qquad \square$

Corollary 3.13. *Let A be a Cohen–Macaulay local ring of dimension n, quotient of a regular local ring B of dimension $n + r$, and let I be a proper ideal of A ($I \neq A$) such that $\text{depth}_I(A) \geq 2$; for example, A could be a normal ring of dimension $n \geq 2$, and I the maximal ideal of A. Put $X = \text{Spec}(A), Y = V(I), U = X \setminus Y$, and $\omega_X = (\text{Ext}_B^r(A, B))\tilde{}$. Then the following conditions are equivalent:*

(i) A is a Gorenstein ring;

(ii) $\omega_X|_U \cong \mathcal{O}_X|_U$.

Proof. If A is Gorenstein, then $\Omega_A \cong A$, so that $\omega_X \cong \mathcal{O}_X$, and therefore condition (ii) is satisfied.

Conversely, assume that (ii) holds. Then by (3.12) we have $\text{depth}_I(\Omega_A) \geq 2$, so that the local cohomology sheaves $\mathcal{H}_Y^0(\omega_X)$ and $\mathcal{H}_Y^1(\omega_X)$ vanish (cf. [Gro4, Theorem 3.8]). But these two sheaves are, respectively, the kernel and cokernel of the canonical map $\omega_X \to i_*(\omega_X|_U)$, and therefore $\omega_X \cong i_*(\omega_X|_U) \cong i_*(\mathcal{O}_X|_U) \cong \mathcal{O}_X$, for $\text{depth}_I(A) \geq 2$ by hypothesis; here i is the open immersion $U \hookrightarrow X$. $\qquad \square$

Corollary 3.14 (Murthy). *Let A be a factorial, Cohen–Macaulay local ring that is the quotient of a regular local ring. Then A is Gorenstein.*

Proof. If $\dim(A) = 0$, the corollary is trivially true, and if $\dim(A) = 1$ then A is a discrete valuation ring—and therefore Gorenstein (in fact, even regular). Therefore we may assume that $\dim(A) \geq 2$. We will prove the corollary by induction on $\dim(A)$. Put $U = X \setminus \{\mathfrak{m}\}$, where $X = \text{Spec}(A)$ and $\mathfrak{m}$ is the maximal ideal of A. Then every invertible $\mathcal{O}_U$-module is free. Indeed, since A is factorial, each such module can be extended to an invertible $\mathcal{O}_X$-module, which is free because A is a local ring. As $\dim(U) < \dim(A)$, we can assume by induction that U is a Gorenstein scheme, and therefore $\omega_X|_U$ is invertible. Hence $\omega_X|_U$ is *free* by the preceding remark. The conclusion follows from (3.13). $\qquad\square$

With these preparations, we will now prove the following remarkable special case of Theorem 3.9.

Theorem 3.15 (M. Artin). *Let X be a (nonsingular, projective) surface and let $E \subset X$ be a connected curve with integral components $E_1, \ldots, E_n$. Then the following conditions are equivalent:*

(1) *There exists a morphism $f : X \to Y$ with the following properties: Y is a normal projective surface, $f(E) = y$ is a Gorenstein point on Y, f is an isomorphism between $X \setminus E$ and $Y \setminus \{y\}$, and $f^*(\omega_Y) \cong \omega_X$.*

(2) *The intersection matrix $\|(E_i \cdot E_j)\|$ is negative definite, the E_i are nonsingular rational curves, and $(E_i^2) = -2, \forall i = 1, \ldots, n$.*

Proof. (1) $\implies$ (2). We only have to show that $p_a(E_i) = 0$ and $(E_i^2) = -2$. Since ω_Y is invertible and $f^*(\omega_Y) \cong \omega_X$, we have $(\omega_X \cdot E_i) = 0$. As $(E_i^2) < 0$ (cf. (2.7)), the genus formula immediately implies $p_a(E_i) = 0$ and $(E_i^2) = -2$.

(2) $\implies$ (1). Let $Z > 0$ be any positive divisor with support contained in E. Then $(\omega_X \cdot E_i) = -2 - (E_i^2) - 2p_a(E_i) = 0$ for all i, so that $(\omega_X \cdot Z) = 0$. We get

$$p_a(Z) = \tfrac{1}{2}(Z^2) + \tfrac{1}{2}(\omega_X \cdot Z) + 1 = \tfrac{1}{2}(Z^2) + 1 \leq 0,$$

because $(Z^2) < 0$ (the intersection matrix $\|(E_i \cdot E_j)\|$ being negative definite). Therefore condition (2) in Theorem 3.9 is satisfied. From that theorem we infer that there exists a projective contraction $f : X \to Y$ such that $\chi(\mathcal{O}_X) = \chi(\mathcal{O}_Y)$. There only remains to prove that ω_Y is invertible and $f^*(\omega_Y) \cong \omega_X$.

Since $(\omega_X \cdot E_i) = 0$ for all i, (3.10) shows that there exists a differential form ω of degree 2 on X whose divisor $\mathrm{div}_X(\omega)$ has support disjoint from E. In other words, there exists an affine open set U in Y, containing the point y, such that $\Omega^2_{X/k}|_{f^{-1}(U)} \cong \mathcal{O}_X|_{f^{-1}(U)}$—and therefore $\Omega^2_{Y/k}|_{U \setminus \{y\}} \cong \mathcal{O}_Y|_{U \setminus \{y\}}$. By (3.13) we see that y is a Gorenstein point on Y, so that ω_Y is invertible. Since on U we have $f^*(\omega_Y) \cong \omega_X$ and f is an isomorphism outside y, we have $f^*(\omega_Y) \cong \omega_X$. $\qquad\square$

Definition 3.16. Let (Y, y) be a normal singularity of an algebraic surface Y (so that, in particular, (Y, y) is an *isolated* singularity). A *desingularization* of (Y, y) is a morphism $f : X \to Y$ such that X is nonsingular in a neighborhood of $f^{-1}(y)$ and f is an isomorphism between $X \setminus f^{-1}(y)$ and $Y \setminus \{y\}$.

By a theorem of Zariski–Abhyankar [Zar4, Zar5, Abh], such a desingularization of a normal singularity of dimension 2 over an algebraically closed field (of any characteristic) always exists. When we study such a singularity (Y, y), we may obviously assume—if necessary—that Y is a normal affine surface, nonsingular outside y (and then the desingularization X will be nonsingular everywhere).

Definition 3.17. A normal singularity (Y, y) of a surface Y is *rational* if there exists a desingularization $f : X \to Y$ of (Y, y) such that $R^1 f_* \mathcal{O}_X = 0$.

Recalling the structure of birational maps between two nonsingular surfaces (as a product of quadratic transformations and their inverses, cf. [Sha1]), and the fact that, given a quadratic transformation $u : X' \to X$ of a nonsingular surface X, we have $R^1 u_* \mathcal{O}_{X'} = 0$, we see immediately that Definition 3.17 is independent of the choice of the desingularization $f : X \to Y$, in the sense that if $R^1 f_* \mathcal{O}_X = 0$, then every other desingularization of (Y, y) has the same property.

Lemma 3.8 gives conditions equivalent to the fact that (Y, y) is a rational singularity, where f is a desingularization of (Y, y) and $E = f^{-1}(y)_{\text{red}}$ is the reduced fiber of f above y. Indeed, if y is not a regular point on Y, then from Zariski's connectedness theorem [EGA III, (4.3.1)] we see that E is a connected closed set with all irreducible components $E_1, \ldots, E_n$ of dimension 1. Corollary 2.7 shows that the intersection matrix $\|(E_i \cdot E_j)\|$ is negative definite. From this we infer that, for any system $a_1, \ldots, a_n$ of integers, there exists a unique $Z \in G$ (G being the free Abelian group with basis $E_1, \ldots, E_n$, as in the proof of (3.9)) such that $(Z \cdot E_i) = da_i$ for all i, where $d = \det \|(E_i \cdot E_j)\|$. In particular, there exist nonzero divisors $Z \in G$ such that $(Z \cdot E_i) \leq 0$ for all i. Each such nonzero divisor Z satisfies $Z \geq E = E_1 + \cdots + E_n$ (cf. the proof of Theorem 3.9).

Lemma 3.18. *With the same hypotheses and notation as earlier, there exists a smallest divisor $Z > 0$, with support equal to E, such that $(Z \cdot E_i) \leq 0$ for all $i = 1, \ldots, n$.*

Proof. Let $Z^1 = \sum_{i=1}^n r_i^1 E_i > 0$, $Z^2 = \sum_{i=1}^n r_i^2 E_i > 0$ such that $(Z^1 \cdot E_i) \leq 0$, $(Z^2 \cdot E_i) \leq 0$ for all i, and let $Z' = \inf(Z^1, Z^2) = \sum_{i=1}^n r_i' E_i$, with $r_i' = \inf\{r_i^1, r_i^2\}$. As $r_i^1 \geq 1$ and $r_i^2 \geq 1$ for every i, we have $r_i' \geq 1$ for every i. On the other hand, if, for example, for a certain i we have $r_i^1 \leq r_i^2$, then $(Z' \cdot E_i) = r_i^1(E_i^2) + \sum_{j \neq i} r_j'(E_i \cdot E_j) \leq r_i^1(E_i^2) + \sum_{j \neq i} r_j^1(E_i \cdot E_j) = (Z^1 \cdot E_i) \leq 0$. $\square$

Remark 3.19. The proof of the existence of the smallest $Z \geq E$ such that $(Z \cdot E_i) \leq 0$ for every i did not use the fact that the morphism $f : X \to Y$ exists; all we need to consider is the curve E on X, with irreducible components $E_1, \ldots, E_n$ having negative definite intersection matrix.

Definition 3.20. In the conditions of (3.18), the smallest (positive) cycle Z such that $(Z \cdot E_i) \leq 0$, $i = 1, \ldots, n$, is called the *fundamental cycle* of E.

Theorem 3.21 (M. Artin). *Let $f : X \to Y$ be a desingularization of a normal two-dimensional singularity (Y, y), $E = f^{-1}(y)_{\text{red}}$, and Z the fundamental cycle of E. Then $p_a(Z) \geq 0$, and moreover $p_a(Z) = 0$ if and only if (Y, y) is a rational singularity.*

Proof. Let $Z' > 0$ ($Z' \in G$) with $Z' \not\geq Z$. Then there exists an index i such that $(Z' \cdot E_i) > 0$ (by definition of the fundamental cycle Z). Then $p_a(Z' + E_i) = p_a(Z') + p_a(E_i) + (Z' \cdot E_i) - 1 \geq p_a(Z')$, for $p_a(E_i) \geq 0$. Therefore

$$p_a(Z' + E_i) \geq p_a(Z') \text{ if } Z' \not\geq Z \text{ and } (Z' \cdot E_i) > 0. \tag{3.21.1}$$

Next assume that $0 < Z' < Z$ and $(Z' \cdot E_i) > 0$. Then the multiplicity of E_i in Z' is necessarily strictly smaller than the multiplicity of the same E_i in Z, and therefore $Z' + E_i \leq Z$. Now the proof of $p_a(Z) \geq 0$ can be completed by induction, noting that $p_a(E_j) \geq 0$ for all j and using (3.21.1)

Now assume that $p_a(Z) = 0$. Let Z_1 be a positive divisor in G. We define inductively a sequence of divisors $\{Z_m\}$, as follows:

(i) if $Z_m \geq Z$, then $Z_{m+1} = Z_m - Z \geq 0$; and

(ii) if $Z_m \not\geq Z$, let i be an index such that $(Z_m \cdot E_i) > 0$. We choose i such that $(Z_m \cdot E_i) > 0$ and the multiplicity of E_i in Z_m is minimal among all such i. Then put $Z_{m+1} = Z_m + E_i$.

The process stops when $Z_m = 0$. In case (i) we have $p_a(Z_m) = p_a(Z_{m+1} + Z) = p_a(Z_{m+1}) + p_a(Z) + (Z \cdot Z_{m+1}) - 1 = p_a(Z_{m+1}) + (Z \cdot Z_{m+1}) - 1 \leq p_a(Z_{m+1}) - 1$ (because $(Z \cdot Z_{m+1}) \leq 0$), and therefore in case (i) we get

$$p_a(Z_{m+1}) > p_a(Z_m). \tag{3.21.2}$$

In case (ii), the inequality (3.21.1) shows that $p_a(Z_{m+1}) \geq p_a(Z_m)$.

Now we remark that, since $(E_i^2) < 0$ for every i, case (ii) cannot occur repeatedly infinitely many times; that is, case (i) will always occur after a finite number of steps of type (ii). On the other hand, we note by induction that, if $Z' = \sum_{i=1}^n s_i E_i$, then $p_a(Z')$ can be expressed as a quadratic polynomial in $s_1, \ldots, s_n$, in which the homogeneous part of degree 2 is $1/2 \sum_{i,j} s_i s_j (E_i \cdot E_j)$. Since the matrix $\|(E_i \cdot E_j)\|$ is negative definite, the set of integers $\{p_a(Z') \mid Z' \in G, Z' > 0\}$ is bounded from above. Thus, from the preceding arguments and (3.21.2) we see that the inductive process described earlier ends after a finite number of steps, that is, there exists an integer $m \geq 1$ such that $Z_{m+1} = 0$. Then $Z_m = Z$, and therefore $p_a(Z_1) \leq p_a(Z) = 0$. This concludes the proof of Theorem 3.21. $\qquad\square$

3.22. Let X be a surface and $I \subset \mathcal{O}_X$ a coherent sheaf of ideals. We define the sheaf

$$I^{-1}(U) = \{a \in \Bbbk(X) \mid a \cdot I(U) \subseteq \mathcal{O}_X(U)\},$$

which is a coherent subsheaf of the constant sheaf $\Bbbk(X)$ of all rational functions on X. The sheaf $(I^{-1})^{-1}$ (defined similarly; note that $I^{-1} \subset \Bbbk(X)$) contains I and is a sheaf of ideals. I is called a *divisorial ideal* if $I = (I^{-1})^{-1}$. By [Bbk], I is divisorial if and only if for every point $x \in X$ and every primary component of $I_x \subset \mathcal{O}_{X,x}$, the associated prime ideal has height one. Since X is nonsingular, this means that I is locally defined by a single equation, that is, that I is invertible.

Lemma 3.23. *Let $f : X \to Y$ be a proper morphism from a nonsingular surface X to an affine normal surface $Y = \operatorname{Spec}(A)$, and let E be a connected curve on X, with integral components $E_1, \ldots, E_n$. Assume that $f(E)$ is a point $y \in Y$, f is an isomorphism between $X \setminus E$ and $Y \setminus \{y\}$, and $R^1 f_* \mathcal{O}_X = H^1(\mathcal{O}_X) = 0$. If $\mathfrak{m}$ is the maximal ideal in A corresponding to the point $y = f(E)$, then the ideal $\mathfrak{m}\mathcal{O}_X$ is invertible.*

Proof. By (3.22), it suffices to show that the ideal $I' = (I^{-1})^{-1}$ coincides with $I = \mathfrak{m}\mathcal{O}_X$. Consider the coherent $\mathcal{O}_X$-module I'/I, whose support can contain only closed points of X—so that $\operatorname{Supp}(I'/I)$ is finite (or empty). In particular, $H^1(X, I'/I) = 0$. In the exact sequence

$$0 \to \Gamma(I'/I) \to \Gamma(\mathcal{O}_X/I) \to \Gamma(\mathcal{O}_X/I') \to H^1(I'/I) = 0, \qquad (3.23.1)$$

we have $\Gamma(\mathcal{O}_X/I) = \Bbbk$. Indeed, since $I = \mathfrak{m}\mathcal{O}_X$, we see that I is generated by (finitely many) global sections, that is, there exists an exact sequence

$$0 \longrightarrow K \longrightarrow \mathcal{O}_X^m \longrightarrow I \longrightarrow 0.$$

From this we get an exact sequence

$$0 = H^1(\mathcal{O}_X^m) \longrightarrow H^1(I) \longrightarrow H^2(K) = 0,$$

and therefore $\Gamma(\mathcal{O}_X/I) = \Gamma(\mathcal{O}_X)/\Gamma(I)$. But $\mathfrak{m} \subset \Gamma(I) \subsetneq \Gamma(\mathcal{O}_X) = A$, whence $\Gamma(\mathcal{O}_X/I) = A/\mathfrak{m} = \Bbbk$.

Then (3.23.1) implies that either $\Gamma(I'/I) = 0$ or $\Gamma(\mathcal{O}_X/I') = 0$. But since f is not an isomorphism, $\dim(\operatorname{Supp}(\mathcal{O}_X/I)) = 1$, so that $I' \neq \mathcal{O}_X$; then $\Gamma(\mathcal{O}_X/I') \neq 0$, and therefore $\Gamma(I'/I) = 0$. Since the support of I'/I is finite (we have already seen this), we get $I' = I$, as required. $\qquad\square$

Lemma 3.24. *With the hypotheses of Lemma 3.23, assume that $A = \mathcal{O}_{Y,y}$. Then for every invertible $\mathcal{O}_X$-module L we have:*

(i) *$(L \cdot E_i) = 0$, $\forall i = 1, \ldots, n$ implies $L \cong \mathcal{O}_X$; and*

(ii) $(L \cdot E_i) \geq 0$, $\forall i = 1, \ldots, n$ *implies that L is generated by its global sections and $H^1(L) = 0$.*

Proof. We will use the following observation: $H^1(L) = 0$ if and only if $H^1(L \otimes \mathcal{O}_Z) = 0$ for every positive divisor Z with support contained in E. Indeed, one implication follows immediately using the fact that $H^2 = 0$ for every coherent $\mathcal{O}_X$-module; the other implication can be proved using Zariski's holomorphic functions theorem (see the proof of (3.8)).

We will show (ii) first. Let $x \in X$ be a closed point. Then $x \in E$. We must show that L has a global section that does not vanish at x. By (3.8) we have $H^1(\mathcal{O}_Z) = 0$ for every $Z > 0$ with support contained in E, and therefore if we set $L_Z = L \otimes \mathcal{O}_Z$, L_Z is generated by its global sections, for $\deg_{E_i}(L_Z|_{E_i}) = (L \cdot E_i) \geq i$ for all i, and (3.6) applies. Since clearly $(\mathfrak{m}\mathcal{O}_X)_Z = \mathfrak{m}\mathcal{O}_X \otimes \mathcal{O}_Z$ is generated by its global sections (and is invertible by (3.23)), we see that $(L \otimes \mathfrak{m}\mathcal{O}_X)_Z$ is an invertible $\mathcal{O}_Z$-module generated by its global sections. Therefore, by (3.6) we get $H^1((L \otimes \mathfrak{m}\mathcal{O}_X)_Z) = 0$. Since Z is arbitrary, the observation made at the beginning of the proof shows that $H^1(L \otimes \mathfrak{m}\mathcal{O}_X) = 0$. The cohomology exact sequence

$$H^0(L) \longrightarrow H^0(L \otimes (\mathcal{O}_X/\mathfrak{m}\mathcal{O}_X)) \longrightarrow H^1(L \otimes \mathfrak{m}\mathcal{O}_X) = 0$$

shows that the first map is surjective. Now let Z' be the curve defined by the invertible ideal $\mathfrak{m}\mathcal{O}_X$. Then $x \in Z'$ and $L_{Z'} = L \otimes (\mathcal{O}_X/\mathfrak{m}\mathcal{O}_X)$ is generated by global sections, and therefore there exists a global section $s' \in H^0(L \otimes \mathcal{O}_{Z'})$ such that $s'(x) \neq 0$. Moreover, s' can be lifted to a global section $s \in H^0(L)$ on X, and we still have $s(x) \neq 0$, so that L is generated by its global sections. The fact that $H^1(L) = 0$ follows immediately from this and the hypothesis that $H^1(\mathcal{O}_X) = 0$. So the proof of (ii) is complete.

Although (i) is essentially contained in (3.10), we can also give the following argument. Just like in the proof of (ii), we can show that in the hypotheses of (i) there exists a section $s' \in H^0(L_{Z'})$ that does not vanish anywhere on Z' (cf. (3.4)). Form this we see that s doesn't vanish at any point of E, and therefore s does not vanish anywhere on X, because A is a local ring. $\qquad \square$

Lemma 3.25. *With the same hypotheses as in Lemma 3.24, the canonical map $\mathfrak{m}^p \to \Gamma(\mathfrak{m}^p\mathcal{O}_X)$ is an isomorphism for every integer $p \geq 1$.*

Proof. Induction on p. The case $p = 1$ has already been considered in the proof of (3.23). All we need to prove is that the natural map (for $p \geq 2$)

$$\Gamma(\mathfrak{m}^{p-1}\mathcal{O}_X) \otimes \Gamma(\mathfrak{m}\mathcal{O}_X) \to \Gamma(\mathfrak{m}^p\mathcal{O}_X) \tag{3.25.1}$$

is surjective. This map factors as

$$\Gamma(\mathfrak{m}^{p-1}\mathcal{O}_X) \otimes \Gamma(\mathfrak{m}\mathcal{O}_X) \xrightarrow{v} \Gamma(\mathfrak{m}^{p-1}\mathcal{O}_X \otimes \mathfrak{m}\mathcal{O}_X) \to \Gamma(\mathfrak{m}^p\mathcal{O}_X), \tag{3.25.2}$$

and since $\mathfrak{m}^i \mathcal{O}_X$ is invertible for every i (cf. (3.23)), the second map in (3.25.2) is an isomorphism, so there only remains to prove the surjectivity of the map v. To show that v is surjective, consider two exact sequences

$$0 \longrightarrow K_1 \longrightarrow \mathcal{O}_X^s \xrightarrow{u_1} \mathfrak{m}^{p-1}\mathcal{O}_X \longrightarrow 0$$

$$0 \longrightarrow K_2 \longrightarrow \mathcal{O}_X^t \xrightarrow{u_2} \mathfrak{m}\mathcal{O}_X \longrightarrow 0$$

(whose existence is clear), inducing cohomology exact sequences

$$\Gamma(\mathcal{O}_X^s) \xrightarrow{\alpha_1} \Gamma(\mathfrak{m}^{p-1}\mathcal{O}_X) \longrightarrow H^1(K_1) \longrightarrow H^1(\mathcal{O}_X^s) = 0$$

$$\Gamma(\mathcal{O}_X^t) \xrightarrow{\alpha_2} \Gamma(\mathfrak{m}\mathcal{O}_X) \longrightarrow H^1(K_2) \longrightarrow H^1(\mathcal{O}_X^t) = 0$$

where α_1 and α_2 may be assumed to be surjective. Then $H^1(K_1) = 0$ and $H^1(K_2) = 0$. Let $N = \mathrm{Ker}(u_1 \otimes u_2)$. The exact sequence

$$K = (K_1 \otimes \mathcal{O}_X^s) \oplus (K_2 \otimes \mathcal{O}_X^t) \to \mathcal{O}_X^s \otimes \mathcal{O}_X^t \to \mathfrak{m}^{p-1}\mathcal{O}_X \otimes \mathfrak{m}\mathcal{O}_X \to 0$$

induces an exact sequence

$$0 \longrightarrow M \longrightarrow K \longrightarrow N \longrightarrow 0.$$

As $H^1(K) = 0$ and $H^2(M) = 0$ (because all the fibers of f have dimension at most one), we have $H^1(N) = 0$, and therefore the natural map $\Gamma(u_1 \otimes u_2) : \Gamma(\mathcal{O}_X^s \otimes \mathcal{O}_X^t) \to \Gamma(\mathfrak{m}^{p-1}\mathcal{O}_X \otimes \mathfrak{m}\mathcal{O}_X)$ is surjective. Now the commutative diagram

$$
\begin{array}{ccc}
\Gamma(\mathcal{O}_X^s) \otimes \Gamma(\mathcal{O}_X^t) & \xrightarrow{\;\cong\;} & \Gamma(\mathcal{O}_X^s \otimes \mathcal{O}_X^t) \\
\Big\downarrow {\scriptstyle \alpha_1 \otimes \alpha_2} & & \Big\downarrow {\scriptstyle \Gamma(u_1 \otimes u_2)} \\
\Gamma(\mathfrak{m}^{p-1}\mathcal{O}_X) \otimes \Gamma(\mathfrak{m}\mathcal{O}_X) & \xrightarrow{\;v\;} & \Gamma(\mathfrak{m}^{p-1}\mathcal{O}_X \otimes \mathfrak{m}\mathcal{O}_X)
\end{array}
$$

shows that v is surjective, as required.   □

Remark 3.26. In fact, Lemma 3.25 shows that all the powers of $\mathfrak{m}$ are *complete*—in Lipman's terminology, cf. [Lip]—and therefore (*loc. cit.*) if (Y, y) is a rational singularity, then the quadratic transformation of Y with center y is again normal. This fact, combined with a simple observation (namely, that the only singularities that can arise on the quadratic transformation of Y with center y are again rational singularities), shows that every rational singularity can be desingularized using only quadratic transformations, that is, no normalizations will be required. A singularity with this property is called an *absolutely isolated* singularity.

Lemma 3.27. *Let (Y, y) be a normal singularity of dimension $r > 1$, and let $f : X \to Y$ be a proper birational morphism such that $L = \mathfrak{m}\mathcal{O}_X$ is invertible. If X is complete, then the multiplicity of the local ring $\mathcal{O}_{Y,y}$ is equal to $-(L^r)_X = (-1)^{r-1} \cdot (D^r)$, where D is the effective divisor defined by the ideal sheaf L.*

Proof. Let $\sigma : X_1 \to Y$ be the quadratic transformation of Y with center y, and set $L_1 = \mathcal{O}_{X_1}(1) = \mathfrak{m}\mathcal{O}_{X_1}$. By the universal property of σ we see that there exists a unique morphism $u : X \to X_1$ such that $\sigma \circ u = f$ and $u^*(L_1) \cong L$. As u is birational, (1.18) shows that $(L_1{}^r)_{X_1} = (L^r)_X$, and therefore it suffices to prove the lemma in the case $f = \sigma$. However, in that case $X = \mathrm{Proj}\left(\oplus_{i=0}^{\infty}\mathfrak{m}^i\right)$, and in the Leray spectral sequence

$$E_2^{pq} = H^p(Y, R^q f_*\mathcal{O}_X(n)) \implies H^{p+q}(X, \mathcal{O}_X(n)) = H^{p+q}(X, L^n)$$

we have $E_2^{pq} = 0$ for all $q > 0$ and $n \gg 0$ ([EGA III, (2.2.1)]). So $H^p(Y, f_*\mathcal{O}_X(n)) = H^p(X, L^n)$ for all $p \geq 0$ and $n \gg 0$. Since $f_*\mathcal{O}_X(n) = \mathfrak{m}^n$ for $n \gg 0$, we get isomorphisms $H^p(Y, \mathfrak{m}^n) \cong H^p(X, L^n)$ for all $p \geq 0$ and $n \gg 0$, whence the equality $\chi(Y, \mathfrak{m}^n) = \chi(X, L^n)$. Put $P(n) = \chi(X, L^n)$. By (1.1) and (1.21), $P(n)$ is a polynomial of the form $(a/r!)n^r + \cdots$, with $a = (L^r)$. Then the difference function has the form $\Delta P(n) = (a/(r-1)!) \cdot n^{r-1} + \cdots$, and on the other hand

$$\Delta P(n) = \Delta \chi(Y, \mathfrak{m}^n) = -\dim_k(\mathfrak{m}^n/\mathfrak{m}^{n+1})$$

(using the exact sequence

$$0 \longrightarrow \mathfrak{m}^{n+1} \longrightarrow \mathfrak{m}^n \longrightarrow \mathfrak{m}^n/\mathfrak{m}^{n+1} \longrightarrow 0$$

and the additivity of the Euler–Poincaré characteristic). $\qquad\square$

We are now ready to prove the following important result.

Theorem 3.28 (M. Artin). *Let (Y, y) be a two-dimensional normal singularity, and let $f : X \to Y$ be a desingularization of (Y, y).*

If (Y, y) is a rational singularity, Z is the fundamental cycle of the fiber $E = f^{-1}(y)_{\mathrm{red}} = E_1 + \cdots + E_n$, and $\mathfrak{m}$ is the maximal ideal of the local ring $\mathcal{O}_{Y,y}$, then $X \times_Y \mathrm{Spec}(\mathcal{O}_{Y,y}/\mathfrak{m}^n) = nZ, H^0(\mathcal{O}_{nZ}) \cong \mathcal{O}_{Y,y}/\mathfrak{m}^n$, and $\dim_k(\mathfrak{m}^n/\mathfrak{m}^{n+1}) = -n \cdot (Z^2) + 1$ for every $n \geq 1$.

Proof. We may obviously assume that $Y = \mathrm{Spec}(A)$, with $A = \mathcal{O}_{Y,y}$, because even though in that case X is no longer a projective surface, the self-intersection of the divisor Z on X still makes sense, as Z is still a projective curve. The first assertion of the theorem is then equivalent to the identity $\mathfrak{m}^n\mathcal{O}_X = \mathcal{O}_X(-nZ)$ for all $n \geq 1$. It is clearly enough to prove that $\mathfrak{m}\mathcal{O}_X = \mathcal{O}_X(-Z)$. Let $g \in \mathfrak{m}, g \neq 0$. Then

$$\mathrm{div}_X(g) = D + Z',$$

with Z' a strictly positive divisor with support equal to E, and D an effective divisor that does not contain any component E_i. As the E_i are projective curves, we have

$$(\mathrm{div}_X(g) \cdot E_i) = 0 \quad \text{for every } i,$$

and therefore $(Z' \cdot E_i) = -(D \cdot E_i) \leq 0$ for every i. By the definition of the fundamental cycle, we get $Z' \geq Z$. In other words, $\mathfrak{m}\mathcal{O}_X \subseteq \mathcal{O}_X(-Z)$.

To prove that $\mathcal{O}_X(-Z) = \mathfrak{m}\mathcal{O}_X$, it suffices to show that for every point $P \in Z$ there is a function $g \in \mathfrak{m}$ whose divisor is equal to Z in a neighborhood of P. To see this, note that $L = \mathcal{O}_X(-Z)$ has the property that $(L \cdot E_i) = -(Z \cdot E_i) \geq 0$ for every i, and therefore, according to (3.24), L is generated by its global sections. Therefore there exists a positive divisor $D \in |L|$ that does not pass through P. From $\mathcal{O}_X(-Z) \cong \mathcal{O}_X(D)$ we get $\mathcal{O}_X(Z + D) \cong \mathcal{O}_X$, so there exists $g \in Q(A)$ (the field of fractions of A) such that $\mathrm{div}_X(g) = Z + D$. As $Z > 0$ and $D > 0$, we must have $g \in \mathfrak{m}$, and the first assertion of the theorem is proved.

Now consider the commutative diagram with exact rows:

$$
\begin{array}{ccccccccc}
0 & \longrightarrow & \mathfrak{m}^n & \longrightarrow & A & \longrightarrow & A/\mathfrak{m}^n & \longrightarrow & 0 \\
& & \downarrow{\scriptstyle \cong} & & \downarrow{\scriptstyle \cong} & & \downarrow & & \\
0 & \longrightarrow & H^0(\mathfrak{m}^n\mathcal{O}_X) & \longrightarrow & H^0(\mathcal{O}_X) & \longrightarrow & H^0(\mathcal{O}_{nZ}) & \longrightarrow & H^1(\mathfrak{m}^n\mathcal{O}_X)
\end{array}
$$

Using (3.24) and the first assertion of the theorem, we get $H^1(\mathfrak{m}^n\mathcal{O}_X) = H^1(\mathcal{O}_X(-nZ)) = 0$, because $(-nZ \cdot E_i) \geq 0$ for every i; this proves the second assertion of the theorem, because the first two vertical arrows in the commutative diagram above are isomorphisms.

Finally, from the exact sequence

$$0 \longrightarrow \mathcal{O}_X(-nZ) \otimes \mathcal{O}_Z \longrightarrow \mathcal{O}_{(n+1)Z} \longrightarrow \mathcal{O}_{nZ} \longrightarrow 0$$

we get the cohomology exact sequence

$$0 \longrightarrow H^0(\mathcal{O}_X(-nZ) \otimes \mathcal{O}_Z) \longrightarrow H^0(\mathcal{O}_{(n+1)Z})$$
$$\longrightarrow H^0(\mathcal{O}_{nZ}) \longrightarrow H^1(\mathcal{O}_X(-nZ) \otimes \mathcal{O}_Z)$$

in which the last term (the H^1) is zero, by (3.6), because $(-nZ \cdot E_i) \geq 0$ for every i. Thus, using the natural isomorphisms $H^0(\mathcal{O}_{(n+1)Z}) \cong A/\mathfrak{m}^{n+1}$ and $H^0(\mathcal{O}_{nZ}) \cong A/\mathfrak{m}^n$, which we established earlier, we get:

$$
\begin{aligned}
\dim \mathfrak{m}^n/\mathfrak{m}^{n+1} &= \dim H^0(\mathcal{O}_X(-nZ) \otimes \mathcal{O}_Z) \\
&= -\sum_i nr_i(Z \cdot E_i) + \dim H^0(\mathcal{O}_Z) \qquad \text{by (3.7)} \\
&= -n \cdot (Z^2) + 1.
\end{aligned}
$$

$\square$

Corollary 3.29. *Under the hypotheses of (3.28), the dimension of the Zariski tangent space $T_{Y,y}$ of the singularity (Y,y) coincides with $-(Z^2)+1$, and the multiplicity of $\mathcal{O}_{Y,y}$ is equal to $-(Z^2)$.*

Theorem 3.30 (Castelnuovo). *Let X be a nonsingular projective surface. Let $E \subset X$ be a connected curve, with integral components $E_1, \ldots, E_n$. Then there exists a morphism $f : X \to Y$ such that $f|_{X \setminus E}$ is an isomorphism between $X \setminus E$ and $Y \setminus f(E)$ and $f(E) = y$ is a smooth point on Y, if and only if the intersection matrix $\|(E_i \cdot E_j)\|$ is negative definite and the fundamental cycle Z of E satisfies the conditions $p_a(Z) = 0$ and $(Z^2) = -1$. If this is the case, then the contraction (Y, f) is uniquely determined up to a canonical isomorphism, and Y is also projective. If E is also an integral curve with $p_a(E) = 0$ and $(E^2) = -1$, then f is the quadratic transformation of Y with center y.*

Proof. Everything but the last assertion follows immediately from Theorem 3.9, Theorem 3.21, and Corollary 3.29. To show that f coincides with the quadratic transformation of Y with center y when E is integral, we note that (3.8) implies that $R^1 f_* \mathcal{O}_X = 0$, and therefore according to (3.23) the ideal $\mathfrak{m}\mathcal{O}_X$ is invertible (using the notation from (3.28)). If $g : X' \to Y$ is the quadratic transformation of Y with center y, by the universal property of g there exists a unique morphism $u : X \to X'$ such that $g \circ u = f$ and $u^* \mathcal{O}_{X'}(1) \cong \mathfrak{m}\mathcal{O}_X$. Since u is a birational morphism with finite fibers and X' is nonsingular, u is an isomorphism, as required. $\qquad\qquad\square$

Theorem 3.31. *Let $f : X \to Y$ be a desingularization of the normal two-dimensional singularity (Y, y), $E = f^{-1}(y)_{\mathrm{red}}$ the reduced fiber of f, with integral components $E_1, \ldots, E_n$, and Z the fundamental cycle of E. Then the following conditions are equivalent:*

(a) $p_a(Z) = 0$ and $(Z^2) = -2$, in other words, (Y, y) is a rational double point (cf. (3.21) and (3.29)); and

(b) $(Z \cdot K) = 0$ and $(Z^2) = -2$.

If E does not contain curves E_i with the properties $p_a(E_i) = 0$ and $(E_i^2) = -1$, then these conditions are also equivalent to:

(c) $p_a(E_i) = 0$ and $(E_i^2) = -2$ for every $i = 1, \ldots, n$.

Proof. The equivalence of (a) and (b) follows immediately from the formula $p_a(Z) = \frac{1}{2}(Z^2) + \frac{1}{2}(Z \cdot K) + 1$. Then we remark that under any one of the conditions (a), (b), or (c), we have $p_a(E_i) = 0$ for every i, and from the hypotheses we get that $(E_i^2) \leq -2$ for every i.

(b) $\implies$ (c). We have $(K \cdot E_i) = 2p_a(E_i) - 2 - (E_i^2) \geq 0$. Let $Z = \sum_{i=1}^n r_i E_i$; then $0 = (Z \cdot K) = \sum_{i=1}^n r_i(K \cdot E_i)$, and since $(K \cdot E_i) \geq 0$ for every i (and all r_i are strictly positive) we get $(K \cdot E_i) = 0$ for every i. From $p_a(E_i) = 0$ we then conclude that $(E_i^2) = -2$ for every i.

(c) $\implies$ (a). From $p_a(E_i) = 0$ and $(E_i^2) = -2$ we get $(K \cdot E_i) = 0$ for every i, and in particular $(Z \cdot K) = 0$. Now (3.21) gives $p_a(Z) \geq 0$, while at the same time $(Z^2) < 0$ and therefore $p_a(Z) = \frac{1}{2}(Z^2) + \frac{1}{2}(Z \cdot K) + 1 = \frac{1}{2}(Z^2) + 1 \leq 0$; therefore $p_a(Z) = 0$ and $(Z^2) = -2$, as stated. $\qquad\square$

Theorem 3.32. *Let X be a nonsingular projective surface, and let E be a connected curve on X, with integral components $E_1, \ldots, E_n$. Assume that the intersection matrix $\|(E_i \cdot E_j)\|$ is negative definite, $p_a(E_i) = 0$ and $(E_i^2) = -2$, $i = 1, \ldots, n$. Then the following are the only possible configuration types for E (corresponding to the Dynkin diagrams (A_n), (D_n), (E_6), (E_7) and (E_8) in the classification of semisimple Lie algebras), in which all the intersections are transversal:*

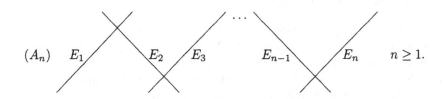

FIGURE 3.1.

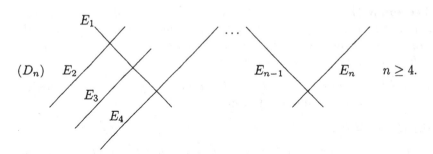

FIGURE 3.2.

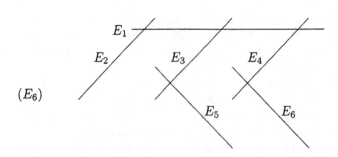

FIGURE 3.3.

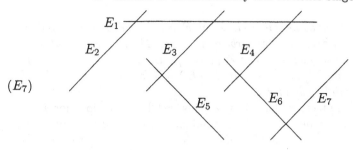

(E_7)

FIGURE 3.4.

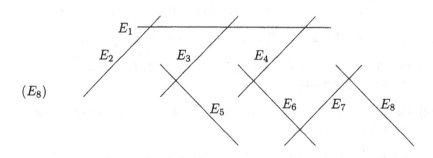

(E_8)

FIGURE 3.5.

To prove this theorem we need some preparation.

(3.32.1) *If in Proposition 2.5 instead of "$z \cdot e_i = 0$ for all i" we have the weaker hypothesis "$z \cdot e_i \leq 0$ for all i," then $x \cdot x \leq 0$ for every $x \in M$. If we also assume that the graph associated to I is connected and $z \cdot z < 0$, then $x \cdot x = 0$ if and only if $x = 0$.*

The proof is almost the same. We may again assume, without loss of generality, that $z = \sum_j e_j$. Let $x = \sum_i c_i e_i$. Then, as in the proof of (2.5), we have:

$$x \cdot x = \sum_i c_i^2 (e_i \cdot e_i) + 2 \sum_{i<j} c_i c_j (e_i \cdot e_j)$$

$$\leq \sum_i c_i^2 (e_i \cdot e_i) + \sum_{i<j} (c_i^2 + c_j^2)(e_i \cdot e_j)$$

$$= \sum_{i,j} c_i^2 (e_i \cdot e_j) = \sum_i c_i^2 (z \cdot e_i) \leq 0.$$

If I is connected, then the first inequality "$\leq$" is an equality if and only if all c_i are equal, that is, if x is proportional to z. Therefore, if $x \cdot x = 0$ and $z \cdot z < 0$, we must have $x = 0$. $\qquad \square$

(3.32.2) *The fundamental cycle Z of E can be calculated as follows: Put $Z_1 = E_{i_0}$, where i_0 is an arbitrary index. Then assume that $Z_j = \sum_{i=1}^{n} a_{ji} E_i$ is already defined. Put $Z_{j+1} = Z_j + E_{i_j}$ if there exists a component E_{i_j} such that $(Z_j \cdot E_{i_j}) > 0$. If $(Z_j \cdot E_i) \leq 0$ for every i, then $Z = Z_j$ is the fundamental cycle of E.*

We prove by induction that $Z_j \leq Z$, this fact being obvious for $j = 1$. If $Z_j < Z$, from the definition of the fundamental cycle we see that there exists an index i_j such that $(Z_j \cdot E_{i_j}) > 0$. Let $Z = \sum_{i=1}^{n} r_i E_i$, $r_i > 0$. Since $(Z \cdot E_{i_j}) \leq 0$, we have $r_{i_j} > a_{ji_j}$, and therefore $Z_{j+1} = Z_j + E_{i_j} \leq Z$. $\qquad\square$

(3.32.3) *If $E = E_1 \cup \cdots \cup E_n$ is the graph of a desingularization $f : X \to Y$ of a rational singularity (Y,y), that is, if $E = f^{-1}(y)_{\mathrm{red}}$, then:*

(a) *for all $i \neq j$ for which $(E_i \cdot E_j) > 0$, we have $(E_i \cdot E_j) = 1$;*

(b) *for all $i \neq j \neq h \neq i$ we have $E_i \cap E_j \cap E_h = \varnothing$, that is, there are no "triple intersections"; and*

(c) *the graph does not contain any loops, that is, E is a tree of nonsingular rational curves.*

Indeed, $0 \geq p_a(E_i + E_j) = p_a(E_i) + p_a(E_j) + (E_i \cdot E_j) - 1 = (E_i \cdot E_j) - 1$; this proves (a).

Analogously, $p_a(E_i + E_j + E_h) = p_a(E_i + E_j) + p_a(E_h) + (E_i + E_j \cdot E_h) - 1 = (E_i \cdot E_j) + (E_j \cdot E_h) + (E_i \cdot E_h) - 2 \leq 0$, and (b) follows.

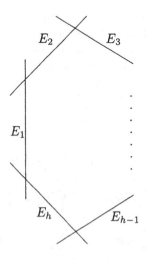

FIGURE 3.6.

If $E_1, \ldots, E_h$ ($h \geq 3$) form a loop (Figure 3.6), let $Z' = E_2 + \cdots + E_h$; by induction on h one shows immediately that $p_a(Z') = 0$. Since $(E_1 \cdot Z') = 2$, we have

$$p_a(E_1 + Z') = p_a(E_1) + p_a(Z') + (E_1 \cdot Z') - 1 = 1,$$

which contradicts the fact that (Y, y) is a rational singularity.

Note. In the hypotheses of Theorem 3.32, the proof of (3.32.3) can be simplified somewhat, using the negative definiteness of the intersection form on the components of E (instead of genus computations). For example, if all $(E_i)^2 = -2$ and $(E_i \cdot E_j) \geq 2$, then $(E_i + E_j \cdot E_i + E_j) \geq 0$, contradiction. Similarly, if E contains a loop consisting of $E_1, \ldots, E_h$, as in Figure 3.6, then $(E_1 + \cdots + E_h)^2 = 0$, contradiction.

However, (3.32.3) was stated—and proved—for *arbitrary* rational singularities, not only for rational double points.

The remaining preliminary results, on the other hand, hold only in the hypotheses of Theorem 3.32.

(3.32.4) *In the hypotheses of (3.32), every component E_i of E meets at most three other components.*

Indeed, if this were not true, then E, which is connected, would contain a subconfiguration of curves like the one shown in Figure 3.7 (after reindexing, if necessary).

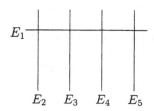

E_1

$E_2 \quad E_3 \quad E_4 \quad E_5$

FIGURE 3.7.

By (3.32.3), $(E_1 \cdot E_j) = 1$, $j = 2, 3, 4, 5$, and E_i does not meet E_j for $i \neq j$, $2 \leq i, j \leq 5$. Let $Z' = 2E_1 + (E_2 + E_3 + E_4 + E_5)$; then a trivial computation gives $(Z')^2 = 0$, contradicting the negative definiteness of the intersection matrix $\|(E_i \cdot E_j)\|$.

(3.32.5) *In the hypotheses of (3.32) there can be at most one component E_i of E that meets three other components.*

Indeed, if this were not true, then E would contain a subconfiguration of curves like the one shown in Figure 3.8, with $h \geq 4$ (after reindexing, if necessary).

By (3.32.3), the only intersections among the E_i, $i = 1, \ldots, h + 2$ are the ones shown in Figure 3.8, and those intersections are transversal. Let $Z' = (E_2 + E_3 + E_{h+1} + E_{h+2}) + 2(E_1 + E_4 + \cdots + E_h)$; then a trivial computation shows that $(Z')^2 = 0$, contradicting negative definiteness.

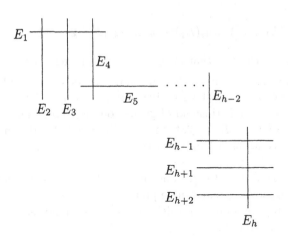

FIGURE 3.8.

Proof of Theorem 3.32.

Case 1. None of the components of E meets three other components.

Then, by (3.32.3), E is necessarily of type (A_n); in that case, one verifies immediately that the fundamental cycle of E is $Z = E_1 + \cdots + E_n$.

Case 2. Exactly one component of E, say, E_1, meets three other components (say, E_2, E_3, and E_4).

In this case, we show first that *at most two of the three curves E_2, E_3, and E_4 meet other curves (besides E_1).*

Indeed, if this were not true, then E would contain a subconfiguration of curves like the one shown in Figure 3.9.

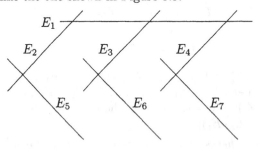

FIGURE 3.9.

Again, the only intersections are those shown in the figure, etc. Put $Z' = 3E_1 + 2(E_2 + E_3 + E_4) + (E_5 + E_6 + E_7)$; then $(Z')^2 = 0$, contradicting negative definiteness.

Subcase (2a). At most one of the components E_2, E_3, and E_4 meets components other than E_1. Since by (3.32.3) E is a tree of smooth rational curves, we see immediately that E is of type D_n. Moreover, if E is of type D_4, then the fundamental cycle of E is $Z = 2E_1 + (E_2 + E_3 + E_4)$, and if E is of type D_n with $n \geq 5$, then the fundamental cycle is $Z = 2E_1 + (E_2 + E_3) + 2(E_4 + \cdots + E_{n-1}) + E_n$ (see Figure 3.2).

Subcase (2b). Two of the curves E_2, E_3 and E_4 intersect curves other than E_1. Let's say that E_2 meets only E_1, while E_3 meets E_5 and E_4 meets E_6 (note that $n \geq 6$ in this case). We will show that the only possible diagrams in this subcase are (E_6), (E_7), and (E_8).

First we show that *at most one of the curves E_5 and E_6 can meet other curves (besides E_3 and E_4, respectively).*

Indeed, if this were not true, then E would contain a subconfiguration of curves like the one shown in Figure 3.10.

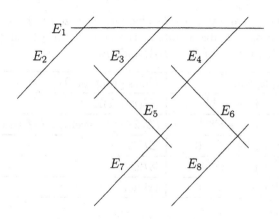

FIGURE 3.10.

Put $Z' = 4E_1 + 2E_2 + 3(E_3 + E_4) + 2(E_5 + E_6) + (E_7 + E_8)$; then $(Z')^2 = 0$, which is a contradiction.

Therefore we may assume that E_3 meets E_5, E_5 meets no other curves, and $E_4, E_6, \ldots, E_n$ form a linear chain of curves (similar to an (A_n)-type configuration); thus the last thing left to prove in this case is that $n \leq 8$. But if n were at least 9, then E would contain a subconfiguration of curves like the one shown in Figure 3.11, and we would get $(Z')^2 = 0$ for $Z' = 6E_1 + 3E_2 + 4E_3 + 2E_5 + 5E_4 + 4E_6 + 3E_7 + 2E_8 + E_9$, which is a contradiction.

Therefore the only possible diagrams in this case are (E_6), with fundamental cycle $Z = 3E_1 + 2(E_2 + E_3 + E_4) + (E_5 + E_6)$; (E_7), with fundamental cycle $Z = 4E_1 + 2E_2 + 3(E_3 + E_4) + 2(E_5 + E_6) + E_7$; and (E_8), with fundamental cycle $Z = 6E_1 + 3E_2 + 4E_3 + 2E_5 + 5E_4 + 4E_6 + 3E_7 + 2E_8$; in all these cases, the fundamental cycle is calculated using (3.32.2).

In all admissible cases (that is, (A_n), (D_n), and $(E_6) \ldots (E_8)$), we can check that the intersection matrix of the components E_i is indeed negative

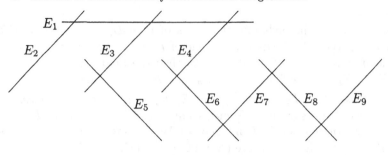

FIGURE 3.11.

definite, using (3.32.1) with $e_i = E_i$ and $z = Z$, because in all cases $z \cdot z = -2$ and $z \cdot e_i \le 0$, as one can easily verify by direct computation.   □

3.33. One can show that (at least in characteristic zero) the following equations define rational double points at the origin corresponding to the Dynkin diagrams (A_n), (D_n), (E_6), (E_7), and (E_8).

Diagram	Equation	Class group of $\hat{O}_{X,x}$
A_n $(n \ge 1)$	$x^2 + y^2 + z^{n+1} = 0$	$\mathbb{Z}/(n+1)\mathbb{Z}$
D_n $(n \ge 4)$	$x^2 + y^2 z + z^{n-1} = 0$	$(\mathbb{Z}/2\mathbb{Z})^2$ (n even), $\mathbb{Z}/4\mathbb{Z}$ (n odd)
E_6	$x^2 + y^3 + z^4 = 0$	$\mathbb{Z}/3\mathbb{Z}$
E_7	$x^2 + y^3 + yz^3 = 0$	$\mathbb{Z}/2\mathbb{Z}$
E_8	$x^2 + y^3 + z^5 = 0$	trivial

EXERCISES

Exercise 3.1. Let $f : X \to Y$ be a desingularization of the rational singularity (Y, y), and let E be a connected curve contained in $f^{-1}(y)$. Show that E can be contracted to a rational singularity, that is, that there exists a morphism $g : X \to Z$ such that Z is a normal projective surface, $z = g(E)$ is a point on Z, $g|_{X \setminus E} : X \setminus E \to Z \setminus g(E)$ is an isomorphism, and (Z, z) is a rational singularity.

Exercise 3.2. Assume that X is a nonsingular surface containing a curve E whose irreducible components E_1, E_2, and E_3 are smooth rational curves with $(E_1^2) = -3$, $(E_2^2) = (E_3^2) = -2$, $(E_1 \cdot E_2) = 1$, $(E_2 \cdot E_3) = 1$, and $(E_1 \cdot E_3) = 0$. Show that E can be contracted to a rational singularity.

Exercise 3.3. Consider the action of the cyclic group G of all complex roots of order n of 1 $(n \ge 2)$ on the polynomial algebra $\mathbb{C}[X, Y]$ given by $\xi \cdot X = \xi X$ and $\xi \cdot Y = \xi Y$, $\forall \xi \in G$. Show that the algebra of invariants $A = \mathbb{C}[X, Y]^G$ is equal to $\mathbb{C}[X^n, X^{n-1}Y, \ldots, XY^{n-1}, Y^n]$ (the coordinate ring of the normal rational curve of degree n in $\mathbb{P}^n$), and prove that $\mathrm{Spec}(A)$

has a rational singularity at the origin. Compute the multiplicity of this singularity.

Exercise 3.4. With the same notation as in Exercise 3.3, consider the action of G on $\mathbb{C}[X,Y]$ given by $\xi \cdot X = \xi X$ and $\xi \cdot Y = \xi^{-1}Y$, $\forall \xi \in G$. Show that the algebra of invariants $B = \mathbb{C}[X,Y]^G$ is equal to $\mathbb{C}[X^n, Y^n, XY]$. Show that the origin is a rational double point of type (A_{n-1}) on $\mathrm{Spec}(B)$.

Exercise 3.5. Consider the binary dihedral subgroup G of order $4n$ of $SL_2(\mathbb{C})$ ($n \geq 2$) generated by the following linear automorphisms of $\mathbb{C}^2$: $(X,Y) \mapsto (-Y,X)$ and $(X,Y) \mapsto (\xi X, \xi^{-1}Y)$ with $\xi^{2n} = 1$. Compute the algebra of invariants $\mathbb{C}[X,Y]^G$ and show that its spectrum has a singularity at the origin, which is a rational double point of type (D_{n+2}).

Exercise 3.6. Prove that the singularity (at the origin) of the affine surface with equation $x^2 + y^3 + z^5 = 0$ ($\mathrm{char}(\Bbbk) \neq 2,3,5$) is a rational double point corresponding to the Dynkin diagram (E_8). [*Hint:* In this case the desingularization process requires only quadratic transformations centered at singular points. In each step write down explicitly the three affine pieces.]

Bibliographic References. The main references for this chapter are [Art1, Art2]. The method of proof of (3.12), (3.13), and (3.14) belongs to the author. Some of the proofs (for example, those of (3.23), (3.24), and (3.25)) were inspired by [Lip].

4
Properties of Rational Singularities

Theorem 4.1 (Laufer–Ramanujam). *Let* $f : X \to Y$ *be a desingularization of a normal singularity* (Y, y), *with* Y *an affine surface,* $E_1, \ldots, E_n$ *the irreducible components of the reduced fiber* $E = f^{-1}(y)_{\mathrm{red}}$, *and* L *an invertible* $\mathcal{O}_X$-*module such that* $(L \cdot E_i) \geq (\omega_X \cdot E_i)$, $i = 1, \ldots, n$. *Then* $H^1(X, L) = 0$.

Proof. Let I_i be the invertible ideal sheaf of the curve E_i; put $J = I_1 \otimes \cdots \otimes I_n$ (the ideal sheaf of $E = E_1 + \cdots + E_n$).

Step 1. For every $h \geq 1$, *the map* $\sigma : H^1(X, J^h L) \to H^1(X, L)$, *induced by the inclusion map* $J^h L \subset L$, *is surjective.*

Proof of Step 1. Since by (2.7) the intersection matrix $\|(E_i \cdot E_j)\|$ is negative definite, we can inductively choose components $E_{i_1}, \ldots, E_{i_{hn}}$, as follows: Put $D_1 = \sum_{i=1}^n h E_i$, and choose E_{i_1} such that $(E_{i_1} \cdot D_1) < 0$. Then put $D_2 = D_1 - E_{i_1}$. In general, assuming that D_j has already been constructed, choose E_{i_j} such that $(E_{i_j} \cdot D_j) < 0$, and then put $D_{j+1} = D_j - E_{i_j}$.

For every j we have an exact sequence

$$0 \longrightarrow \mathcal{O}_X(-D_j) \cdot L \longrightarrow \mathcal{O}_X(-D_{j+1}) \cdot L \longrightarrow M_j \longrightarrow 0$$

where M_j is supported on E_{i_j} and $\deg(M_j) = -(E_{i_j} \cdot D_{j+1}) + (L \cdot E_{i_j}) = -(E_{i_j} \cdot D_j) + (E_{i_j}^2) + (L \cdot E_{i_j})$. By hypothesis, $(L \cdot E_{i_j}) \geq (\omega_X \cdot E_{i_j}) = \deg(\omega_{E_{i_j}}) - (E_{i_j}^2)$, and therefore we get $\deg(M_j) > \deg(\omega_{E_{i_j}})$. Therefore by Serre duality on E_{i_j} we have $H^1(X, M_j) = H^1(E_{i_j}, M_j) = H^0(E_{i_j}, M_j^{-1} \otimes \omega_{E_{i_j}}) = 0$, for $\deg(M_j^{-1} \otimes \omega_{E_{i_j}}) < 0$. Therefore the homomorphism $\sigma_j :$

$H^1(X, \mathcal{O}_X(-D_j) \cdot L) \to H^1(X, \mathcal{O}_X(-D_{j+1}) \cdot L)$ is surjective. Since $D_{hn+1} = 0$, the homomorphism σ is surjective, because it is equal to the composite $\sigma_{hn} \circ \cdots \circ \sigma_1$.

Step 2. There exists an integer $h \geq 1$ such that the map σ from Step 1 is identically zero.

Proof of Step 2. Let $\tau_i : H^1(X, \mathfrak{m}^i L) \to H^1(X, L)$ be the map induced by the inclusion $\mathfrak{m}^i L \subset L$ (where $\mathfrak{m}$ is the maximal ideal corresponding to the point y on Y), and let V_i denote the image of τ_i. Then we get a filtration $V_1 \supseteq V_2 \supseteq \cdots$ of $H^1(X, L)$. As $H^1(X, L)$ is finite-dimensional $(H^1(X, L) = R^1 f_*(L)$, because Y is affine, and $R^1 f_*(L)$ is a coherent $\mathcal{O}_Y$-module concentrated at most at the point y, because f is a desingularization), there exists an $h_1 \geq 1$ such that $V_{h_1} = V_{h_1+1} = \cdots = V$.

On the other hand, by [EGA III, (3.3.2)] there exists an $h_2 \geq 1$ such that $V_{h_2+r} = \mathfrak{m}^r V_{h_2}$ for all $r \geq 1$ (each $a \in \mathfrak{m}^r$ induces a multiplication map $\mathfrak{m}^i L \to \mathfrak{m}^{i+r} L$, and therefore a map $H^1(X, \mathfrak{m}^i L) \to H^1(X, \mathfrak{m}^{i+r} L))$. In particular, we get $V = \mathfrak{m} V$, and therefore $V = 0$ by Nakayama's lemma. In other words, if we put $h' = \max(h_1, h_2)$, the map $\tau_{h'}$ is identically zero. Finally, since $\mathfrak{m} \mathcal{O}_X$ is the ideal sheaf defining the fiber $f^{-1}(y)$ (note that in general $\mathfrak{m} \mathcal{O}_X$ is not invertible), there exists an $s \geq 1$ such that $J^s \subseteq \mathfrak{m} \mathcal{O}_X$. To conclude the proof of Step 2, take $h = sh'$.

The proof of Theorem 4.1 is obtained by combining Steps 1 and 2.   □

This proof belongs to Laufer [Lau], except the algebraic argument in Step 2 (which makes Laufer's proof valid in any characteristic). Another proof of Theorem 4.1 can be found in [Ram].

Lemma 4.2. *Let (Y, y) be a normal singularity, with Y an affine surface. Then the $\mathfrak{m}$-adic completion $\hat{\mathcal{O}}_{Y,y}$ of the local ring $\mathcal{O}_{Y,y}$ of (Y, y) is normal. Moreover, if $f : X \to Y$ is a desingularization of (Y, y), then $\hat{f} : \hat{X} = X \times_Y \operatorname{Spec} \hat{\mathcal{O}}_{Y,y} \to \operatorname{Spec} \hat{\mathcal{O}}_{Y,y} = \hat{Y}$ is a desingularization of $(\hat{Y}, \hat{y})$, with $\hat{y}$ the closed point of $\hat{Y}$, the inclusions $f^{-1}(y) \hookrightarrow X$ and $\hat{f}^{-1}(\hat{y}) \hookrightarrow \hat{X}$ are formally equivalent, and $H^1(X, \mathcal{O}_X) \otimes \hat{\mathcal{O}}_{Y,y} \cong H^1(\hat{X}, \mathcal{O}_{\hat{X}})$.*

Proof. By the Zariski–Abhyankar theorem, there exists a desingularization $f : X \to Y$ of (Y, y). Since the commutative square

$$
\begin{array}{ccc}
\hat{X} & \xrightarrow{\;v\;} & X \\
{\scriptstyle \hat{f}} \downarrow & & \downarrow {\scriptstyle f} \\
\hat{Y} & \xrightarrow{\;u\;} & Y
\end{array}
$$

is Cartesian and u is a faithfully flat morphism, we get the isomorphism $H^1(X, \mathcal{O}_X) \otimes \hat{\mathcal{O}}_{Y,y} \cong H^1(\hat{X}, \mathcal{O}_{\hat{X}})$ (see [EGA III, (1.4.15)]). Moreover, the fibers $f^{-1}(y)$ and $\hat{f}^{-1}(\hat{y})$ are canonically identified via the morphism v, and the embeddings $f^{-1}(y) \hookrightarrow X$ and $\hat{f}^{-1}(\hat{y}) \hookrightarrow \hat{X}$ are formally equivalent. Assume for the time being that we have proved that $\hat{X}$ is a nonsingular scheme. Then, since $H^0(X, \mathcal{O}_X) \otimes \hat{\mathcal{O}}_{Y,y} = H^0(\hat{X}, \mathcal{O}_{\hat{X}})$ (loc. cit.), we have $H^0(\hat{X}, \mathcal{O}_{\hat{X}}) = \hat{\mathcal{O}}_{Y,y}$, and therefore $\hat{\mathcal{O}}_{Y,y}$ is normal.

There only remains to prove that $\hat{f} : \hat{X} \to \hat{Y}$ is a desingularization of $(\hat{Y}, \hat{y})$. Since f is an isomorphism between $X \setminus f^{-1}(y)$ and $Y \setminus \{y\}$, $\hat{f}$ is also an isomorphism between $\hat{X} \setminus \hat{f}^{-1}(\hat{y})$ and $\hat{Y} \setminus \{\hat{y}\}$. Now let $\hat{x} \in \hat{X}$ be an arbitrary closed point. If we show that the local ring $\mathcal{O}_{\hat{X},\hat{x}}$ is regular, then it will be clear that $\hat{X}$ is a nonsingular scheme. As f is a proper morphism, $\hat{f}$ is also proper, hence $\hat{f}(\hat{x}) = \hat{y}$. In particular, $\Bbbk(\hat{x}) = \Bbbk(\hat{y}) = \Bbbk(y) = \Bbbk$ ($\Bbbk$ being algebraically closed). Let $x = v(\hat{x})$. Then, since $v^{-1}(x) = \hat{X} \otimes_X \Bbbk(x) = \hat{Y} \otimes_Y \Bbbk(y) = \mathrm{Spec}(\hat{\mathcal{O}}_{Y,y}/\mathfrak{m}_y\hat{\mathcal{O}}_{Y,y}) = \mathrm{Spec}(\Bbbk(y)) = \mathrm{Spec}(\Bbbk)$, we have $\mathfrak{m}_{X,x}\mathcal{O}_{\hat{X},\hat{x}} = \mathfrak{m}_{\hat{X},\hat{x}}$. Since v is flat and the local ring $\mathcal{O}_{X,x}$ is regular, we see that the local ring $\mathcal{O}_{\hat{X},\hat{x}}$ is also regular. $\qquad\square$

Definition 4.3. Let X be a nonsingular surface (not necessarily projective), and let C be an integral curve on X. We say that C is an *exceptional curve of the first kind* if C is projective, $p_a(C) = 0$ and $(C^2) = -1$ (in general the intersection number $(C_1 \cdot C_2)$ of two curves C_1 and C_2 on a surface that is not necessarily projective makes sense if all the components of at least one of the curves are projective).

If C is an exceptional curve of the first kind on a nonsingular surface X, then there exists a unique nonsingular surface Y and a morphism denoted $\mathrm{cont}_C : X \to Y$ with the following properties: the morphism cont_C is proper, it contracts the curve C to a point, and it is an isomorphism outside C. This fact is an immediate consequence of Theorem 3.30, the fact that X is not projective not being an obstruction (we can complete the surface X to a projective surface $\tilde{X}$ and then apply Theorem 3.30 to $\tilde{X}$; indeed, even if $\tilde{X}$ has singularities at infinity, that wouldn't matter).

Definition 4.4. Let $f : X \to Y$ be a desingularization of the normal singularity (Y, y) of dimension 2. We will say that f is a *minimal desingularization* if there are no exceptional curves of the first kind among the components of the reduced fiber $E = f^{-1}(y)_{\mathrm{red}}$.

We remark that every such singularity (Y, y) has at least one minimal desingularization. Indeed, if $f : X \to Y$ is an arbitrary, nonminimal desingularization of (Y, y), let E_1 be a component of E that is an exceptional curve of the first kind. Then the nonsingular surface $\mathrm{cont}_{E_1}(X)$ dominates Y, and therefore we get a desingularization $f_1 : \mathrm{cont}_{E_1}(X) = X_1 \to Y$. If f_1 is minimal we are done. If not, notice that the fiber $f_1^{-1}(y)_{\mathrm{red}}$ has $n-1$ components, where n is the number of components of E. Repeating the process described earlier, one necessarily arrives at a minimal desingularization after a finite number of steps.

The importance of minimal desingularizations is illustrated by the following proposition.

Proposition 4.5. *If (Y, y) is a singularity of the normal surface Y, then there exists a minimal desingularization $f : X \to Y$ of (Y, y). Furthermore, for any other desingularization $f' : X' \to Y$ there exists a unique morphism $u : X' \to X$ such that $f \circ u = f'$. In particular, any two minimal desingularizations are canonically isomorphic. Finally, if f is minimal and its reduced fiber E has irreducible components $E_1, \ldots, E_n$, then $(\omega_X \cdot E_i) \geq 0$ for all $i = 1, \ldots, n$.*

Proof. The existence of f has already been shown. Now let f' be any other desingularization. Let $u = f^{-1} \circ f'$; then u is a birational map. By the structure theorem of birational maps between the two nonsingular projective surfaces X' and X (X' and X may indeed be assumed to be projective, because Y may be assumed, without loss of generality, to be projective, and the morphisms f' and f are also projective), there exists a morphism $v : X'' \to X'$ that is a composite of finitely many quadratic transformations, such that $u \circ v$ is a morphism. Choose v such that it is a composite of a *minimal* number of quadratic transformations. We must show that this minimal number is zero, that is, that u is a morphism. Assume by way of contradiction that the number of quadratic transformations involved in v is strictly positive. Then X'' contains an exceptional curve of the first kind, say, E'', contained in the fiber $(f' \circ v)^{-1}(y)$. If $(u \circ v)(E'')$ would be a point, then $\text{cont}_{E''}(X'')$ would dominate X and X', and that would contradict the minimality of v. Hence $(u \circ v)(E'')$ is a component, say, E_1, of the fiber $f^{-1}(y)$. Since $u \circ v$ is a birational morphism, $u \circ v$ is a composite of a finite number of quadratic transformations. Since by hypothesis f is minimal, we have $(E_1^2) \leq -2$; and because the proper transform of a curve C by a quadratic transformation has self-intersection less than or equal to (C^2), we get $(E''^2) \leq -2$, contradicting the hypothesis that E'' is an exceptional curve of the first kind.

Finally, from the adjunction formula:

$$(\omega_X \cdot E_i) = 2p_a(E_i) - 2 - (E_i^2)$$

and from the fact that $(E_i^2) < 0$ (cf. (2.7)) combined with the inequality $p_a(E_i) \geq 0$ we see that $(\omega_X \cdot E_i)$ could be < 0 if and only if $p_a(E_i) = 0$ and $(E_i^2) = -1$, that is, if and only if E_i were an exceptional curve of the first kind. □

Theorem 4.6. *Let (Y, y) be a normal singularity, with Y an affine surface. Then (Y, y) is a rational singularity if and only if the divisor class group of the $\mathfrak{m}_y$-adic completion $\hat{\mathcal{O}}_{Y,y}$ of the local ring $\mathcal{O}_{Y,y}$ is finite. In that case the order of the divisor class group is equal to the absolute value d of the determinant of the intersection matrix $\|(E_i \cdot E_j)\|$.*

Proof. By (4.2) we may replace (Y, y) with $(\hat{Y}, \hat{y})$, where $\hat{Y} = \operatorname{Spec} \hat{\mathcal{O}}_{Y,y}$; hence we may assume without loss of generality that $Y = \operatorname{Spec}(A)$, with A a normal local Noetherian ring, of dimension 2 and complete in the topology of the maximal ideal, having a desingularization $f : X \to Y$. Moreover, by (4.5) we may assume that the reduced fiber $E = f^{-1}(y)_{\mathrm{red}}$ does not contain exceptional curves of the first kind, that is, that f is a minimal desingularization.

Let Z be the unique cycle with support contained in E such that $(Z \cdot E_i) = -(\omega_X \cdot E_i) \cdot d$, $i = 1, \ldots, n$ (recall that the intersection matrix $\|(E_i \cdot E_j)\|$ is negative definite). Since by (4.5) we have $(\omega_X \cdot E_i) \geq 0$ for every i, we get $Z > 0$ (as in the proof of (3.9)), and therefore Z is greater than or equal to the fundamental cycle of E. In particular, the support of Z is equal to E.

Since A is complete, the natural restriction homomorphism $\operatorname{Pic}(X) \to \operatorname{Pic}(X_E)$ is an isomorphism, where X_E denotes the formal completion of X along E (see [EGA III, (5.1.4)]). But $\operatorname{Pic}(X_E)$ is naturally identified with $\operatorname{proj lim}_m \operatorname{Pic}(mZ)$, where the homomorphisms in the projective system are the restriction homomorphisms (see [Har3, Ch. IV] and [EGA III, Ch. 0, (13.3.1)]).

Considering the standard exact sequence (for $m \geq 1$):

$$0 \longrightarrow \mathcal{O}_X(-mZ) \otimes \mathcal{O}_Z \longrightarrow \mathcal{O}^*_{(m+1)Z} \longrightarrow \mathcal{O}^*_{mZ} \longrightarrow 1,$$

we get the cohomology exact sequence

$$H^1(\mathcal{O}_X(-mZ) \otimes \mathcal{O}_Z) \longrightarrow \operatorname{Pic}((m+1)Z)$$
$$\longrightarrow \operatorname{Pic}(mZ) \longrightarrow H^2(\mathcal{O}_X(-mZ) \otimes \mathcal{O}_Z).$$

But, since $(\mathcal{O}_X(-mZ) \cdot E_i) = -m(Z \cdot E_i) = md(\omega_X \cdot E_i) \geq (\omega_X \cdot E_i)$ for every i, (4.1) gives $H^1(\mathcal{O}_X(-mZ)) = 0$, and therefore the first cohomology group in the exact sequence is zero. Since every H^2 vanishes on Z, we see that all the homomorphisms in the projective system are in fact isomorphisms. In conclusion, we see that the natural restriction homomorphism $\operatorname{Pic}(X) \to \operatorname{Pic}(Z)$ is an isomorphism.

On the other hand, since X is a nonsingular scheme, we have equivalence between Weil divisors and Cartier divisors on X, and therefore we see immediately that the restriction homomorphism $\operatorname{Pic}(X) \to \operatorname{Pic}(U)$, with $U = X \setminus E = Y \setminus \{y\}$, is surjective and its kernel is generated by the classes of the invertible sheaves $\mathcal{O}_X(E_1), \ldots, \mathcal{O}_X(E_n)$. Considering all these facts and the isomorphisms discussed earlier, we see that $\operatorname{Pic}(U)$ can be identified with $\operatorname{Pic}(mZ)/H$, where $m \geq 1$ is arbitrary, and H is the subgroup of $\operatorname{Pic}(mZ)$ generated by the classes of the invertible sheaves $\mathcal{O}_X(E_i) \otimes \mathcal{O}_{mZ}$. But because A is normal and of dimension > 1, $\operatorname{Pic}(U)$ can also be identified naturally with the divisor class group of A. In the commutative diagram with exact rows

$$
\begin{array}{ccccccccc}
0 & \longrightarrow & H & \longrightarrow & \mathrm{Pic}(mZ) & \longrightarrow & \mathrm{Pic}(U) & \longrightarrow & 0 \\
& & \downarrow{\scriptstyle\delta'}\,\cong & & \downarrow{\scriptstyle\delta} & & \downarrow{\scriptstyle\delta''} & & \\
0 & \longrightarrow & \delta(H) & \longrightarrow & \mathbb{Z}^n & \longrightarrow & \mathbb{Z}^n/\delta(H) & \longrightarrow & 0
\end{array}
$$

the homomorphism δ' is an isomorphism, because the intersection matrix $\|(E_i \cdot E_j)\|$ is negative definite. Therefore $\mathrm{Ker}(\delta) = \mathrm{Ker}(\delta'')$.

Assuming now that $\mathrm{Pic}(U)$ is finite, we get $\mathrm{Ker}(\delta)$ finite, and by (3.4) and (3.3), $H^1(\mathcal{O}_{mZ}) = 0$ for every $m \geq 1$, and therefore, by Zariski's holomorphic functions theorem, $H^1(\mathcal{O}_X) = 0$. In other words, in this case (Y, y) is a rational singularity.

Conversely, assuming that (Y, y) is a rational singularity, we have that δ is an isomorphism, by (3.4) and (3.3), and therefore δ'' is also an isomorphism. In other words, $\mathrm{Pic}(U)$ is identified with the cokernel of the following group homomorphism between free Abelian groups of rank n:

$$
\mathbb{Z}E_1 \oplus \cdots \oplus \mathbb{Z}E_n = H \to \mathbb{Z}^n, \quad E_i \mapsto \big((E_i \cdot E_1), \ldots, (E_i \cdot E_n) \big).
$$

Form this we see that $\mathrm{Pic}(U)$ is finite, of order equal to the absolute value of the determinant of the intersection matrix $\|(E_i \cdot E_j)\|$. $\qquad\square$

Corollary 4.7. *Let (Y, y) be a normal singularity of an affine surface Y, such that the $\mathfrak{m}_y$-adic completion of $\mathcal{O}_{Y,y}$ is factorial. Then (Y, y) is a rational singularity.*

4.8. Now let (Y, y) be an isolated normal singularity of dimension $d \geq 2$ of an affine variety $Y = \mathrm{Spec}(A)$, nonsingular outside y. Assume that (Y, y) has a desingularization $f : X \to Y$, that is, a projective morphism f such that:

(a) the reduced fiber $E = f^{-1}(y)_{\mathrm{red}}$ is a (closed) subset of pure codimension 1 in X, with irreducible components $E_1, \ldots, E_n$;

(b) X is a nonsingular variety; and

(c) f is an isomorphism between $X \setminus E$ and $Y \setminus \{y\}$.

By Hironaka's results [Hir], such a desingularization always exists in characteristic zero. If $d = 2$, then, as we have already noted, (Y, y) admits a desingularization over any algebraically closed field, of arbitrary characteristic. Let $\mathfrak{m}$ be the maximal ideal of A corresponding to the point y. If M is an A-module, we denote the $\mathfrak{m}$-adic completion of M by $\hat{M}$. If N is a $\Bbbk$-vector space, then we denote the algebraic dual of N by N'; that is, $N' = \mathrm{Hom}_{\Bbbk}(N, \Bbbk)$. Also, if L is a locally free $\mathcal{O}_Z$-module of finite rank (with Z an arbitrary scheme), then $\check{L}$ denotes the dual of L, that is, $\check{L} = \mathbf{Hom}_{\mathcal{O}_Z}(L, \mathcal{O}_Z)$.

Theorem 4.9. *With the notation introduced earlier, assume that the isolated normal singularity (Y, y) of dimension $d \geq 2$ admits a desingularization $f : X \to Y$. Then for every locally free $\mathcal{O}_X$-module L of finite rank and for every integer $p \geq 0$ there exists a natural $\Bbbk$-vector space isomorphism between $H^p_E(X, L)$ (the local cohomology group with support in $E = f^{-1}(y)_{\mathrm{red}}$) and $(\widehat{H^{d-p}(X, \check{L} \otimes \omega_X)}^{\,\hat{}})'$. For $p < d$ the spaces $H^{d-p}(X, \check{L} \otimes \omega_X)$ are finite-dimensional, and in this case the natural isomorphism becomes $H^p_E(X, L) \cong H^{d-p}(X, \check{L} \otimes \omega_X)'$.*

Proof. Since L is locally free we have $H^0_E(X, L) = 0$, and since $d > \dim(E)$ we have $H^d(X, \check{L} \otimes \omega_X) = 0$ as well. We may therefore assume that $p > 0$. Since Y is affine and f is a proper morphism and an isomorphism outside y, $H^{d-p}(X, \check{L} \otimes \omega_X)$ is finite-dimensional if $p < d$ (because it coincides with $R^{d-p} f_*(\check{L} \otimes \omega_X)$), so that $H^{d-p}(X, \check{L} \otimes \omega_X)^{\hat{}} = H^{d-p}(X, \check{L} \otimes \omega_X)$.

Assuming now that $p > 0$, we have by [Gro4, Theorem 2.8]:

$$H^p_E(X, L) \cong \operatorname*{inj\,lim}_m \operatorname{Ext}^p_{\mathcal{O}_X}(\mathcal{O}_{mE}, L), \tag{4.9.1}$$

where $\mathcal{O}_{mE} = \mathcal{O}_X / I^m$ with $I = \mathcal{O}_X(-E)$, the ideal of E in $\mathcal{O}_X$. We will use the following lemma.

Lemma 4.10. *Let D be an effective divisor on X, defined by the ideal $J \subset \mathcal{O}_X$. Then for every locally free $\mathcal{O}_X$-module L of finite rank on X and for every integer $p > 0$ there exists a canonical isomorphism*

$$H^{p-1}(D, L \otimes J^{-1} \otimes \mathcal{O}_D) \xrightarrow{\;\cong\;} \operatorname{Ext}^p_{\mathcal{O}_X}(\mathcal{O}_D, L).$$

Assume for the time being that Lemma 4.10 is true. Then the isomorphism (4.9.1) becomes

$$H^p_E(X, L) = \operatorname*{inj\,lim}_m H^{p-1}(mE, L \otimes I^{-m} \otimes \mathcal{O}_{mE}). \tag{4.9.2}$$

But $\omega_X \otimes I^{-m} \otimes \mathcal{O}_{mE}$ is isomorphic to the dualizing sheaf ω_{mE} (by (3.11.5)). Applying the Grothendieck–Serre duality theorem on the Cohen–Macaulay projective scheme mE, we get:

$$H^p_E(X, L) = \operatorname*{inj\,lim}_m H^{p-1}(mE, L \otimes \omega_X^{-1} \otimes \omega_{mE})$$

$$= \operatorname*{inj\,lim}_m H^{(d-1)-(p-1)}(mE, \check{L} \otimes \omega_X \otimes \mathcal{O}_{mE})'$$

$$= \left(\operatorname*{proj\,lim}_m H^{d-p}(mE, \check{L} \otimes \omega_X \otimes \mathcal{O}_{mE}) \right)'.$$

Finally, by Zariski's holomorphic functions theorem ([EGA III, (4.2.1)]), this last vector space coincides with the dual of the $\mathfrak{m}$-adic completion of $H^{d-p}(X, \check{L} \otimes \omega_X)$. The proof of Theorem 4.9 is therefore complete, modulo the proof of Lemma 4.10.

Proof of Lemma 4.10. Consider the spectral sequence associated with the composite of Ext (cf. [AK, IV(2.9.1)]):

$$E_2^{pq} = \mathrm{Ext}^p_{\mathcal{O}_D}(\mathcal{O}_D, \mathbf{Ext}^q_{\mathcal{O}_X}(\mathcal{O}_D, L)) \implies \mathrm{Ext}^{p+q}_{\mathcal{O}_X}(\mathcal{O}_D, L).$$

Since $\mathcal{O}_D$ admits the locally free resolution

$$0 \longrightarrow J \longrightarrow \mathcal{O}_X \longrightarrow \mathcal{O}_D \longrightarrow 0,$$

for every $q \geq 2$ we have $\mathbf{Ext}^q_{\mathcal{O}_X}(\mathcal{O}_D, L) = 0$. On the other hand, since L is locally free, we have $\mathbf{Ext}^0_{\mathcal{O}_X}(\mathcal{O}_D, L) = \mathbf{Hom}_{\mathcal{O}_X}(\mathcal{O}_D, L) = 0$ as well, and therefore $E_2^{pq} = 0$ for every $q \neq 1$. In other words, the spectral sequence considered earlier degenerates, and therefore we have canonical isomorphisms:

$$H^{p-1}(D, \mathbf{Ext}^1_{\mathcal{O}_X}(\mathcal{O}_D, L))$$
$$= \mathrm{Ext}^{p-1}_{\mathcal{O}_D}(\mathcal{O}_D, \mathbf{Ext}^1_{\mathcal{O}_X}(\mathcal{O}_D, L)) \xrightarrow{\cong} \mathrm{Ext}^p_{\mathcal{O}_X}(\mathcal{O}_D, L).$$

Finally, $\mathbf{Ext}^1_{\mathcal{O}_X}(\mathcal{O}_D, L) \cong L \otimes J^{-1} \otimes \mathcal{O}_D$ by an identification similar to (3.11.5) (see [AK, I(2.4)]); this completes the proof of Lemma 4.10.   □

Corollary 4.11. *With the same notation as in Theorem 4.1, if Y is a surface, we have $H^1(X, \omega_X) = 0$ and $H^1_E(X, M) = 0$ for every invertible $\mathcal{O}_X$-module M such that $(M \cdot E_i) \leq 0$ for all $i = 1, \ldots, n$.*

Proof. The first claim follows directly from Theorem 4.1. On the other hand, from (4.9) we infer that $\dim H^1_E(X, M) = \dim H^1(X, M^{-1} \otimes \omega_X)$. But, because $(M \cdot E_i) \leq 0$, we have $(M^{-1} \otimes \omega_X \cdot E_i) \geq (\omega_X \cdot E_i)$ for all $i = 1, \ldots, n$, and therefore $H^1(X, M^{-1} \otimes \omega_X) = 0$, again by Theorem 4.1.   □

Corollary 4.12. *With the same notation as in Corollary 4.11, for every invertible $\mathcal{O}_X$-module L such that $(L \cdot E_i) \geq (\omega_X \cdot E_i)$, $i = 1, \ldots, n$, there is an exact sequence*

$$0 \longrightarrow H^0(X, L) \longrightarrow H^0(X \setminus E, L) \longrightarrow H^1(X, L^{-1} \otimes \omega_X)' \longrightarrow 0.$$

In particular, $\dim H^1(X, \mathcal{O}_X) = \dim \left[H^0(X \setminus E, \omega_X)/H^0(X, \omega_X) \right]$, and therefore (Y, y) is a rational singularity if and only if every section of ω_X over $X \setminus E$ can be extended to a section over all of X.

Proof. The exact sequence is obtained from the local cohomology exact sequence

$$0 \longrightarrow H^0(X, L) \longrightarrow H^0(X \setminus E, L) \longrightarrow H^1_E(X, L) \longrightarrow H^1(X, L),$$

because $H^1(X, L) = 0$ by (4.1) and $H^1_E(X, L) = H^1(X, L^{-1} \otimes \omega_X)'$ by (4.9). The last part of the corollary follows by taking $L = \omega_X$.   □

4.13. With the same notation as in Corollary 4.12, if L is a locally free $\mathcal{O}_X$-module of finite rank, then the exact sequence

$$0 \to \Gamma(X, L) \to \Gamma(X \setminus E, L) \to H^1(X, \check{L} \otimes \omega_X)' \to H^1(X, L)$$

(obtained from the local cohomology exact sequence combined with Theorem 4.9) shows that $\Gamma(X \setminus E, L)/\Gamma(X, L)$ is a finite-dimensional space, and

$$\dim\left[\Gamma(X \setminus E, L)/\Gamma(X, L)\right] = \dim \operatorname{Ker}\left[H^1(X, \check{L} \otimes \omega_X)' \to H^1(X, L)\right]$$
$$= \dim \operatorname{Ker}\left[R^1 f_*(\check{L} \otimes \omega_X)' \to R^1 f_*(L)\right].$$

From this we infer that $\dim\left[\Gamma(X \setminus E, L)/\Gamma(X, L)\right]$ is independent of Y, in the following sense: if Y' is an affine open subset of Y containing y, and if $X' = f^{-1}(Y')$, then

$$\dim\left[\Gamma(X \setminus E, L)/\Gamma(X, L)\right] = \dim\left[\Gamma(X' \setminus E, L)/\Gamma(X', L)\right]. \qquad (4.13.1)$$

On the other hand, let $g : \tilde{X} \to X$ be the blow-up of X centered at the (reduced) point $u \in E$, and let C be the exceptional curve $g^{-1}(u)$, defined by the invertible ideal $J \subset \mathcal{O}_{\tilde{X}}$. For each $s \geq 0$ put $L_s = g^*(L) \otimes J^{-s}$. Then by the projection formula we have $g_*(L_s) \cong L \otimes g_*(J^{-s}) \cong L$, because for every $s \geq 0$ we have $g_*(J^{-s}) \cong \mathcal{O}_X$. (*Proof:* The problem being local around u, we may assume that X is affine. We argue by induction on s, the case $s = 0$ being trivial. In general, the exact sequence

$$0 \longrightarrow J^{-(s-1)} \longrightarrow J^{-s} \longrightarrow \mathcal{O}_C(-s) \longrightarrow 0$$

shows that the map $\Gamma(\tilde{X}, J^{-(s-1)}) \to \Gamma(\tilde{X}, J^{-s})$ is an isomorphism, because $\Gamma(C, \mathcal{O}_C(-s)) = 0$ for $s > 0$.)

Therefore we get $\Gamma(X \setminus E, L) \cong \Gamma(\tilde{X} \setminus \tilde{E} \cup C, L_s)$ and $\Gamma(X, L) \cong \Gamma(\tilde{X}, L_s)$, where $\tilde{E}$ is the proper transform of E by g. In other words, we have:

$$\dim\left[\Gamma(X \setminus f^{-1}(y), L)/\Gamma(X, L)\right]$$
$$= \dim\left[\Gamma(\tilde{X} \setminus (f \circ g)^{-1}(y), L_s)/\Gamma(\tilde{X}, L_s)\right]. \qquad (4.13.2)$$

If $L = \omega_X$, then we have the well-known formula $g^*(\omega_X) \otimes J^{-1} = \omega_{\tilde{X}}$, and therefore for every $q \geq 1$ we have $g^*(\omega_X^q) \otimes J^{-q} = \omega_{\tilde{X}}^q$. From the preceding discussion we see that for each $q \geq 1$ the integer

$$r_y(q) = \dim\left[\Gamma(X \setminus E, \omega_X^q)/\Gamma(X, \omega_X^q)\right]$$

is a numerical invariant of the singularity (Y, y) (to show that it does not depend on the desingularization, we use the fact that any other desingularization of (Y, y) is obtained from f by a composition of quadratic transformations and their inverses). In the complex-analytic case the invariants $r_y(q)$ ($q \geq 1$) were introduced and studied by Laufer [Lau] and Knöller [Knö].

Here are several other consequences of Theorems 4.1 and 4.9.

Corollary 4.14. *Let (Y, y) be a normal singularity, with Y an affine surface, and let $f : X \to Y$ be the minimal desingularization of (Y, y), cf. (4.5). Then $H^1(X, \omega_X^q) = 0$ and $r_y(q) = \dim H^1(X, \omega_X^{1-q})$ for every $q \geq 1$.*

Proof. By (4.5) we have $(\omega_X \cdot E_i) \geq 0$ for all i, and therefore $(\omega_X^q \cdot E_i) \geq (\omega_X \cdot E_i)$ for all i. Then (4.1) gives $H^1(X, \omega_X^q) = 0$. Then the formula for $r_y(q)$ follows from (4.12). $\qquad\square$

Corollary 4.15. *With the same notation as in the previous corollary, (Y, y) is a rational singularity if and only if $r_y(1) = 0$. If (Y, y) is a rational double point, then $r_y(q) = 0$ for all $q \geq 1$. Conversely, if y is a normal, nonregular point on Y and $r_y(1) = r_y(q_0) = 0$ for at least one $q_0 \geq 3$, then (Y, y) is a rational double point.*

Proof. Let $f : X \to Y$ be the minimal desingularization of (Y, y). The first statement follows from 4.14. If (Y, y) is a rational double point, then by (3.29) and (3.31) we have $p_a(E_i) = 0$ and $(E_i^2) = -2$, and therefore $(\omega_X \cdot E_i) = 0$, for every $i = 1, \ldots, n$. From this and (4.1) we get $H^1(X, \omega_X^{1-q}) = 0$, and therefore $r_y(q) = 0$, for all $q \geq 1$.

Conversely, $r_y(1) = 0$ means exactly that (Y, y) is a rational singularity. The exact sequence

$$0 \to I_i \otimes \omega_X^{1-q} \to \omega_X^{1-q} \to (\omega_{E_i} \otimes I_i / I_i^2)^{1-q} = L_{i,q} \to 0$$

$(q \geq 1, i = 1, \ldots, n)$, where I_i is the invertible ideal of E_i in X, gives the cohomology exact sequence:

$$H^1(X, \omega_X^{1-q}) \to H^1(E_i, L_{i,q}) \to H^2(X, I_i \otimes \omega_X^{1-q}) = 0.$$

From the hypothesis and (4.14) we see that the first cohomology group in the last exact sequence is zero for some $q_0 \geq 3$; therefore $H^1(E_i, L_{i,q_0}) = 0$. On the other hand, $\deg(L_{i,q_0}) = (1 - q_0) \cdot (-2 - (E_i^2)) \leq 0$, because (Y, y) being a rational singularity we have $p_a(E_i) = 0$, and f being a minimal desingularization of (Y, y) we also have $(E_i^2) \leq -2$ for every i. If (Y, y) were not a rational double point, then $-2 - (E_i^2) > 0$ for some i, and therefore $\deg(L_{i,q_0}) \leq -2$, because $q_0 \geq 3$. As E_i is isomorphic to the projective line, this would contradict the vanishing of $H^1(E_i, L_{i,q_0})$. $\qquad\square$

4.16. Let (Y, y) be a normal singularity of the affine surface Y, and let $f : X \to Y$ be the minimal desingularization of (Y, y). If (Y, y) is a Gorenstein singularity, then the dualizing sheaf ω_Y is invertible in a neighborhood of y; then let $L = \omega_X \otimes f^*(\omega_Y)^{-1}$. Since L is trivial on $X \setminus E$, L is necessarily of the form $L \cong I_1^{\alpha_1} \otimes \cdots \otimes I_n^{\alpha_n}$, where as before $I_i = \mathcal{O}_X(-E_i)$, and $\alpha_i \in \mathbb{Z}$, $i = 1, \ldots, n$. As $(L \cdot E_i) = (\omega_X \cdot E_i) - (f^*(\omega_Y) \cdot E_i) = (\omega_X \cdot E_i) \geq 0$

for all i (cf. (4.5), since f is minimal), we have $\alpha_i \geq 0$ for all i (see the proof of Theorem 3.9). Thus we have:

$$\omega_X \cong f^*(\omega_Y) \otimes I_1^{\alpha_1} \otimes \cdots \otimes I_n^{\alpha_n}, \qquad \alpha_i \geq 0. \qquad (4.16.1)$$

Replacing Y with a smaller affine set containing y and on which ω_Y is trivial, formula (4.16.1) becomes:

$$\omega_X \cong I_1^{\alpha_1} \otimes \cdots \otimes I_n^{\alpha_n}, \qquad \alpha_i \geq 0. \qquad (4.16.1)$$

Consider then the effective divisor $D = \sum_{i=1}^{n} \alpha_i E_i$. If $D = 0$, then ω_X is trivial, and the adjunction formula, together with (3.31), shows that (Y, y) is a rational double point.

Assume now that $D > 0$. Then, as one can see immediately, $\alpha_i > 0$ for all i. By (3.11.5), the dualizing sheaf ω_D of the curve D is given by the formula:

$$\omega_D \cong \omega_X \otimes \omega_X^{-1} \otimes \mathcal{O}_D \cong \mathcal{O}_D. \qquad (4.16.3)$$

The exact sequence

$$0 \longrightarrow \omega_X \longrightarrow \mathcal{O}_X \longrightarrow \mathcal{O}_D \longrightarrow 0$$

induces the cohomology exact sequence:

$$0 \to \Gamma(X, \omega_X) \to \Gamma(X, \mathcal{O}_X) \to \Gamma(D, \mathcal{O}_D)$$
$$\to H^1(X, \omega_X) = 0 \to H^1(X, \mathcal{O}_X) \to H^1(D, \mathcal{O}_D) \to 0, \qquad (4.16.4)$$

which shows that $r_y(1) = \dim H^1(D, \mathcal{O}_D) = \dim \Gamma(D, \mathcal{O}_D) > 0$ (using duality on D).

Conversely, assume that (Y, y) is a normal singularity such that, if $f : X \to Y$ is its minimal desingularization, then there exists an effective divisor $D > 0$ with support equal to E and whose dualizing sheaf ω_D is trivial. We will then show that (Y, y) is a Gorenstein singularity. Indeed, the short exact sequence

$$0 \to \omega_X \to \omega_X \otimes I^{-1} \to \omega_D \cong \mathcal{O}_D \to 0$$

(with $I = \mathcal{O}_X(-D)$) induces the cohomology exact sequence:

$$\Gamma(X, \omega_X \otimes I^{-1}) \to \Gamma(D, \mathcal{O}_D) \to H^1(X, \omega_X) = 0 \quad \text{(by Corollary 4.11).}$$

Therefore there exists a section $s \in \Gamma(X, \omega_X \otimes I^{-1})$ such that $s|_D = 1$. Since $\mathrm{Supp}(D) = E$, we see that in a neighborhood U of E of the form $U = f^{-1}(V)$, with V a suitable affine open neighborhood of y in Y, we have $\omega_X|_U \cong \mathcal{O}_X(-D)|_U$. Since we may assume that $V \setminus \{y\}$ is nonsingular and $U \setminus E \cong V \setminus \{y\}$, we get $\Omega^2_{Y/k}|_{V \setminus \{y\}} \cong \mathcal{O}_Y|_{V \setminus \{y\}}$. By 3.13, (Y, y) is then a Gorenstein singularity.

If $D' > 0$ is another strictly positive divisor with $\text{Supp}(D') = E$ whose dualizing sheaf is trivial, then the earlier argument shows that, at least in some neighborhood of E, we have $\omega_X \cong \mathcal{O}_X(-D')$, and therefore $\mathcal{O}_X(D) \cong \mathcal{O}_X(D')$. Since the supports of D and D' are equal to E and the intersection matrix $\|(E_i \cdot E_j)\|$ is negative definite, this already implies that $D = D'$. Finally, in view of (3.31), the preceding discussion leads to the following theorem.

Theorem 4.17. *Let (Y,y) be a normal singularity of dimension 2 such that y is not a regular point on Y, and let $f : X \to Y$ be the minimal desingularization of (Y, y), with $E = f^{-1}(y)_{\text{red}}$ having irreducible components $E_1, \ldots, E_n$. Then (Y, y) is a Gorenstein singularity if and only if one of the following two conditions holds:*

(i) *(Y, y) is a rational double point; or*

(ii) *there exists a strictly positive divisor $D > 0$ such that $\text{Supp}(D) = E$ whose dualizing sheaf ω_D is trivial.*

Furthermore, if (ii) holds, then the divisor D is uniquely determined by the fact that ω_D is trivial, and we have $r_y(1) = \dim \Gamma(D, \mathcal{O}_D)$ and $\omega_X = f^*(\omega_Y) \otimes \mathcal{O}_X(-D)$.

The divisor D in case (ii) of Theorem 4.17 could reasonably be called *the Gorenstein divisor of the singularity (Y, y).* In case (i) we may consider the Gorenstein divisor to be zero.

Corollary 4.18. (a) *If (Y,y) is a Gorenstein rational singularity, then the multiplicity of (Y, y) is ≤ 2.*

(b) *If (Y, y) is a rational singularity whose local ring $\mathcal{O}_{Y,y}$ is factorial, then the multiplicity of (Y, y) is ≤ 2.*

Proof. The first statement follows directly from Theorem 4.17, for in case (ii) we have $r_y(1) = \dim \Gamma(D, \mathcal{O}_D) > 0$. The second statement follows from the first via Murthy's theorem (Theorem 3.14). $\qquad\square$

The following result is a characterization of rational double points.

Corollary 4.19. *If $f : X \to Y$ is the minimal desingularization of a normal singularity (Y, y) of dimension 2 (with y a nonregular point on Y), then the following conditions are equivalent:*

(a) *(Y, y) is a rational double point;*

(b) *(Y, y) is a Gorenstein singularity and $f_*(\omega_X) = \omega_Y$; and*

(c) *(Y, y) is a Gorenstein rational singularity.*

Proof. The equivalence of (a) and (c) is already contained in Corollary 4.18. On the other hand, according to (4.12), (Y, y) is a rational singularity if and only if $\Gamma(X \setminus E, \omega_X) = \Gamma(X, \omega_X)$, which by (3.12) can be rewritten as $f_*(\omega_X) = \omega_Y$ (namely, $\Gamma(X \setminus E, \omega_X) \cong \Gamma(Y \setminus \{y\}, \omega_Y) = \Gamma(Y, \omega_Y)$, the last equality being due to (3.12)). This proves the equivalence of (b) and (c). $\qquad\square$

Corollary 4.20. *With the same hypotheses and notation as in Theorem 4.17, assume that case* (ii) *holds. Then for every $q \geq 1$ we have the formula:*

$$r_y(q) = -(D^2) \cdot \frac{q(q-1)}{2} + r_y(1).$$

Proof. Shrinking Y if necessary, we may assume that $\omega_X = \mathcal{O}_X(-D)$, where $f : X \to Y$ is the minimal desingularization of (Y, y). The exact sequence

$$0 \to \omega_X \to \omega_X \otimes \mathcal{O}_X(qD) = \omega_X^{1-q} \to \omega_{qD} \to 0,$$

together with (4.14), shows that $\dim H^1(X, \omega_X^{1-q}) = \dim H^1(qD, \omega_{qD})$, whence by duality on the curve qD we get $r_y(q) = \dim \Gamma(qD, \mathcal{O}_{qD})$. On the other hand, the exact sequence

$$0 \longrightarrow \omega_X^q \longrightarrow \mathcal{O}_X \longrightarrow \mathcal{O}_{qD} \longrightarrow 0$$

and (4.14) show that $\dim H^1(qD, \mathcal{O}_{qD}) = r_y(1)$. Computing the arithmetic genus of qD and noting that $-D$ is a canonical divisor on X, we have

$$1 - \chi(\mathcal{O}_{qD}) = \frac{1}{2}(qD \cdot (q-1)D) + 1,$$

which implies the formula given in the corollary. $\qquad\square$

4.21. *Examples of Nonrational Gorenstein Singularities.* A normal singularity (Y, y) of dimension 2 is *simply elliptic* if in its minimal desingularization $f : X \to Y$ the reduced fiber $E = f^{-1}(y)_{\mathrm{red}}$ is a nonsingular elliptic curve. Such singularities can be obtained as follows: let E be an arbitrary elliptic curve, and let L be an arbitrary invertible $\mathcal{O}_E$-module with $\deg(L) > 0$. Let $X = V(L)$, the geometric vector bundle of rank 1 associated to L (see [EGA II]), and embed E in X as the zero section, so that—as one can readily see—the conormal sheaf of E in X is isomorphic to L. By Grauert's contractability criterion ([EGA II, §8]) there exists a proper morphism $f : X \to Y$ that contracts E to a normal point $y \in Y$ and such that f is an isomorphism between $X \setminus E$ and $Y \setminus \{y\}$. Then (Y, y) is a simply elliptic singularity, and $f : X \to Y$ is its minimal desingularization.

Theorem 4.22. *Let (Y, y) be a normal singularity of an affine surface Y, such that the $\mathfrak{m}_y$-adic completion of the local ring $\mathcal{O}_{Y,y}$ is a factorial, nonregular ring. Then (Y, y) is a rational double point, and if $f : X \to Y$ is* its minimal desingularization, then the graph associated to the reduced fiber $E = f^{-1}(y)_{\text{red}}$, with irreducible components $E_1, \ldots, E_n$ (with $p_a(E_i) = 0$ and $(E_i^2) = -2$, $i = 1, \ldots, n$) corresponds to the Dynkin diagram (E_8) in Theorem 3.32.

Proof. This follows immediately from Theorems 4.6 and 3.32, Corollary 4.18, and the fact (easy to verify) that from among the five Dynkin diagrams in Theorem 3.32, (E_8) is the only one to have the absolute value of the determinant of the matrix $\|(E_i \cdot E_j)\|$ equal to 1.   □

EXERCISES

Exercise 4.1. Let (Y, y) be a normal surface singularity such that, if $f : X \to Y$ is the minimal desingularization of (Y, y) and $E = f^{-1}(y)_{\text{red}}$, then $E \cong \mathbb{P}^1$ and $(E^2) = -n$, with $n \geq 2$. Prove that (Y, y) is analytically equivalent to the vertex (Y', y') of the cone Y' over the normal rational curve of degree n in $\mathbb{P}^n$, that is, $\hat{\mathcal{O}}_{Y,y} \cong \hat{\mathcal{O}}_{Y',y'}$.

Exercise 4.2. In Exercise 4.1, compute the divisor class group of $\hat{\mathcal{O}}_{Y,y}$.

Exercise 4.3. Generalize Exercises 4.1 and 4.2 to the case of an n-dimensional isolated normal singularity (Y, y) that admits a desingularization $f : X \to Y$ such that $f^{-1}(y)_{\text{red}} \cong \mathbb{P}^{n-1}$ and $\mathcal{O}_X(-Y)/\mathcal{O}_X(-2Y) \cong \mathcal{O}_{\mathbb{P}^{n-1}}(s)$ with $n \geq 2$ and $s \geq 2$. Prove that (Y, y) is analytically equivalent to the vertex of the cone over the s-fold Veronese embedding of $\mathbb{P}^{n-1}$, and show that the divisor class group of $\hat{\mathcal{O}}_{Y,y}$ is isomorphic to $\mathbb{Z}/s\mathbb{Z}$. [*Hint:* There are several ways of proving the assertion about the analytic equivalence. One way, in the complex-analytic case, is to use a criterion of Grauert, cf. [Gra*]. The algebraic way, which works in arbitrary characteristic, is in the spirit of Lemma 4.2, using Grothendieck's existence theorem [EGA III, (5.2)].]

Exercise 4.4. Let $\Bbbk$ be a field of characteristic different from 2, 3, and 5. By Exercise 3.6, the singularity (at the origin) of the affine surface with equation $x^2 + y^3 + z^5 = 0$ is a rational double point of type (E_8). Check that in this case the absolute value of the determinant of the intersection matrix $\|(E_i \cdot E_j)\|_{i,j=1,\ldots,8}$ equals 1. Then, using Theorem 4.6, deduce the following result of Mumford, cf. [Mum5]: the ring $\Bbbk[[X, Y, Z]]/(X^2 + Y^3 + Z^5)$ is a UFD (unique factorization domain).

Exercise 4.5. The fact that $\Bbbk[[X, Y, Z]]/(X^2 + Y^3 + Z^5)$ is a UFD implies that $A = \Bbbk[X, Y, Z]/(X^2 + Y^3 + Z^5)$ is also a UFD (char($\Bbbk$) $\neq 2, 3, 5$). Prove this latter fact directly, by methods of elementary algebra. [*Hint:* Let x, y, and z be the images of X, Y, and Z in the quotient ring A. Show that z is a prime element in A and that $A[z^{-1}]$ is a UFD. Then A is a UFD by an elementary lemma of Nagata.]

Exercise 4.6. Let E be the diagonal of the product $X = C \times C$, where C is a nonsingular projective curve of genus $g \geq 2$. Show that $\omega_{2E} \cong \mathcal{O}_{2E}$. (*Note:* By Exercise 1.2, $(E^2) = 2 - 2g < 0$; assuming that E can be contracted projectively—which is, in fact, true, cf. Exercise 14.2—it follows by Theorem 4.17 that the singularity obtained by contraction is Gorenstein.)

Exercise 4.7. Let $C \subset \mathbb{P}^2$ be an irreducible, reduced, singular curve of degree 3, and blow up $\mathbb{P}^2$ at $n \geq 10$ different nonsingular points of C to get the surface X. Let E be the proper transform of C under the blow-up morphism $X \to \mathbb{P}^2$. Then $(E^2) = 9 - n$. Assume that E contracts projectively to a normal singularity (Y, y) (by a result of Grauert [Gra*], if $k = \mathbb{C}$, then there exists at least an analytic contraction $X \to Y$ of E to a point $y \in Y$, because $(E^2) < 0$). Prove that the singularity (Y, y) is Gorenstein.

Bibliographic References. Theorem 4.1 was proved by Laufer [Lau] in the case $k = \mathbb{C}$ and independently by Ramanujam [Ram] in the general case. The proof given here follows Lipman's idea, with a modification made by the author to make the proof valid in any characteristic. Lemma 4.2, Theorem 4.6, and Corollary 4.7 are from [Lip]. Theorem 4.9 and its consequences and Theorem 4.17 were proved by the author, cf. [Băd1] (see also [Rei3*] for similar ideas). The idea of using the numerical invariants $r_y(q)$ was originated by Laufer [Lau] and Knöller [Knö].

5

Noether's Formula, the Picard Scheme, the Albanese Variety, and Plurigenera

From this point on by *surface* we mean a nonsingular projective surface X defined over an algebraically closed field $\Bbbk$ of arbitrary characteristic. When we have to deal with surfaces with singularities, we state that explicitly (for example: let X be a normal surface...).

We have already seen that if X is a surface and L is an invertible $\mathcal{O}_X$-module, then the Riemann–Roch theorem is written as:

$$\chi(L) = \frac{1}{2}(L \cdot L \otimes \omega_X^{-1}) + \chi(\mathcal{O}_X).$$

In a more precise form, the Riemann–Roch theorem expresses $\chi(\mathcal{O}_X)$ in terms of the Chern classes of the surface X. Namely, one has Noether's formula:

$$\chi(\mathcal{O}_X) = \frac{1}{12}(c_1^2 + c_2) = \frac{1}{12}((K^2) + c_2), \qquad (*)$$

where K is the canonical class of X (corresponding to the invertible $\mathcal{O}_X$-module ω_X). Moreover, c_2 is equal to the Euler characteristic of X, and is given by the formula

$$c_2 = b_0 - b_1 + b_2 - b_3 + b_4, \qquad (**)$$

where b_i $(i = 0, 1, 2, 3, 4)$ are the Betti numbers of the surface X. If the ground field $\Bbbk$ is the complex field $\mathbb{C}$, then $b_i = \dim_{\mathbb{C}} H^i(X, \mathbb{C})$, where $H^i(X, \mathbb{C})$ is the singular cohomology space of X in dimension i, with complex coefficients. If $\Bbbk$ is an arbitrary field, then the Betti numbers can be defined via l-adic cohomology (where l is a prime integer, distinct from the characteristic of $\Bbbk$). These numbers do not depend on the choice of l, and for $\Bbbk = \mathbb{C}$ they coincide with the classical Betti numbers. As in classical algebraic topology, the Poincaré duality theorem holds: $b_i = b_{4-i}$ for $i = 0, 1$. Thus equation $(**)$ becomes (as $b_0 = b_4 = 1$):

$$c_2 = 2 - 2b_1 + b_2.$$

If we denote by

$$q = \dim H^1(X, \mathcal{O}_X) \quad \text{and}$$
$$p_g = \dim H^2(X, \mathcal{O}_X) = \dim H^0(X, \omega_X)$$

the irregularity and, respectively, the geometric genus of X, then formula $(*)$ becomes (using the equation derived earlier):

$$12 - 12q + 12p_g = (K^2) + 2 - 2b_1 + b_2,$$

or

$$10 - 8q + 12p_g = (K^2) + b_2 + 2\Delta, \quad \text{where } \Delta = 2q - b_1. \qquad (***)$$

We will use Noether's formula in the form $(***)$ very often in the classification of surfaces.

For more details on Noether's formula the reader is referred to [BS] (together with [Gro2]), and for details on the Betti numbers, to [Igu] and [SGA $4\frac{1}{2}$] (= [Del]).

Let $\text{Pic}(X)$ denote the Picard group of the surface X, and let $\text{Pic}^0(X)$ (resp. $\text{Pic}^\tau(X)$) denote the subgroup of $\text{Pic}(X)$ consisting of all isomorphism classes of invertible $\mathcal{O}_X$-modules algebraically (resp. numerically) equivalent to zero. Recall that two invertible $\mathcal{O}_X$-modules L_1 and L_2 are *algebraically equivalent* (notation: $L_1 \approx L_2$) if there exists a connected scheme T, two closed points $t_1, t_2 \in T$, and an invertible $\mathcal{O}_{X \times T}$-module L such that $L|_{X \times \{t_1\}} \cong L_1$ and $L|_{X \times \{t_2\}} \cong L_2$. An invertible $\mathcal{O}_X$-module L' is algebraically equivalent to zero if $L' \approx \mathcal{O}_X$. On the other hand, L_1 and L_2 are *numerically equivalent* (notation: $L_1 \equiv L_2$) if $(L_1 \cdot C) = (L_2 \cdot C)$ for every integral curve C on X, and L' is numerically equivalent to zero if $L' \equiv \mathcal{O}_X$. Clearly, $\text{Pic}^0(X) \subseteq \text{Pic}^\tau(X)$.

Given the surface (or any nonsingular projective variety) X, we may consider the Picard functor

$$\mathbf{Pic}_X : \text{Sch}_{\Bbbk} \to \text{Ens},$$

defined on the category $\text{Sch}_{\Bbbk}$ of all algebraic $\Bbbk$-schemes (i.e., of finite type over $\Bbbk$), with values in the category Ens of sets, as follows: if $T \in \text{Sch}_{\Bbbk}$ is an algebraic $\Bbbk$-scheme, then $\mathbf{Pic}_X(T)$ is the set of all T-isomorphism classes of invertible $\mathcal{O}_{X \times T}$-modules, two invertible $\mathcal{O}_{X \times T}$-modules L and L' being T-isomorphic if there exists an invertible $\mathcal{O}_T$-module M such that $L \cong L' \otimes p_2^*(M)$, where p_2 is the canonical projection of $X \times T$ onto T. If $f : T' \to T$ is a morphism of k-schemes, then $\mathbf{Pic}_X(f) : \mathbf{Pic}_X(T) \to \mathbf{Pic}_X(T')$ is defined by $\mathbf{Pic}_X(f)(\text{class of } L) = \text{class of } (\mathrm{id}_X \times f)^*(L)$ (thus $\mathbf{Pic}_X$ is a contravariant functor). It is easy to see that $\mathbf{Pic}_X$ has a subfunctor $\mathbf{Pic}_X^0$, defined by

$$\mathbf{Pic}_X^0(T) = \{L \in \mathbf{Pic}_X(T) \mid L_t = L|_{X \times \{t\}} \approx \mathcal{O}_X \text{ for all } t \in T\}$$

(with slightly abusive notation). Then the functors $\mathbf{Pic}_X$ and $\mathbf{Pic}_X^0$ are representable, and their representing pairs, denoted by $(\underline{\mathrm{Pic}}(X), L)$ and $(\underline{\mathrm{Pic}}^0(X), L')$ have the following properties:

- $\underline{\mathrm{Pic}}(X)$ is a (topological) disjoint union of an infinite family of proper k-schemes.

- L is an invertible sheaf on $X \times \underline{\mathrm{Pic}}(X)$, and the k-rational points of $\underline{\mathrm{Pic}}(X)$ are in a natural bijective correspondence with the elements of the Picard group of X, $\mathrm{Pic}(X)$.

- $\underline{\mathrm{Pic}}^0(X)$ is the connected component of the origin 0 in $\underline{\mathrm{Pic}}(X)$, where 0 is the point of $\underline{\mathrm{Pic}}(X)$ corresponding to the class of $\mathcal{O}_X$ in the bijective correspondence mentioned earlier, and $\underline{\mathrm{Pic}}^0(X)$ is a proper scheme over k.

- The sheaf L' is the restriction of L to $X \times \underline{\mathrm{Pic}}^0(X)$, and L' is uniquely determined up to a $\underline{\mathrm{Pic}}^0(X)$-isomorphism.

If $M = L'|_{\{x_0\} \times \underline{\mathrm{Pic}}^0(X)}$, with x_0 a fixed closed point on X, then replacing L' with $L' \otimes p_2^*(M)^{-1}$ we get a canonical isomorphism class, which represents the $\underline{\mathrm{Pic}}^0(X)$-isomorphism class of L', and a representative L' from this class is called the *Poincaré invertible sheaf* on $X \times \underline{\mathrm{Pic}}^0(X)$ (or the *Poincaré line bundle* on $X \times \underline{\mathrm{Pic}}^0(X)$). Thus we see that the Poincaré line bundle L' is uniquely determined up to isomorphism, and it is completely characterized by the following properties (which in particular express the fact that the pair $(\underline{\mathrm{Pic}}^0(X), L')$ represents the functor $\mathbf{Pic}_X^0$):

(i) $L'|_{\{x_0\} \times \underline{\mathrm{Pic}}^0(X)} \cong \mathcal{O}_{\underline{\mathrm{Pic}}^0(X)}$, and for every closed (i.e., k-rational) point $a \in \underline{\mathrm{Pic}}^0(X)$ the invertible $\mathcal{O}_X$-module $L_a' = L'|_{X \times \{a\}}$ is algebraically equivalent to zero. More precisely, the association $a \mapsto$ class of (L_a') defines a bijection between the set of closed points of $\underline{\mathrm{Pic}}^0(X)$ and the set $\mathrm{Pic}^0(X)$ of all isomorphism classes of invertible $\mathcal{O}_X$-modules that are algebraically equivalent to zero (the bijection already mentioned earlier).

(ii) If T is an arbitrary algebraic k-scheme and L'' is an invertible $\mathcal{O}_{X \times T}$-module such that $L''|_{\{x_0\} \times T} \cong \mathcal{O}_T$ and the invertible $\mathcal{O}_X$-module

$L''_t = L''|_{X \times \{t\}}$ is algebraically equivalent to zero for every $t \in T$, then there is a unique morphism of k-schemes $f : T \to \underline{\mathrm{Pic}}^0(X)$ such that L'' is isomorphic to $(\mathrm{id}_X \times f)^*(L')$.

Another remarkable property of the scheme $\underline{\mathrm{Pic}}^0(X)$ is that the Zariski tangent space $T_{\underline{\mathrm{Pic}}^0(X),0}$ of $\underline{\mathrm{Pic}}^0(X)$ at the origin is canonically isomorphic to the k-vector space $H^1(X, \mathcal{O}_X)$.

Since the functor $\mathbf{Pic}^0_X$ factors through the category of Abelian groups (the Abelian group structure on $\mathbf{Pic}^0_X(T)$ being induced by the tensor product of invertible $\mathcal{O}_{X \times T}$-modules), the scheme $\underline{\mathrm{Pic}}^0(X)$ is in fact a commutative group scheme. If $\underline{\mathrm{Pic}}^0(X)$ is a reduced scheme, then it is an Abelian variety. Note that a theorem of Cartier (see [Mum1, Lecture 25]) says that every commutative group scheme is reduced when the characteristic of k is equal to zero.

In what follows the group scheme $\underline{\mathrm{Pic}}^0(X)$ will be called the *Picard scheme* of the variety X, and the commutative algebraic group $\underline{\mathrm{Pic}}^0(X)_{\mathrm{red}}$ will be called the *(classical) Picard variety* of X (on which the operation is induced by the group scheme operation on $\underline{\mathrm{Pic}}^0(X)$). As we have already mentioned, if X is a nonsingular projective variety, then the scheme $\underline{\mathrm{Pic}}^0(X)$ is proper over k, so that in this case the classical Picard variety of X is an Abelian variety. The functor $\mathbf{Pic}^0_X$ is representable even if X is not regular (but is still projective), but in that case $\underline{\mathrm{Pic}}^0(X)$ may be a nonproper scheme over k. This is the case, for example, for algebraic curves with singularities.

Some references for the facts we have stated here without proofs are: [Gro3] for the general theory of the Picard scheme, [Mum1] for the Picard scheme of a surface, [Mum2] for the Picard scheme of an Abelian variety, and [Ser3] and [Oor] for the Picard scheme of a curve with singularities.

Theorem 5.1. *If X is a surface, then the following formula (Noether's formula) holds:*

$$10 - 8q + 12p_g = (K^2) + b_2 + 2\Delta,$$

where $\Delta = 2q - b_1 = 2(q - s)$, with $q = \dim H^1(X, \mathcal{O}_X)$ and $s = $ the dimension of the classical Picard variety of X. Moreover, $\Delta = 0$ if the characteristic of k is zero, and when the characteristic of k is positive we have the inequalities $0 \leq \Delta \leq 2p_g$.

Proof. By the definition of b_1 we have

$$(\mathbb{Z}/l\mathbb{Z})^{b_1} = H^1_{\text{ét}}(X, \mathbb{Z}/l\mathbb{Z}) = \{a \in \mathrm{Pic}(X) \mid l \cdot a = 0\}$$
$$= \{a \in \mathrm{Pic}^0(X) \mid l \cdot a = 0\} = (\mathbb{Z}/l\mathbb{Z})^{2s},$$

because $\mathrm{Pic}^0(X)$ is the underlying Abelian group of the Picard variety of X, which is an Abelian variety of dimension s, and we can use the following elementary property of Abelian varieties: the set of points of order l (with l relatively prime to $p = \text{char}(\Bbbk)$) of an Abelian variety A of dimension s is in a natural bijection with $(\mathbb{Z}/l\mathbb{Z})^{2s}$ (see [Mum2, §6, Proposition 2]). Therefore $b_1 = 2s$ (this is precisely the definition given by Igusa in [Igu]). Thus

$$\Delta = 2q - b_1 = 2(q - s) \geq 0,$$

for q is the dimension of the tangent space to the s-dimensional scheme $\underline{\text{Pic}}^0(X)$ at the origin. The preceding equation can be rewritten as:

$$\Delta = 2(\dim T_{\underline{\text{Pic}}^0(X),0} - \dim T_{\underline{\text{Pic}}^0(X)_{\text{red}},0}).$$

But these two tangent spaces are related by the Bockstein operations (see [Mum1, Lecture 27]), and we have:

$$\begin{aligned} q - s &= \dim T_{\underline{\text{Pic}}^0(X),0} - \dim T_{\underline{\text{Pic}}^0(X)_{\text{red}},0} \\ &= \dim H^1(X, \mathcal{O}_X) - \dim \left(\cap_{i=1}^{\infty} \text{Ker}(\beta_i) \right) \\ &\leq \dim \left(\cup_{i=1}^{\infty} \text{Im}(\beta_i) \right) \\ &\leq \dim H^2(X, \mathcal{O}_X) \\ &= p_g, \end{aligned}$$

where the β_i are the Bockstein operators on X; thus we get the inequalities for Δ stated in the theorem. The fact that $\Delta = 0$ in characteristic zero follows from Cartier's theorem, and Noether's formula follows from formula $(***)$ on page 70. □

Definition 5.2. Let X be a nonsingular projective variety and let $x_0 \in X$ be a fixed closed point. A pair (A, α) consisting of an Abelian variety A and a morphism $\alpha : X \to A$ such that $\alpha(x_0) = 0$ (the zero element of A) is called the *Albanese variety* of the variety X if for every morphism $f : X \to B$ such that B is an Abelian variety and $f(x_0) = 0$, there exists a unique homomorphism $g : A \to B$ of Abelian varieties such that $g \circ \alpha = f$.

In connection with this definition we should note the elementary result which says that every morphism $g' : A \to B$ between two Abelian varieties A and B has the form $g' = g + b$, where $g : A \to B$ is a homomorphism of Abelian varieties and $b = g'(0) \in B$. Therefore we see easily that Definition 5.2 could also be formulated without reference to a point x_0, as follows: the pair (A, α) consisting of an Abelian variety A and a morphism $\alpha : X \to A$ is the Albanese variety of X if for every morphism $g : X \to B$ to an Abelian variety B there exists a unique morphism $g : A \to B$ such that $g \circ \alpha = f$. From this we conclude that any two Albanese varieties of X are canonically isomorphic, and in particular, that the Albanese variety of X is independent (up to isomorphism) of the choice of the base point x_0.

By a slight abuse of notation, we will denote the Albanese variety of the (nonsingular and projective) variety X by $(\mathrm{Alb}(X), \alpha)$.

The existence of the Albanese variety of a nonsingular projective variety X is proved in [Ser4] using relatively elementary methods. Here we give a quick proof, which shows not only the existence of the Albanese variety, but also its relation to the (classical) Picard variety of X. But we need to use the theory of the Picard scheme.

Let B be an Abelian variety. By [Mum2] we have that its Picard scheme $\hat{B} = \underline{\mathrm{Pic}}^0(B)$ is always reduced, and therefore an Abelian variety. Hence $\dim(\hat{B}) = \dim T_{B,0} = \dim H^1(B, \mathcal{O}_B) = \dim(B)$. In this case $\hat{B}$ is called the *dual Abelian variety* of the Abelian variety B (*loc. cit.*).

If X is an arbitrary nonsingular and projective variety, and if we denote the Picard variety of X by $P(X)$, that is—$P(X) = \underline{\mathrm{Pic}}^0(X)_{\mathrm{red}}$, then there exists a canonical morphism

$$u_X : X \to (P(X))\hat{} = \underline{\mathrm{Pic}}^0(P(X)),$$

defined as follows. Fix a base point $x_0 \in X$. Let L' be the Poincaré line bundle on $X \times \underline{\mathrm{Pic}}^0(X)$, L'' the Poincaré line bundle on $P(X) \times (P(X))\hat{}$, L'_{red} the restriction of L' to the subscheme $X \times P(X)$ of $X \times \underline{\mathrm{Pic}}^0(X)$, and $v : P(X) \times X \to X \times P(X)$ the isomorphism $(a, b) \mapsto (b, a)$. By the universal property of the dual Abelian variety $(P(X))\hat{}$ we see that there exists a (unique) canonical morphism $u_X : X \to (P(X))\hat{}$ such that

$$v^*(L'_{\mathrm{red}}) = (\mathrm{id}_{P(X)} \times u_X)^*(L'').$$

A notable fact is that if X is an Abelian variety, then the canonical morphism $u_X : X \to (\hat{X})\hat{} = X\hat{}\hat{}$ is an isomorphism (this justifies the name *dual Abelian variety* given to $\hat{X}$, in the sense that the bidual of an Abelian variety must be (canonically) isomorphic to the original variety; see [Mum2]).

If X is a nonsingular projective variety and x_0 is a base point in X, then we put $\mathrm{Alb}(X) = (P(X))\hat{}$ and $\alpha = u_X$, the morphism described earlier. By construction we have $\alpha(x_0) = 0$. It is now very easy to check (combining various universal properties) that if $f : X \to B$ is a morphism to an Abelian variety B such that $f(x_0) = 0$, then the homomorphism $g = (P(f))\hat{} : \mathrm{Alb}(X) \to \mathrm{Alb}(B) = B\hat{}\hat{}$ has the property that $g \circ u_X = u_B \circ f$, and is unique with this property, where $P(f) : P(B) \to P(X)$ is the morphism induced by f on the Picard varieties. Since we have already remarked that u_B is an isomorphism, we have proved indeed that $((P(X))\hat{}, u_X)$ is the Albanese variety of X.

In other words, we have proved the following.

Theorem 5.3. *If X is a nonsingular projective variety, then the Albanese variety $(\mathrm{Alb}(X), \alpha)$ exists and is unique up to an isomorphism, and $\mathrm{Alb}(X)$ coincides with the dual of the Picard variety $P(X)$ of X. In particular, if*

X is a surface, then the dimension of the variety $\mathrm{Alb}(X)$ is equal to $s = \dim(P(X))$ (in the notation of Theorem 5.1), and therefore $\dim(\mathrm{Alb}(X)) \leq q$. Moreover, we have equality if and only if $\Delta = 0$ (e.g., if $\mathrm{char}(\Bbbk) = 0$ or if $p_g = 0$). $\square$

We note that if X is a nonsingular projective curve then the Picard scheme $\underline{\mathrm{Pic}}^0(X)$ is always reduced, and therefore it coincides with the Picard variety of X. Furthermore, in this case the Abelian variety $P(X)$ is isomorphic to its dual, and therefore the Albanese variety $\mathrm{Alb}(X)$ (which in this case is also called the *Jacobian* of the curve X) is isomorphic to $P(X)$.

We will use the following additional notation. $NS(X)$ will denote the quotient group $\mathrm{Pic}(X)/\mathrm{Pic}^0(X)$, which is called the *Néron–Severi group* of the nonsingular projective variety X. A theorem first proved by Severi in characteristic zero and then by Néron in arbitrary characteristic says that this group is finitely generated, and its rank ρ is called the *base number* of X. An inequality proved by Igusa says that $\rho \leq b_2$ (cf. [Gro8]). Over the complex field this result is elementary, because it is easy to prove that $\rho \leq \dim H^1(X, \Omega^1_{X/\Bbbk})$ if $\mathrm{char}(\Bbbk) = 0$—and then one can use the Hodge decomposition of $H^2(X, \mathbb{C})$. See [Har1, p. 368, Exercise 1.8] or Exercise 5.5 below for the case when X is a surface.

5.4. Now let X be a nonsingular projective variety and let L be an invertible $\mathcal{O}_X$-module. To L we can associate the graded $\Bbbk$-algebra $\Gamma_*(L) = \oplus_{n=0}^{\infty} H^0(X, L^n)$, where the homogeneous part of degree n is $H^0(X, L^n)$, and the product of $s \in H^0(X, L^m)$ and $s' \in H^0(X, L^n)$ is given by $s \cdot s' = s \otimes s' \in H^0(X, L^{m+n})$. It is clear that if $s \otimes s' = 0$ then either $s = 0$ or $s' = 0$, so that $\Gamma_*(L)$ is an integral domain. Moreover, $\Gamma_*(L)$ contains $\Bbbk = H^0(X, L^0) = H^0(X, \mathcal{O}_X)$. Thus it makes sense to talk about the transcendence degree $\lambda(L)$ of $\Gamma_*(L)$ over $\Bbbk$.

Lemma 5.5. *The following inequalities hold:*

$$0 \leq \lambda(L) \leq \dim(X) + 1.$$

If in addition L is ample, then $\lambda(L) = \dim(X) + 1$, and for n sufficiently large $\Gamma_*(L^n)$ is a $\Bbbk$-algebra of finite type.

Proof. Let H be a very ample effective divisor on X such that $L' = L \otimes \mathcal{O}_X(H)$ is ample (cf. [EGA II]). Multiplication by a global section of $\mathcal{O}_X(H)$ corresponding to H induces an injective homomorphism $L' \hookrightarrow L$, which in turn induces an injective homomorphism of $\Bbbk$-algebras, $\Gamma_*(L) \to \Gamma_*(L')$. Thus $\lambda(L) \leq \lambda(L')$. Therefore it suffices to prove the second claim of the lemma.

Assume now that L is ample. We remark that for every integer $n > 0$, the ring extension $\Gamma_*(L^n) \subset \Gamma_*(L)$ has the property that the nth power of every homogeneous element of $\Gamma_*(L)$ is in $\Gamma_*(L^n)$; hence $\lambda(L) = \lambda(L^n)$.

Thus we may assume that L is very ample. Then let $X \hookrightarrow \mathbb{P}^m$ be an embedding of X into a projective space such that $L \cong \mathcal{O}_X(1)$. Then the Serre homomorphism

$$\alpha : S(X) \to \Gamma_*(L)$$

is a TN-isomorphism, where $S(X)$ is the homogeneous coordinate ring of X in $\mathbb{P}^m$ (see [EGA III, (2.3.1)]). Therefore for n sufficiently large the composite of the immersion $X \hookrightarrow \mathbb{P}^m$ with the Veronese immersion $\mathbb{P}^m \hookrightarrow \mathbb{P}^{N(n)}$, with $N(n) = \binom{m+n}{n} - 1$, has the property that the associated Serre homomorphism is in fact an isomorphism. Thus the second claim of the lemma is proved, as the homogeneous coordinate ring of X in $\mathbb{P}^{N(n)}$ is a $\Bbbk$-algebra of finite type, of transcendence degree equal to $\dim(X) + 1$. $\square$

Definition 5.6. If X is a nonsingular projective variety of dimension $d \geq 1$, then the ring $\Gamma_*(\omega_X)$, with $\omega_X = \Omega^d_{X/\Bbbk}$, is called the *canonical ring* of X; it will be denoted by $R(X)$ throughout this book.

The number $\kappa(X) = \lambda(\omega_X) - 1$ if $\lambda(\omega_X) > 0$, resp. $\kappa(X) = -\infty$ if $\lambda(\omega_X) = 0$, is called the *canonical dimension* (or *Kodaira dimension*) of X, and $p_n = p_n(X) = \dim H^0(X, \omega_X^n)$ (for $n \geq 1$) is called the *n-genus* (or n^{th} *plurigenus*) of X.

From Lemma 5.5 we have $-\infty \leq \kappa(X) \leq \dim(X)$, with $\kappa(X) = -\infty$ if and only if $p_n = 0$ for all $n \geq 1$, and otherwise $\kappa(X) \geq 0$. Of course, p_1 is just the geometric genus p_g of X.

Proposition 5.7. *If X and Y are two surfaces and $u : X \to Y$ is a birational map, then u induces an isomorphism of graded $\Bbbk$-algebras, $u^* : R(Y) \to R(X)$. In particular, for each $n \geq 1$, the n-genus p_n is a birational numerical invariant.*

Proof. Considering the structure of birational maps of surfaces (as composites of quadratic transformations and their inverses), we immediately reduce the proof to the case when u is the quadratic transformation of Y with center y. Let $E = u^{-1}(y)$. We have the formula

$$\omega_X = u^*(\omega_Y) \otimes \mathcal{O}_X(E),$$

and therefore, for every $n \geq 1$:

$$\omega_X^n = u^*(\omega_Y^n) \otimes \mathcal{O}_X(nE).$$

By the projection formula we have $u_*(\omega_X^n) \cong \omega_Y^n \otimes u_*\mathcal{O}_X(nE) \cong \omega_Y^n$, since $u_*\mathcal{O}_X(nE) \cong \mathcal{O}_Y$ (easy to check by induction on n; see (4.13)). Therefore $H^0(Y, u_*(\omega_X^n)) = H^0(Y, \omega_Y^n)$, or equivalently, $H^0(X, \omega_X^n) = H^0(Y, \omega_Y^n)$. $\square$

Examples 5.8.   (i) Let X be a nonsingular projective curve of genus g.
    Then:

(a) $g = 0$ implies $X = \mathbb{P}^1$, $\omega_X = \mathcal{O}_X(-2)$, $R(X) = \Bbbk$, $\kappa(X) = -\infty$, $p_n = 0$ for all $n \geq 1$.

(b) $g = 1$ implies $\omega_X = \mathcal{O}_X$, $\kappa(X) = 0$, $p_n = 1$ for all $n \geq 1$.

(c) $g \geq 2$ implies $\deg(\omega_X) = 2g - 2 > 0$, and therefore ω_X is ample, and $\kappa(X) = 1$.

(ii) Let $X = C_1 \times C_2$ be the product of two nonsingular projective curves of genera g_1 and g_2, respectively. Then $\omega_X \cong p_1^*(\omega_{C_1}) \otimes p_2^*(\omega_{C_2})$, where p_1 and p_2 are the projections of X onto C_1 and C_2, respectively. Thus $H^0(X, \omega_X^n) = H^0(C_1, \omega_{C_1}^n) \otimes H^0(C_2, \omega_{C_2}^n)$, $n > 0$, and we have:

(a) $[g_1 = 0$ or $g_2 = 0] \implies [\kappa(X) = -\infty]$.

(b) $[g_1 = g_2 = 1] \implies [\kappa(X) = 0]$.

(c) $[g_1 \geq 2$ and $g_2 = 1$, or $g_1 = 1$ and $g_2 \geq 2] \implies [\kappa(X) = 1]$.

(d) $[g_1 \geq 2$ and $g_2 \geq 2] \implies [\kappa(X) = 2]$.

All these assertions can be easily proved using the following elementary lemma.

Lemma 5.9. *Let B be a nonsingular projective curve, and let M be an invertible $\mathcal{O}_B$-module. Then:*

(i) $\dim H^0(B, M^n) = n \deg(M) + const.$, *if* $\deg(M) \geq 1$ *and* $n \gg 0$.

(ii) $\dim H^0(B, M^n) = 1$ *if* $M^n \cong \mathcal{O}_B$.

(iii) $\dim H^0(B, M^n) = 0$ *if* $\deg(M) < 0$, *and also if* $\deg(M) = 0, n \geq 1$, *and* $M^n \not\cong \mathcal{O}_B$. □

Remarks 5.10. Let X be a surface.

(a) If $p_n \geq 2$ for some $n \geq 1$, then $\kappa(X) \geq 1$.

(b) If $p_n \leq 1$ for all $n \geq 1$, then $\kappa(X) \leq 0$. Moreover, $\kappa(X) = 0$ if and only if $p_n \leq 1$ for all $n \geq 1$ and $p_n = 1$ for at least one $n \geq 1$.

Proof. If we prove (a), then the remainder of (b) is clear. We prove (a).

Let $\alpha, \beta \in H^0(X, \omega_X^n)$ be two $\Bbbk$-linearly independent sections. Then for each $m \geq 1$ the sections $\alpha^m, \alpha^{m-1} \cdot \beta, \ldots, \beta^m \in H^0(X, \omega_X^{mn})$ are also linearly independent over $\Bbbk$; indeed, if

$$\lambda_0 \alpha^m + \lambda_1 \alpha^{m-1} \beta + \cdots + \lambda_m \beta^m = 0, \qquad \lambda_i \in \Bbbk, \qquad (*)$$

then the polynomial $P(U, V) = \lambda_0 U^m + \lambda_1 U^{m-1} V + \cdots + \lambda_m V^m$ has the form $P(U, V) = \prod_{i=1}^m (\gamma_i U + \mu_i V)$ with $\gamma_i, \mu_i \in \Bbbk$ (because $\Bbbk$ is algebraically closed), and therefore $(*)$ becomes $\prod_{i=1}^m (\gamma_i \alpha + \mu_i \beta) = 0$; since $R(X)$ is an integral domain, we get $\gamma_i \alpha + \mu_i \beta = 0$ for some i, and therefore $\gamma_i = \mu_i = 0$, and finally $P(U, V) = 0$.

Therefore the $\Bbbk$-subalgebra generated by α and β in $R(X)$ is isomorphic to the polynomial algebra $\Bbbk[U, V]$; thus $\operatorname{tr.deg} R(X) \geq \operatorname{tr.deg} \Bbbk[U, V] = 2$, that is, $\kappa(X) \geq 1$. $\square$

EXERCISES

We recall the convention made in this chapter: from now on by the term *surface* we shall mean a nonsingular projective surface, unless stated otherwise.

Exercise 5.1. Let $f : X' \to X$ be the quadratic transformation of a surface X with center x, and let $E = f^{-1}(x)$. Prove the formula $\omega_{X'} \cong f^*(\omega_X) \otimes \mathcal{O}_{X'}(E)$. Use this to deduce that $(K'^2) = (K^2) - 1$, where K and K' are canonical divisors of X and X', respectively.

Exercise 5.2. With the notation of Exercise 5.1, show that if ω_X^{-1} is not ample, then $\omega_{X'}^{-1}$ is not ample either. [*Hint:* Use Exercise 5.1 and the Nakai–Moishezon criterion.]

Exercise 5.3. Let X be a surface over $\Bbbk$, with $p_g = 0, q > 0$, and $\operatorname{char}(\Bbbk) = 0$. If $\alpha : X \to \operatorname{Alb}(X)$ is the Albanese morphism, prove that $\alpha(X)$ is a curve. [*Hint:* Assuming that $Y = \alpha(X)$ is a surface, then the morphism $\alpha' : X \to Y$ deduced from α is generically étale (because $\operatorname{char}(\Bbbk) = 0$). Pick a nonsingular point $y \in Y$ above which α' is étale and produce an invariant differential form ω of degree 2 on $\operatorname{Alb}(X)$ that is nonzero at y. Then $\alpha^*(\omega)$ would be a nonzero element in $H^0(X, \omega_X)$, contradicting $p_g = 0$.]

Exercise 5.4. Let $\alpha : X \to \operatorname{Alb}(X)$ be the Albanese morphism of a surface X. Assume that $C = \alpha(X)$ is a curve and $\operatorname{char}(\Bbbk) = 0$. Prove that C is nonsingular and that the fibers of α are all connected. [*Hint:* Consider the Stein factorization $\alpha = u \circ f$, with $u : B \to C$ finite, B nonsingular, and $f : X \to B$ with connected fibers. Using the universal property of the Albanese variety, show that $\operatorname{Alb}(u) : \operatorname{Alb}(B) \to \operatorname{Alb}(X)$ is an isomorphism (here $\operatorname{char}(\Bbbk) = 0$ is essential!). Finally, use the fact that the morphism $B \to \operatorname{Alb}(B) = \operatorname{Jac}(B)$ is injective.]

Exercise 5.5. For any surface X, the homomorphism of sheaves of groups $\operatorname{dlog} : \mathcal{O}_X^* \to \Omega^1_{X|\Bbbk}$ defined by $\operatorname{dlog}(f) = df/f$ induces a group homomorphism $c : H^1(X, \mathcal{O}_X^*) = \operatorname{Pic}(X) \to H^1(X, \Omega^1_{X|\Bbbk})$ (which will also play an important role in the proof of Theorem 6.2 below). Then, using c, to every divisor D on X we associate a cohomology class $c(D) \in H^1(X, \Omega^1_{X|\Bbbk})$. On the other hand, by Serre duality we have a nondegenerate bilinear map

$$\langle \, , \, \rangle : H^1(X, \Omega^1_{X|\Bbbk}) \times H^1(X, \Omega^1_{X|\Bbbk}) \to \Bbbk.$$

(i) Prove that this bilinear map is compatible with the intersection pairing, that is, for every two divisors D and D' on X we have

$$\langle c(D), c(D') \rangle = (D \cdot D') \cdot 1 \text{ in } \Bbbk.$$

[*Hint:* Reduce to the case when D and D' are two nonsingular curves meeting transversally. Then consider the analogous map $c : \mathrm{Pic}(D) \to H^1(D, \Omega^1_{D|\Bbbk})$ and show that the cohomology class of a point of D goes to 1 under the natural identification $H^1(D, \Omega^1_{D|\Bbbk}) \cong \Bbbk$.]

(ii) If $\mathrm{char}(\Bbbk) = 0$, use the fact that $H^1(X, \Omega^1_{X|\Bbbk})$ is a finite-dimensional $\Bbbk$-vector space to show that $\mathrm{Num}(X)$ is a finitely generated free Abelian group.

Bibliographic References. Noether's formula may be found in [BS]. The Betti numbers of an abstract algebraic surface were originally defined by Igusa (cf. [Igu]); it was later observed that they can also be introduced using étale cohomology, cf. [Del]. Details about the theory of the Picard scheme can be found in [Gro3], [Mum1, Mum2], [Ser3], [Oor]. The construction and elementary properties of the Albanese variety can be found in [Ser4]. The Igusa–Severi inequality is proved in [Igu] and [Gro8]. The canonical ring of a surface was first considered by Mumford [Mum3].

6

Existence of Minimal Models

Definition 6.1. Let X be a surface. X is a *minimal model* if every birational morphism $f : X \to Y$, with Y surface (nonsingular and projective, just like X), is an isomorphism.

From the structure theorem for birational morphisms of surfaces and Castelnuovo's contractability criterion (Theorem 3.30), we see that X is a minimal surface if and only if X doesn't contain any exceptional curves of the first kind.

Our goal in this chapter is to prove the following theorem and its corollary.

Theorem 6.2. *Let $\sigma : X' \to X$ be the quadratic transformation of the surface X with center x. Then*

$$\dim H^1(X, \Omega^1_{X/\mathrm{k}}) < \dim H^1(X', \Omega^1_{X'/\mathrm{k}}).$$

Corollary 6.3. *Every surface X dominates a minimal surface. In particular, every birational isomorphism class of surfaces contains at least one minimal surface.*

Proof of Corollary 6.3. If X is a minimal surface, there is nothing to prove. Otherwise X contains an exceptional curve of the first kind, say, E. By Theorem 3.30 there exists a contraction $f_1 : X \to X_1$, with X_1 a (nonsingular) surface and f_1 the quadratic transformation of X_1 with its center at some point $x_1 \in X_1$. If X_1 is a minimal surface, then we are done. Otherwise there exists a contraction $f_2 : X_1 \to X_2$ of an exceptional curve E_1 of the first kind on X_1, and so on. Thus we get a tower of quadratic transformations

$$X \longrightarrow X_1 \longrightarrow X_2 \longrightarrow \cdots .$$

We must show that this sequence is finite. But this follows from Theorem 6.2, because we have

$$\dim H^1(X, \Omega^1_{X/\Bbbk}) > \dim H^1(X_1, \Omega^1_{X_1/\Bbbk}) > \dim H^1(X_2, \Omega^1_{X_2/\Bbbk}) > \cdots$$

while the $\Bbbk$-vector space $H^1(X, \Omega^1_{X/\Bbbk})$ is finite-dimensional.   □

Proof of Theorem 6.2.

Step 1. Let $U = \operatorname{Spec}(A) \subset X$ be an affine open neighborhood of the point x such that, if we denote by $\mathfrak{m}$ the maximal ideal in A corresponding to the point x, then $\mathfrak{m}$ is generated by two elements $u, v \in \mathfrak{m}$ (in other words, u and v form a regular system of parameters around the point x). Then $\sigma^{-1}(U)$ has a covering consisting of two affine open sets, $U_1 = \operatorname{Spec}(A[v/u])$ and $U_2 = \operatorname{Spec}(A[u/v])$. Let $w_1 = v/u$ and $w_2 = u/v$. Then $U_1 \cap U_2 = \operatorname{Spec}(B)$ with $B = A[w_1, w_2]$ ($w_1 w_2 = 1$). We claim that *the surjective A-algebra homomorphism $\phi : A[T] \to A[w_1]$ that maps T to w_1 (T being a variable over A) induces an isomorphism $A[T]/(uT - v) \cong A[w_1]$ (and similarly for w_2).*

Indeed, the inclusion $(uT - v) \subset \operatorname{Ker}(\phi)$ is obvious. Conversely, let $f = a_0 + a_1 T + \cdots + a_n T^n \in A[T]$ with $a_n \neq 0$, $n \geq 0$ and $f(w_1) = 0$. If $n = \deg(f) = 0$ then $f = 0$, so we may assume that $n \geq 1$. The equation $f(w_1) = 0$ can be rewritten as $a_n v^n = -u(a_0 u^{n-1} + a_1 u^{n-2} v + \cdots + a_{n-1} v^{n-1})$, and therefore u divides $a_n v^n$ in the regular local ring $A_{\mathfrak{m}} = \mathcal{O}_{X,x}$. As $A_{\mathfrak{m}}$ is a unique factorization domain and u does not divide v, we see that u divides a_n (note that u and v are prime elements in $A_{\mathfrak{m}}$), that is, $a_n/u \in A_{\mathfrak{m}}$. Now let $\mathfrak{m}' \neq \mathfrak{m}$ be another maximal ideal of A. Then at least one of u and v is invertible in the local ring $A_{\mathfrak{m}'}$. If u is invertible in $A_{\mathfrak{m}'}$, then $a_n/u \in A_{\mathfrak{m}'}$; and if v is invertible in $A_{\mathfrak{m}'}$, then the equation $a_n/u = -(v^n)^{-1}(a_0 u^{n-1} + a_1 u^{n-2} v + \cdots + a_{n-1} v^{n-1})$ again shows that $a_n/u \in A_{\mathfrak{m}'}$ (for any maximal ideal $\mathfrak{m}' \neq \mathfrak{m}$). Therefore $a_n/u \in A$. Thus there exists $a' \in A$ such that $a_n = u a'$. Hence there exists $g \in A[T]$ such that $f = a' T^{n-1}(uT - v) + g$, with $\deg(g) < \deg(f) = n$ and $g(w_1) = 0$. By the inductive hypothesis g is divisible by $(uT - v)$, and therefore f is also divisible by $(uT - v)$.

Step 2. The canonical homomorphism $\Omega^1_X \to \sigma_* \Omega^1_{X'}$ is surjective and the support of its kernel is contained in $\{x\}$ (we simplified the notation $\Omega^1_{X/\Bbbk}$ to Ω^1_X). In fact this homomorphism is an isomorphism, but in the proof of Theorem 6.2 we will not need this additional bit of information.

Proof of Step 2. As σ is an isomorphism outside $E = \sigma^{-1}(x)$, we see immediately that the support of the kernel of the homomorphism considered in Step 2 is contained in $\{x\}$. There only remains to prove the surjectivity of that homomorphism. The problem being local around the point x, we may assume that $X = U = \text{Spec}(A)$ as in Step 1. Then $X' = U_1 \cup U_2$. To give a differential form $\omega \in \Gamma(X, \sigma_* \Omega^1_{X'}) = \Gamma(X', \Omega^1_{X'})$ amounts to giving two differential forms $\omega_1 \in \Gamma(U_1, \Omega^1_{X'})$ and $\omega_2 \in \Gamma(U_2, \Omega^1_{X'})$ such that $\omega_1|_{U_1 \cap U_2} = \omega_2|_{U_1 \cap U_2}$. Since $U_1 = \text{Spec}(A_1)$, $U_2 = \text{Spec}(A_2)$, $U_1 \cap U_2 = \text{Spec}(B)$, with $A_1 = A[w_1]$, $A_2 = A[w_2]$ and $B = A[w_1, w_2]$, we can write:

$$\begin{cases} \omega_1 = \sum_i b'_i \, da'_i + f_1(w_1) \, dw_1 \\ \omega_2 = \sum_j b''_j \, da''_j + f_2(w_2) \, dw_2 \end{cases} \qquad (*)$$

with $a'_i, a''_j \in A$, $b'_i \in A_1$, $b''_j \in A_2$, $f_1, f_2 \in A[T]$. Moreover, ω_1 and ω_2 have the same image in $\Omega_B = \Omega^1_{B/k}$.

If $\bar{f}_1$ (resp. $\bar{f}_2$) is the reductions of f_1 (resp. f_2) modulo the ideal $\mathfrak{m}$, then $\omega_1|_{E \cap U_1} = \bar{f}_1(T) \, dT$, $\omega_2|_{E \cap U_2} = \bar{f}_2(T') \, dT'$, with T and T' nonhomogeneous coordinates on $E = \mathbb{P}^1$ such that $T \cdot T' = 1$. Since the two restrictions to E coincide on the intersection $E \cap U_1 \cap U_2$, they define a differential form $\omega' \in \Gamma(E, \Omega^1_E)$. However, $\Omega^1_E \cong \mathcal{O}_E(-2)$, and $\mathcal{O}_E(-2)$ has no nonzero global sections; therefore $\bar{f}_1 = \bar{f}_2 = 0$. That is, all the coefficients of f_1 and f_2 are in the ideal $\mathfrak{m}$. But as $\mathfrak{m} A_1 = u A_1$, we can write $f_1(w_1) \, dw_1 = g_1(w_1) \cdot u \cdot dw_1 = g_1(w_1) \cdot dv - g_1(w_1) \cdot w_1 \cdot du$, with $g_1 \in A_1[T]$. In other words, we have shown that, without loss of generality, we may assume in $(*)$ that $f_1 = f_2 = 0$, and therefore that

$$\begin{cases} \omega_1 = \sum_i b'_i \, da'_i & a'_i \in A, \; b'_i \in A_1 \\ \omega_2 = \sum_j b''_j \, da''_j & a''_j \in A, \; b''_j \in A_2. \end{cases} \qquad (**)$$

Now recall the following general elementary facts:

(a) If C is a $\Bbbk$-algebra and T is a variable over C, then one has the following exact sequence:

$$0 \longrightarrow \Omega_C \otimes_C C[T] \longrightarrow \Omega_{C[T]} \longrightarrow \Omega_{C[T]/C} \longrightarrow 0$$

($\Omega_C = \Omega^1_{C/\Bbbk}$), and since $\Omega_{C[T]/C} = C[T] \cdot dT$, we get an isomorphism:

$$\Omega_{C[T]} \cong (\Omega_C \otimes_C C[T]) \oplus C[T] \cdot dT.$$

(b) If I is an ideal in C such that all the localizations of C and C/I are regular, then there is a canonical exact sequence:

$$0 \longrightarrow I/I^2 \longrightarrow \Omega_C \otimes_C C/I \longrightarrow \Omega_{C/I} \longrightarrow 0.$$

(c) If S is a multiplicative system in C, then there is a canonical isomorphism $\Omega_{S^{-1}C} = S^{-1}\Omega_C$.

Using Step 1 and these general facts (with $A[w_1] = A[T]/I$, $I = (uT - v)A[T]$ and $B = A[w_1, w_1^{-1}]$), we get an isomorphism:

$$\Omega_B \cong \frac{(\Omega_A \otimes_A B) \oplus B \cdot dw_1}{B \cdot (w_1 \cdot du + u \cdot dw_1 - dv)},$$

where $B \cdot dw_1$ is a free B-module of rank one. From this we see that the natural homomorphism $\Omega_A \otimes_A B \to \Omega_B$ is injective.

Since $\sigma_* \mathcal{O}_{X'} = \mathcal{O}_X$, we have an exact sequence of A-modules:

$$0 \longrightarrow A \xrightarrow{\phi} A_1 \oplus A_2 \xrightarrow{\psi} B \longrightarrow 0,$$

where $\phi(a) = (a, a)$ and $\psi(a_1, a_2) = a_1 - a_2$ (actually the equality $\sigma_* \mathcal{O}_{X'} = \mathcal{O}_X$ gives the inclusion $\mathrm{Ker}(\psi) \subseteq \mathrm{Im}(\phi)$, and since the opposite inclusion is obvious, we have $\mathrm{Ker}(\phi) = \mathrm{Im}(\psi)$; and the surjectivity of ψ follows immediately from the fact that $B = A[w_1, w_2]$ with $w_1 w_2 = 1$). Tensoring this exact sequence with Ω_A we get an exact sequence:

$$\Omega_A \xrightarrow{\phi \otimes \mathrm{id}} (A_1 \otimes_A \Omega_A) \oplus (A_2 \otimes_A \Omega_A) \xrightarrow{\psi \otimes \mathrm{id}} B \otimes_A \Omega_A \longrightarrow 0.$$

Since $(**)$ shows that $(\omega_1, \omega_2) \in \mathrm{Ker}(\psi \otimes \mathrm{id}) = \mathrm{Im}(\phi \otimes \mathrm{id})$ (using the injectivity of $\Omega_A \otimes_A B \to \Omega_B$), we see that there exists $\omega \in \Omega_A$ such that $(\phi \otimes \mathrm{id})(\omega) = (\omega_1, \omega_2)$, and the proof of Step 2 is complete.

Step 3. Consider the Leray spectral sequence:

$$E_2^{pq} = H^p(X, R^q \sigma_* \Omega_{X'}^1) \implies H^{p+q}(X', \Omega_{X'}^1).$$

Writing down part of the exact sequence in small degrees associated to this spectral sequence, we have an exact sequence:

$$0 \longrightarrow H^1(X, \sigma_* \Omega_{X'}^1) \longrightarrow H^1(X', \Omega_{X'}^1) \xrightarrow{\alpha} H^0(X, R^1 \sigma_* \Omega_{X'}^1).$$

From Step 2 we know that the canonical homomorphism $H^1(X, \Omega_X) \to H^1(X, \sigma_* \Omega_{X'}^1)$, induced by the homomorphism $\Omega_X^1 \to \sigma_* \Omega_{X'}^1$ (which is surjective and has kernel with at most finite support), is an isomorphism.

Therefore the proof of Theorem 6.2 is complete if we show that:

Step 4. The homomorphism α is nontrivial. First we remark that α is the map from a member of the inductive system $\{H^1(\sigma^{-1}(U), \Omega_{X'}^1) \mid U \ni x\}$, namely, the one with $U = X$, to the direct limit of this inductive system, the stalk $R^1 f_*(\Omega_{X'}^1)_x$. Since every element of $H^0(X, R^1 \sigma_* \Omega_{X'}^1) = (R^1 \sigma_* \Omega_{X'}^1)_x$ is represented by an element of $H^1(\sigma^{-1}(U), \Omega_{X'}^1)$ for a suitable neighborhood U of x in X, and since for each such U we have a natural restriction map

$$\mathrm{res} : H^1(\sigma^{-1}(U), \Omega_{X'}^1) \longrightarrow H^1(E, \Omega_E^1),$$

we get a natural homomorphism

$$\beta : H^0(X, R^1 \sigma_* \Omega_{X'}^1) \longrightarrow H^1(E, \Omega_E^1).$$

To finish the proof of Step 4, it would suffice to show that the natural homomorphism $\beta \circ \alpha$ is nontrivial. To that end, consider the commutative diagram:

$$\begin{array}{ccc}
\mathrm{Pic}(X') & \xrightarrow{\ \gamma\ } & H^1(X',\Omega^1_{X'}) \\
{\scriptstyle \mathrm{res}}\downarrow & & \downarrow{\scriptstyle \beta\circ\alpha} \\
\mathrm{Pic}(E) & \xrightarrow{\ \gamma'\ } & H^1(E,\Omega_E)
\end{array}$$

where the top horizontal map is given by $\gamma(\xi) =$ the cohomology class of the 1-cocycle $\{d\xi_{ij}/\xi_{ij}\}_{i,j} \in Z^1(\mathfrak{U},\Omega^1_{X'})$, where $\mathfrak{U} = \{U_i\}_i$ is an open cover of X' in which ξ is represented by the 1-cocycle $\{\xi_{ij}\}_{i,j} \in Z^1(\mathfrak{U},\mathcal{O}^*_{X'})$. Now let $\xi \in \mathrm{Pic}(X')$ be the cohomology class associated to the exceptional curve E on X'. Then $(\beta \circ \alpha)(\gamma(\xi)) = \gamma'(\mathrm{res}(\xi))$. But the restriction of ξ to E is the (unique) line bundle of degree -1 in $\mathrm{Pic}(E)$ (since we have $(E^2) = -1$). The proof of the theorem will be completed if we can show that $\gamma'(\mathrm{res}(\xi)) \neq 0$. But we can see immediately from the definitions that $\gamma'(\mathrm{res}(\xi))$ is the cohomology class in $H^1(E,\Omega_E)$ represented by the 1-cocycle $dz/z = \eta_{12} \in Z^1(\{U'_1,U'_2\},\Omega^1_E)$, with U'_1, U'_2 the canonical open cover of the projective line, and z a nonhomogeneous coordinate on $\mathbb{P}^1$. As $U'_1 \cong$ the affine line $\cong U'_2$, if η_{12} were a coboundary then there would exist differential forms $\omega_1 \in \Gamma(U'_1,\Omega^1_E)$ and $\omega_2 \in \Gamma(U'_2,\Omega^1_E)$, given by

$$\omega_1 = \left(\sum_{i\geq 0} a_i z^i\right)\cdot dz, \qquad a_i \in \Bbbk, \qquad \text{and}$$

$$\omega_2 = \left(\sum_{j\geq 0} b_j z'^j\right)\cdot dz', \qquad b_j \in \Bbbk, \qquad z' = 1/z,$$

such that $dz/z = \omega_1 - \omega_2$ on $U'_1 \cap U'_2$. This implies, however, the impossible equation

$$\left(\sum_{i\geq 0} a_i z^i + \sum_{j\geq 0} b_j/z^{2+j}\right)\cdot dz = dz/z.$$

The proof of Theorem 6.2 is now complete.

Remarks 6.4. (a) The significance of Corollary 6.3 for the classification of surfaces (which is our main interest in the next chapters) is that, since the classification is "birational," we may always assume that the surface X is a minimal model (because by Corollary 6.3 every surface dominates a minimal model). Furthermore, we will see that, with the exception of ruled surfaces, every birational isomorphism class of nonsingular projective surfaces contains a *unique* minimal model. We will also classify all the minimal models of ruled surfaces. The classification of surfaces is by means of certain numerical invariants, such as the plurigenera p_n and in particular the geometric genus p_g, cf. (5.7). Another important birational numerical invariant is the irregularity q. (Indeed, using the structure theorem for birational morphisms of surfaces, we reduce easily to showing that $\dim H^1(X, \mathcal{O}_X)$ is invariant under a quadratic transformation $\sigma : X' \to X$. But since X is nonsingular—so that $E = \mathbb{P}^1$—and using (3.8), we have $R^i\sigma_*\mathcal{O}_{X'} = 0$ for all $i > 0$, or equivalently $\chi(\mathcal{O}_{X'}) = \chi(\mathcal{O}_X)$. Since p_g is a birational invariant, we see that q is also a birational invariant.)

(b) Corollary 6.3 can also be proved using the isomorphism

$$\text{Pic}(X') \cong \text{Pic}(X) \oplus \mathbb{Z}\mathcal{O}_{X'}(E)$$

(the proof is left to the reader), whence one immediately proves the isomorphism

$$\text{Num}(X') \cong \text{Num}(X) \oplus \mathbb{Z}\mathcal{O}_{X'}(E).$$

The conclusion follows from the fact that the rank of $\text{Num}(X)$ is finite, by the Néron–Severi theorem.

EXERCISES

Exercise 6.1. Let $f : X' \to X$ be the quadratic transformation of the surface X with center x, and let $E = f^{-1}(x)$. Show that $\text{Pic}(X') \cong f^*(\text{Pic}(X)) \oplus \mathbb{Z}[\mathcal{O}_{X'}(E)]$.

Exercise 6.2. Let $f : X' \to X$ be the quadratic transformation of a surface X with center x. Prove that the canonical homomorphisms

$$H^i(X, \mathcal{O}_X) \to H^i(X', \mathcal{O}_{X'})$$

are isomorphisms for all $i \geq 0$. In particular, $\chi(X, \mathcal{O}_X) = \chi(X', \mathcal{O}_{X'})$.

Exercise 6.3. Prove that $\text{Pic}(B \times \mathbb{P}^1) \cong p_1^*(\text{Pic}(B)) \oplus p_2^*(\text{Pic}(\mathbb{P}^1))$, where B is an arbitrary algebraic variety and p_1 and p_2 are the canonical projections of $B \times \mathbb{P}^1$. [*Hint:* Use the base change theorems, cf. [Har1, III.12].]

Exercise 6.4. Let Δ be the diagonal of the product $C \times C$, where C is a nonsingular projective curve of genus $g > 0$. Prove that for every $(m, n, s) \in \mathbb{Z}^3 \setminus \{(0, 0, 0)\}$, $m\Delta$ cannot be numerically equivalent to a divisor of the form $n(\{x_1\} \times C) + s(C \times \{x_2\})$, with $x_1, x_2 \in C$. In particular, $\text{rank}(\text{Num}(C \times C)) \geq 3$, and $\text{Pic}(C \times C) \neq p_1^*(\text{Pic}(C)) \oplus p_2^*(\text{Pic}(C))$, where $p_1, p_2 : C \times C \to C$ are the canonical projections.

Bibliographic References. The proof of Theorem 6.2 is presented after [Sha2].

7

Morphisms from a Surface to a Curve. Elliptic and Quasielliptic Fibrations

Theorem 7.1. *Let $f : X \to Y$ be a dominant morphism from an irreducible, nonsingular algebraic variety X to an algebraic variety Y, such that the field extension $f^* : \Bbbk(Y) \to \Bbbk(X)$ between the rational function fields is separable and $\Bbbk(Y)$ is algebraically closed in $\Bbbk(X)$. Then there exists a nonempty open subset $V \subset Y$ such that for every $y \in V$ the fiber $f^{-1}(y)$ is geometrically integral.*

Proof. Let ξ be the generic point of X and let η be the generic point of Y. Then it is clear that the fiber $f^{-1}(\eta)$ contains ξ (because f is dominant); it is irreducible, because its closure is equal to X, and reduced: for each $x \in X$, the local ring $\mathcal{O}_{f^{-1}(\eta),x}$ coincides with $\mathcal{O}_{X,x}$. In other words, the scheme $f^{-1}(\eta)$ is integral over $\Bbbk(Y)$.

In fact, the hypotheses of the theorem imply that $f^{-1}(\eta)$ is even *geometrically* integral over $\Bbbk(Y)$, that is, for any field extension $\Bbbk(Y) \subset K$, the scheme $f^{-1}(\eta) \otimes K$ is still integral. Indeed, this follows by [EGA IV, (4.5.9) and (4.6.1)] from the fact that the field extension $\Bbbk(X)/\Bbbk(Y)$ is separable and $\Bbbk(Y)$ is algebraically closed in $\Bbbk(X)$.

Therefore there only remains to show that, if the generic fiber is geometrically integral, then there exists a nonempty open subset V in Y such that the fiber $f^{-1}(y)$ is geometrically integral for every $y \in V$. First, the problem is clearly local on Y, so we may assume that $Y = \mathrm{Spec}(A)$ is a nonsingular affine variety, with A an integral $\Bbbk$-algebra of finite type. Now let U be a nonempty affine open subset of X, and let $g : U \to Y$ be the restriction of f to U. We show that if the theorem holds for g, then it also holds for f.

Indeed, let V be a nonempty affine open subset of Y such that for every $y \in V$ the fiber $g^{-1}(y) = f^{-1}(y) \cap U$ is geometrically integral. Let $F_1, \ldots, F_m$ be the irreducible components of $F = X \setminus U$. Since for each i we have $\xi \notin F_i$, we see that $F \cap f^{-1}(\eta)$ is a proper closed subset of $f^{-1}(\eta)$. Then the theorem on the dimension of fibers [EGA IV, (13.1.1)] shows that there exists a nonempty open subset $V' \subset Y$ such that for every $y \in V'$ and every $x \in f^{-1}(y)$, $\dim_x f^{-1}(y) = \dim(f^{-1}(y) - F) = d$, where $d = \dim(X) - \dim(Y) = \dim f^{-1}(\eta)$. As U is dense in X, we have that $f^{-1}(y)$ is geometrically irreducible if and only if $g^{-1}(y)$ is geometrically irreducible. Moreover, we have that every geometric fiber of f over a point $y \in V \cap V'$ is irreducible and generically reduced. To see that this fiber is (globally) reduced there only remains to note that it cannot have any embedded components [AK, Chap. VII, (2.2)]. But this is automatically true, for all the local rings of the scheme $f^{-1}(y)$ are local complete intersection rings (because X and Y are nonsingular).

We have therefore reduced the proof to the case when $X = \mathrm{Spec}(B)$ and $Y = \mathrm{Spec}(A)$ are affine, with $A \subset B$ via f^*, and the corresponding extension $\Bbbk(X)/\Bbbk(Y)$ of rational function fields satisfies the hypotheses of the theorem. Then there exists a separating transcendence basis $t_1, \ldots, t_d$ of $\Bbbk(X)/\Bbbk(Y)$ and a primitive element $t_{d+1} \in \Bbbk(X)$ for the finite and separable extension $\Bbbk(X)/\Bbbk(Y)(t_1, \ldots, t_d)$, with t_{d+1} being a root of an irreducible and separable polynomial

$$\phi = T^r + a_1 T^{r-1} + \cdots + a_0, \quad a_i \in \Bbbk(Y)(t_1, \ldots, t_d).$$

Multiplying t_{d+1} by a suitable element of $\Bbbk(Y)$ if necessary, we may further assume that $a_i \in A[t_1, \ldots, t_d]$. As X is birationally isomorphic to the hypersurface $\phi = \phi(T_1, \ldots, T_d, T_{d+1}) = 0$ in $\mathbb{A}^{d+1}$, the earlier argument allows us to reduce the proof to the case when $B = A[T_1, \ldots, T_{d+1}]/(\phi)$, with $\phi \in A[T_1, \ldots, T_{d+1}]$.

The hypothesis that the generic fiber of f is geometrically integral means that ϕ remains irreducible over the algebraic closure $\overline{\Bbbk(Y)}$ of $\Bbbk(Y)$. For any point $y \in Y = \mathrm{Spec}(A)$, denote by $\phi_y(T_1, \ldots, T_{d+1})$ the image of ϕ by the canonical homomorphism $A[T_1, \ldots, T_{d+1}] \to \Bbbk(y)[T_1, \ldots, T_{d+1}]$. Then we must show that the set $\{y \in Y \mid \phi_y \text{ is irreducible over } \overline{\Bbbk(y)}\}$ contains an open subset of Y.

Let n be the total degree of ϕ, and let $\alpha(p)$ ($p \geq 0$) be the number of monomials of degree $\leq p$ in $d + 1$ variables. For all pairs $p, q \geq 0$ with $p + q = n$, we consider the morphism

$$u_{p,q} : \mathbb{A}^{\alpha(p)}(A) \times_A \mathbb{A}^{\alpha(q)}(A) \to \mathbb{A}^{\alpha(n)}(A)$$

given by

$$((a_{i_0, \ldots, i_d}), (b_{j_0, \ldots, j_d})) \mapsto (c_{h_0, \ldots, h_d}), \qquad \text{where}$$

$$c_{h_0, \ldots, h_d} = \sum_{0 \leq i_j \leq h_j} a_{i_0, \ldots, i_d} b_{h_0 - i_0, \ldots, h_d - i_d},$$

which corresponds to multiplication of polynomials of degrees p and q; $\mathbb{A}^n(A)$ denotes the affine space Spec $A[T_1, \ldots, T_n]$ over A.

Clearly, for $y \in Y$, ϕ_y is a product of factors of degrees p and q if and only if the point in $\mathbb{A}^{\alpha(n)}(\overline{\mathbb{k}(y)}) = \mathbb{A}^{\alpha(n)}(A) \otimes_A \overline{\mathbb{k}(y)}$ corresponding to ϕ_y is in the image of the morphism

$$u_{p,q} \otimes_A \overline{\mathbb{k}(y)} : \mathbb{A}^{\alpha(p)}(\overline{\mathbb{k}(y)}) \times_{\overline{\mathbb{k}(y)}} \mathbb{A}^{\alpha(q)}(\overline{\mathbb{k}(y)}) \to \mathbb{A}^{\alpha(n)}(\overline{\mathbb{k}(y)}).$$

But ϕ defines a morphism $\Phi : Y = \text{Spec}(A) \to \mathbb{A}^{\alpha(n)}(A)$. Let $Z = \cup_{p+q=n}[\text{Image of } u_{p,q}]$. By a theorem of Chevalley, Z is a constructible set, and therefore $\Phi^{-1}(Z)$ is also constructible. Since the generic point of Y is not in $\Phi^{-1}(Z)$, there exists a nonempty open subset V of Y that is disjoint from $\Phi^{-1}(Z)$. This concludes the proof of the theorem.   □

If, in the hypotheses of the theorem, Y is an algebraic curve, then the separability of the extension $\mathbb{k}(X)/\mathbb{k}(Y)$ is a consequence of $\mathbb{k}(Y)$ being algebraically closed in $\mathbb{k}(X)$. This follows from the next lemma.

Lemma 7.2. *Let K be an algebraic function field in one variable over a perfect field $\mathbb{k}$ (that is, K is a finitely generated extension of $\mathbb{k}$ of transcendence degree 1), and let $L \supset K$ be an extension of K such that K is algebraically closed in L. Then L/K is a separable extension.*

Proof. Let $p = \text{char}(\mathbb{k})$. If $p = 0$ there is nothing to prove; assume that $p > 0$.

Since $\mathbb{k}$ is perfect, there exists an element $f \in K$ such that the extension $K/\mathbb{k}(f)$ is finite and separable. Then $K^{1/p} = K(f^{1/p})$. Indeed, since $K \subseteq K^{1/p}$, the inclusion $K(f^{1/p}) \subseteq K^{1/p}$ is obvious. On the other hand, let $g \in K$ be a primitive element of the finite and separable extension $K/\mathbb{k}(f)$. Then $K^{1/p} = \mathbb{k}(f^{1/p})(g^{1/p})$, and since g is separable over $\mathbb{k}(f)$, $g^{1/p}$ is also separable over $\mathbb{k}(f^{1/p})$, and therefore $K^{1/p} = \mathbb{k}(f^{1/p})(g) \subseteq K(f^{1/p})$; thus $K^{1/p} = K(f^{1/p})$, as stated.

To prove that L/K is separable, we must show that L and $K^{1/p}$ are linearly disjoint over K. But $f^{1/p} \notin L$ (because K is algebraically closed in L), and since $f \in K \subseteq L$ we have

$$\left[L(f^{1/p}) : L \right] = p = \left[K(f^{1/p}) : K \right];$$

the proof of the lemma is complete.   □

Corollary 7.3. *Let $f : X \to Y$ be a dominant morphism from an irreducible nonsingular variety X of dimension at least 2 to an irreducible curve Y, such that $\mathbb{k}(Y)$ is algebraically closed in $\mathbb{k}(X)$. Then the fiber $f^{-1}(y)$ is geometrically integral for all but a finite number of closed points $y \in Y$.*

This follows directly from Theorem 7.1 and Lemma 7.2.

Proposition 7.4. *Let $f : X \to Y$ be a dominant morphism of nonsingular irreducible varieties over an algebraically closed field $\Bbbk$ of characteristic zero. Then there exists a nonempty open set $V \subseteq Y$ such that $f|_{f^{-1}(V)} : f^{-1}(V) \to V$ is a smooth morphism, that is, such that for every $y \in V$ the fiber $f^{-1}(y)$ is geometrically smooth of pure dimension $\dim(X) - \dim(Y)$.*

Proof. Step 1. Let $g : U \to V$ be a dominant morphism of irreducible varieties over $\Bbbk$ (with $\operatorname{char}(\Bbbk) = 0$). Then there exists a nonempty open set $U' \subseteq U$ such that $g' = g|_{U'} : U' \to V$ is smooth.

Proof of Step 1. By replacing U and V with suitable open subsets, if necessary, we may assume that U and V are nonsingular varieties. Since $\Bbbk(U)$ is separably generated over $\Bbbk(V)$, we have that $\Omega_{U/V}$ is locally free of rank $n = \dim(U) - \dim(V)$ at the generic point of U, by a well-known commutative algebra result and property (c) in the proof of (6.2). Therefore $\Omega_{U/V}$ is locally free of rank n over a nonempty open subset U' of U. Since U and V are nonsingular, we get that $g' = g|_{U'}$ is smooth.

Step 2. Let $f : X \to Y$ be a morphism of algebraic schemes over $\Bbbk$, with $\operatorname{char}(\Bbbk) = 0$. For each $r \geq 0$ let X_r denote the set of closed points $x \in X$ such that $\operatorname{rank}(T_{f,x}) \leq r$, where $T_{f,x} : T_{X,x} \to T_{Y,f(x)}$ is the tangent map. Then $\dim(\overline{f(X_r)}) \leq r$.

Proof of Step 2. Let V be an irreducible component of $f(X_r)$, and let U be an irreducible component of X_r such that $f(U) \subseteq V$ and $\overline{f(U)} = V$. Consider U and V with the corresponding reduced subscheme structures, and let $g = f_U : U \to V$. By Step 1 there exists a nonempty subset U' of U such that $g' = g|_{U'} : U' \to V$ is a smooth morphism. By construction there exists a point $x \in U' \cap X_r$. In the commutative diagram

$$
\begin{array}{ccc}
T_{U',x} & \longrightarrow & T_{X,x} \\
{\scriptstyle T_{g',x}}\Big\downarrow & & \Big\downarrow{\scriptstyle T_{f,x}} \\
T_{V,f(x)} & \longrightarrow & T_{Y,f(x)}
\end{array}
$$

the horizontal maps are injective (because U' and V are locally closed subschemes in X and Y, respectively). But $x \in X_r \implies \operatorname{rank}(T_{f,x}) \leq r$, and $T_{g',x}$ is surjective (because g' is a smooth morphism); therefore $\dim(T_{V,f(x)}) \leq r$, so that $\dim(V) \leq r$.

Step 3. Conclusion.

Let $r = \dim(Y)$ and let X_{r-1} be the set defined in Step 2. By Step 2, $\dim(\overline{f(X_{r-1})}) \leq r - 1$; removing $\overline{f(X_{r-1})}$ from Y, we may assume that $\operatorname{rank}(T_{f,x}) \geq r$ for every closed point x of X. Since Y is nonsingular and r-dimensional, we get that $T_{f,x}$ is surjective for every closed point $x \in X$. In other words, f is smooth. $\qquad\square$

Definition 7.5. Let $f : X \to B$ be a surjective morphism from a surface X to a nonsingular (projective) curve B. Then f is called a *fibration* over the curve B.

f is a *minimal fibration* if $\forall b \in B$, the fiber $F_b = f^{-1}(b) = X \times_B \mathrm{Spec}(\Bbbk(b))$ does not contain any exceptional curves of the first kind among its components.

Note that if $f : X \to B$ is a fibration, then f is necessarily flat (because B is a nonsingular curve), and therefore, by the base change theorems, cf. [Mum2, §5] or [EGA III, (2)], the map $b \to p_a(F_b)$ is constant. (In fact, this is why f is called a fibration.)

Let $f : X \to B$ be a fibration such that $f_* \mathcal{O}_X = \mathcal{O}_B$. Then, by Zariski's connectedness theorem [EGA III, (4.3.2)], all the fibers of f are connected. Since the condition $f_* \mathcal{O}_X = \mathcal{O}_B$ is equivalent (when B is a nonsingular curve) to $\Bbbk(B)$ being algebraically closed in $\Bbbk(X)$, Corollary 7.3 shows that all but finitely many fibers of f are integral curves.

Definition 7.6. A fibration f is *elliptic* if f is minimal, $f_* \mathcal{O}_X = \mathcal{O}_B$, and almost all the fibers of f (i.e., except finitely many closed fibers) are nonsingular elliptic curves.

f is *quasielliptic* if f is minimal, $f_* \mathcal{O}_X = \mathcal{O}_B$ and almost all the fibers of f are singular integral curves of arithmetic genus one.

Therefore a minimal fibration f with $f_* \mathcal{O}_X = \mathcal{O}_B$ is elliptic or quasielliptic if $p_a(F_b) = 1$ for some $b \in B$. If all the fibers of f are singular curves, then f is quasielliptic. If at least one fiber F_b is nonsingular, then the morphism f is generically smooth and f is an elliptic fibration. Proposition 7.4 shows that quasielliptic fibrations cannot exist in characteristic zero.

In the final part of this chapter we will show that quasielliptic fibrations exist only in characteristic 2 or 3, and almost every fiber of a quasielliptic fibration is a rational curve with a unique ordinary cuspidal point as singularity. Elliptic and quasielliptic fibrations play a central role in the classification of surfaces.

Let $f : X \to B$ be an elliptic or quasielliptic fibration, and let $D = F_b = \sum_{i=1}^{p} n_i E_i$, with $b \in B$ a closed point, $p \geq 1$, $n_i \geq 1$ and $E_1, \ldots, E_p$ the integral components of D such that $E_i \neq E_j$ for $i \neq j$.

If K is a canonical divisor on X, then $(D \cdot E_i) = (K \cdot E_i) = 0$ for all $i = 1, \ldots, p$. Indeed, if F is a fiber of f, distinct from D, then $F \equiv D$, so that $(D \cdot E_i) = (F \cdot E_i) = 0$ for all i, because $\mathrm{Supp}(F) \cap E_i = \varnothing$. On the other hand, the genus formula gives $2p_a(D) - 2 = (D^2) + (D \cdot K)$; as $p_a(D) = 0$ and $(D^2) = 0$, we get that $(D \cdot K) = 0$, and therefore $\sum_{i=1}^{p} n_i(K \cdot E_i) = 0$. If $p = 1$ and $n_1 \geq 1$, we get $(K \cdot E_1) = 0$. If $p \geq 2$, then by (2.6) we have $(E_i^2) < 0$ for all $i = 1, \ldots, p$. But then we must have $(K \cdot E_i) \geq 0$ for all i, because if $(K \cdot E_i) < 0$ for some i, then—as $(E_i^2) < 0$—it would immediately follow that E_i is an exceptional curve of the first kind, contradicting the minimality of f. Finally, from $\sum_{i=1}^{p} n_i(K \cdot E_i) = 0$ and $(K \cdot E_i) \geq 0$ for all

i we get that $(K \cdot E_i) = 0$ for all i, as stated. This observation motivates the following definition.

Definition 7.7. Let X be a minimal surface and let $D = \sum_{i=1}^{p} n_i E_i > 0$ be an effective divisor on X. Then D is a *divisor* (or a *curve*) *of canonical type* if $(K \cdot E_i) = (D \cdot E_i) = 0$ for all $i = 1, \ldots, p$. If D is also connected and the greatest common divisor of the integers $n_1, \ldots, n_p$ is equal to 1, then we say that D is an *indecomposable* divisor (or curve) of canonical type.

From the preceding discussion we see that if a minimal surface X has an elliptic or quasielliptic fibration structure $f : X \to B$, then X contains at least one indecomposable curve of canonical type.

Theorem 7.8. *Let* $D = \sum_{i=1}^{p} n_i E_i > 0$ *be an indecomposable curve of canonical type on a minimal surface* X, *and let* L *be an invertible* $\mathcal{O}_D$-*module. If* $\deg(L \otimes \mathcal{O}_{E_i}) = 0$ *for every* $i = 1, \ldots, p$, *then* $H^0(D, L) \neq 0$ *if and only if* $L \cong \mathcal{O}_D$, *and* $H^0(D, \mathcal{O}_D) = \Bbbk$.

Proof. It is enough to show that every nonzero section $s \in H^0(D, L)$ generates L, that is, it defines an isomorphism of $\mathcal{O}_D$ onto L. This would show also that $H^0(D, \mathcal{O}_D)$ is a field, and therefore that $H^0(D, \mathcal{O}_D) \cong \Bbbk$, because $H^0(D, \mathcal{O}_D)/\Bbbk$ is a finite extension and $\Bbbk$ is algebraically closed.

Let $s_i = s|_{E_i} \in H^0(E_i, L \otimes \mathcal{O}_{E_i})$. Since $\deg(L \otimes \mathcal{O}_{E_i}) = 0$, we have that either s_i is identically zero on E_i or s_i does not vanish anywhere on E_i (in which case s_i generates $L \otimes \mathcal{O}_{E_i}$). If s_i is identically zero on a component E_i, then s_j is also identically zero for every j, for D is a connected curve. Therefore, if s_i does not vanish identically on E_i, then s does not vanish anywhere on D, so that s generates L; in this case the proof is complete.

Now assume that s_i is identically zero on E_i for every i. We shall show that this assumption leads to a contradiction.

Let k_i be the order of vanishing of s along E_i. This means that

$$s \in \mathrm{Ker}\left[H^0(D, L) \to H^0(k_i E_i, L_{k_i E_i})\right],$$

and if $k_i < n_i$, then

$$s \notin \mathrm{Ker}\left[H^0(D, L) \to H^0((k_i + 1)E_i, L_{(k_i+1)E_i})\right],$$

where for any closed subscheme $Z \subset D$ we write L_Z for the restriction of L to Z: $L_Z = L \otimes \mathcal{O}_Z$. We have $1 \leq k_i \leq n_i$ for every i.

First assume that $k_i < n_i$. From the commutative diagram

$$H^0(L)$$

$$0 \longrightarrow H^0(E_i, L \otimes (\mathcal{O}_X(-k_i E_i) \otimes \mathcal{O}_{E_i})) \longrightarrow H^0(L_{(k_i+1)E_i}) \longrightarrow H^0(L_{k_i E_i})$$

$$(7.8.1)$$

where the horizontal sequence is exact, we see that s defines a nonzero section in $H^0(E_i, L \otimes (\mathcal{O}_X(-k_i E_i) \otimes \mathcal{O}_{E_i}))$. This section vanishes at every point $P \in E_i$ with order at least equal to the intersection multiplicity $(E_i \cdot \sum_{j \neq i} k_j E_j; P)$.

Indeed, this assertion is local around P. For simplicity assume that E_i intersects exactly one component E_j with $j \neq i$ at P (the general case is similar). Let $A = \mathcal{O}_{X,P}$, and let $t_i = 0$ (resp. $t_j = 0$) be a local equation of E_i (resp. E_j) at P. We have a commutative diagram similar to (7.8.1):

$$s \in A/(t_i^{n_i} \cdot t_j^{n_j})$$

$$0 \longrightarrow A/(t_i) \xrightarrow{t_i^{k_i}} A/(t_i^{k_i+1}) \longrightarrow A/(t_i^{k_i}) \longrightarrow 0$$

Write $s = t_i^{k_i} \cdot \lambda = t_j^{k_j} \cdot \mu$, with $\lambda, \mu \in A$; since t_i, t_j is an A-regular sequence, we have $\mu = t_i^{k_i} \cdot \mu'$ for some $\mu' \in A$, and then $\lambda = t_j^{k_j} \cdot \mu'$. The section s is represented by $\bar{\lambda} = \lambda \mod (t_i) = t_j^{k_j} \cdot \mu' \mod (t_i)$. Then $\mathrm{ord}_P(\bar{\lambda}) = \dim(A/(t_i, \lambda)) = \dim(A/(t_i, t_j^{k_j} \cdot \mu')) \geq \dim(A/(t_i, t_j^{k_j})) = $ intersection multiplicity $(E_i \cdot k_j E_j; P)$, and the inequality claimed earlier is proved.

Therefore if $k_i < n_i$ we have

$$\left(E_i \cdot \sum_{j \neq i} k_j E_j\right) \leq \deg_{E_i} \left(L \otimes (\mathcal{O}_X(-k_i E_i) \otimes \mathcal{O}_{E_i})\right)$$

$$= k_i \cdot \deg(\mathcal{O}_X(-E_i)/\mathcal{O}_X(-2E_i))$$

$$= -k_i \cdot (E_i^2), \tag{7.8.2}$$

because $\deg(L_{E_i}) = 0$.

On the other hand, if $k_i = n_i$, then $(E_i \cdot D) = 0$ implies

$$\left(E_i \cdot \sum_j k_j E_j\right) = -\left(E_i \cdot \sum_{j \neq i}(n_j - k_j)E_j\right) \leq 0, \tag{7.8.3}$$

for $n_i - k_i = 0$ and $(E_i \cdot E_j) \geq 0$ if $i \neq j$.

Therefore, if we put $D_1 = \sum_j k_j E_j$, from (7.8.2) and (7.8.3) we have $(D_1 \cdot E_i) \leq 0$ for all i. But $\sum_i n_i(D_1 \cdot E_i) = (D \cdot D_1) = \sum_i k_i(D \cdot E_i) = 0$, and therefore $(D_1 \cdot E_i) = 0$ for all i. In particular, $(D_1^2) = 0$.

By (2.5) we have that D_1 is a rational multiple of D, because the support of D is connected. That is, the rational numbers k_i/n_i are equal for all i: $D_1 = rD$, $r \in \mathbb{Q}$, $0 < r \leq 1$, and $k_i/n_i = r$ for all i.

But the greatest common divisor of $n_1, \ldots, n_p$ is 1, and therefore there exist integers $a_1, \ldots, a_p$ such that $\sum_i a_i n_i = 1$. Then $r = a_1 k_1/a_1 n_1 = \cdots = a_p k_p/a_p n_p = (\sum_i a_i k_i)/(\sum_i a_i n_i) \in \mathbb{Z}$, and therefore $r = 1$, that is, $D_1 = D$. In other words, $s = 0$ in $H^0(D, L)$, which is a contradiction.   □

Corollary 7.9. *If D is an indecomposable curve of canonical type, then $\omega_D \cong \mathcal{O}_D$, where ω_D is the dualizing sheaf of D.*

Proof. By Serre duality, $\dim H^1(\omega_D) = \dim H^0(\mathcal{O}_D) = 1$. From the exact sequence

$$0 \to \mathcal{O}_X(K) \to \mathcal{O}_X(K+D) \to \omega_D \to 0$$

and the Riemann–Roch theorem we have $\chi(\omega_D) = \chi(\mathcal{O}_X(K+D)) - \chi(\mathcal{O}_X(K)) = \frac{1}{2}(K+D \cdot D) = 0$. Thus $\dim H^0(\omega_D) = 1$. Then Theorem 7.8 shows that $\omega_D \cong \mathcal{O}_D$, for $\deg(\omega_D|_{E_i}) = (K+D \cdot E_i) = 0$ for all i. $\square$

Corollary 7.10. *If $D = \sum_{i=1}^p n_i E_i$ is an indecomposable curve of canonical type and D' is an effective divisor on X such that $(D' \cdot E_i) = 0$ for all $i = 1, \ldots, p$, then $D' = nD + D''$, where $n \geq 0$ and D'' is an effective divisor with $\mathrm{Supp}(D'') \cap \mathrm{Supp}(D) = \varnothing$.*

Proof. Let n be the nonnegative integer defined by the conditions $D' - nD \geq 0$ and $D' - (n+1)D \not\geq 0$. Put $D'' = D' - nD$ and $L = \mathcal{O}_D(D'')$. Then we have an exact sequence

$$0 \to \mathcal{O}_X(D'' - D) \to \mathcal{O}_X(D'') \to \mathcal{O}_D(D'') = L \to 0.$$

Let $s \in H^0(X, \mathcal{O}_X(D''))$ with $\mathrm{div}_X(s) = D''$. Since $D'' - D = D' - (n+1)D \not\geq 0$, the image $\tilde{s}$ of s in $H^0(D, L)$ is nonzero. On the other hand, as $\deg(L|_{E_i}) = (D'' \cdot E_i) = (D' \cdot E_i) - n(D \cdot E_i) = 0$ for all i, Theorem 7.8 implies $L \cong \mathcal{O}_D$. Therefore $s(x) \neq 0$ for all $x \in D$, that is, $\mathrm{Supp}(D'') \cap \mathrm{Supp}(D) = \varnothing$. $\square$

Theorem 7.11. *Let X be a minimal surface with $(K^2) = 0$ and $(K \cdot C) \geq 0$ for all curves C on X. If D is an indecomposable curve of canonical type on X, then there exists an elliptic or quasielliptic fibration structure $f : X \to B$ on X. (Of course, the converse is also true.) More precisely, there exists an integer $n > 0$ such that the Stein factorization of the morphism $\phi_{|nD|} : X \to \mathbb{P}(H^0(\mathcal{O}_X(nD))\check{})$ is a pencil of curves of canonical type (that is, an elliptic or quasielliptic fibration).*

Proof. Case $p_g = 0$.

By (7.9), for each integer $n \geq 0$ we have an exact sequence

$$0 \to \mathcal{O}_X(nK + (n-1)D) \to \mathcal{O}_X(nK + nD) \to \mathcal{O}_D \to 0. \qquad (7.11.1)$$

We have also $H^2(\mathcal{O}_X(nK + (n-1)D)) = 0$ for all $n \geq 2$. Indeed, by Serre duality it is enough to show that $H^0(\mathcal{O}_X(-p(K+D))) = 0$, or $|-p(K+D)| = \varnothing$, for all $p \geq 1$. By way of contradiction, assume that there exists an effective divisor $\Delta \in |-p(K+D)|$. Then there are two possibilities:

(a) $\Delta = 0$, so that $p(K + D) \sim 0$, or $pK \sim -pD$. Then let C be an irreducible curve such that $(D \cdot C) > 0$. Then $(K \cdot C) = 1/p(pK \cdot C) = -1/p(pD \cdot C) = -(D \cdot C) < 0$, and this contradicts the hypotheses of the theorem.

(b) $\Delta > 0$. Then let C be an irreducible curve such that $(\Delta \cdot C) > 0$. On the other hand, $(\Delta \cdot C) = -p(K \cdot C) - p(D \cdot C) \leq 0$, because by hypothesis we have $(K \cdot C) \geq 0$ and $(D \cdot C) \geq 0$ (for C is a curve of canonical type), and again we have reached a contradiction.

From $H^2(\mathcal{O}_X(nK + (n-1)D)) = 0$ and the exact sequence (7.11.1) we get a cohomology exact sequence

$$0 \to H^2(\mathcal{O}_X(nK + nD)) \to H^2(\mathcal{O}_D) = 0,$$

and therefore, for every $n \geq 2$, we have $H^2(\mathcal{O}_X(nK + nD)) = 0$. As $H^1(\mathcal{O}_D) = H^0(\omega_D) = H^0(\mathcal{O}_D) \neq 0$ (by (7.9)), we get $H^1(\mathcal{O}_X(nK+nD)) \neq 0$. The Riemann–Roch theorem gives:

$$\chi(\mathcal{O}_X(nK + nD)) = \chi(\mathcal{O}_X) = 1 - q.$$

Using (5.1), Noether's formula becomes $10 - 8q = b_2$, whence $q \leq 1$. Therefore we have proved that

$$\chi(\mathcal{O}_X(nK + nD)) = 0 \text{ or } 1 \text{ for } n \geq 2.$$

As we have seen, $H^1(\mathcal{O}_X(nK + nD)) \neq 0$ and $H^2(\mathcal{O}_X(nK + nD)) = 0$, so we conclude that $H^0(\mathcal{O}(nK + nD)) \neq 0$ for $n \geq 2$. In other words, if $n \geq 2$ then there exists a $D_n \in |nK + nD|$. As in (a), we see that $D_n \neq 0$. We claim that D_n is of canonical type.

Indeed, let $D = \sum_i n_i E_i$; then $(D_n \cdot E_i) = n(K \cdot E_i) + n(D \cdot E_i) = 0$ for all i, and therefore by Corollary 7.10 we can write $D_n = aD + \sum_j k_j F_j$, $a \geq 0$, $k_j > 0$, and the F_j are pairwise distinct irreducible curves that do not intersect D. But $(K \cdot F_j) \geq 0$ for all j, and $\sum_j k_j (K \cdot F_j) = (K \cdot \sum_j k_j F_j + aD) = (K \cdot nK + nD) = 0$; therefore $(K \cdot F_j) = 0$ for all j. Finally, $(D_n \cdot F_j) = n(K \cdot F_j) + n(D \cdot F_j) = 0$ for all j, because $\mathrm{Supp}(D) \cap F_j = \varnothing$, and the claim is proved.

Now we remark that we cannot have $D_n = a_n D$ for integers $a_n \geq 0$ for all $n \geq 2$: otherwise $nK \sim \lambda_n D$ with $\lambda_n \in \mathbb{Z}$ and $n \geq 2$; in particular, $K = 3K - 2K \sim (\lambda_3 - \lambda_2)D = \lambda D$. If $\lambda < 0$, take an irreducible curve C such that $(D \cdot C) > 0$; then $(K \cdot C) < 0$, which is a contradiction. If $\lambda \geq 0$, then $|K| = |\lambda D| \neq \varnothing$, which is another contradiction.

Therefore we have proved that there exists at least one indecomposable curve of canonical type D' on X, which is disjoint from D. Using Corollary 7.9 again, we get an exact sequence

$$0 \to \mathcal{O}_X(2K + D + D') \to \mathcal{O}_X(2K + 2D + 2D') \to \mathcal{O}_D \oplus \mathcal{O}_{D'} \to 0.$$

Arguing as before, we see that $H^2(\mathcal{O}_X(2K+D+D')) = 0$, and therefore $H^2(\mathcal{O}_X(2K+2D+2D')) = 0$, and since $H^1(\mathcal{O}_D \oplus \mathcal{O}_{D'})$ is two-dimensional, we have $\dim H^1(\mathcal{O}_X(2K+2D+2D')) \geq 2$, and finally by Riemann–Roch we have $\chi(\mathcal{O}_X(2K+2D+2D')) = \chi(\mathcal{O}_X) = 0$ or 1, whence

$$\dim H^0(\mathcal{O}_X(2K+2D+2D')) \geq 2.$$

Let $\Delta \in |2K+2D+2D'|$. We have $\Delta > 0$, $(\Delta^2) = 0$ and $\dim |\Delta| \geq 1$. Moreover, since D and D' are both of canonical type, Δ is also of canonical type. Therefore the complete linear system is composed with a pencil.

Indeed, if C is the fixed part of $|\Delta|$, then from (2.6) and the fact that Δ is of canonical type we infer that $((\Delta - C)^2) \leq 0$. In other words, the rational map $\phi_{|\Delta|} : X \to \phi_{|\Delta|}(X) = B \subseteq |\Delta|$ is defined everywhere, because otherwise we would have $((\Delta - C)^2) > 0$. Since $\dim |\Delta| \geq 1$, B cannot be a point. B cannot be a surface either; indeed, if B is a surface, then $\Delta - C = \phi^*(H)$ with H a hyperplane section on B, and then $((\Delta - C)^2) = [\Bbbk(X) : \Bbbk(B)](H^2) > 0$, which is a contradiction. Thus $|\Delta|$ is indeed composed with a pencil, and $\phi_{|\Delta|}$ is a morphism.

From $(D \cdot \Delta) = (D \cdot 2K+2D+2D') = 0$, $(D \cdot \Delta - C) \geq 0$ and $(D \cdot C) \geq 0$ we get that $(D \cdot \Delta - C) = 0$. Since D is connected, we see that D is contained in one of the fibers of $\phi_{|\Delta|}$. Using (2.6) and the fact that $(D^2) = 0$, we get that a rational multiple of D is equal to one of the fibers of the morphism $f : X \to B'$, where f is the morphism induced by $\phi_{|\Delta|}$ by Stein factorization (that is, B' is the normalization of B in the field $\Bbbk(X)$). Since the greatest common divisor of the coefficients of D is equal to 1, one of the fibers of f is in fact a (positive) integral multiple of D. In particular, the arithmetic genus of that fiber of f is 1, so that f is an elliptic or quasielliptic fibration. Theorem 7.11 is proved if $p_g = 0$.

Case $p_g > 0$.

As in the previous case, it will suffice to show that the dimension of the $\Bbbk$-linear space $H^0(\mathcal{O}_X(\Delta))$ is greater than or equal to 2 for some divisor Δ of canonical type. Specifically, we shall prove that there exists an $n > 0$ such that $\dim H^0(\mathcal{O}_X(nD)) \geq 2$.

Let $F_n = \mathcal{O}_X(nD)/\mathcal{O}_X$. The exact sequence

$$H^0(\mathcal{O}_X(nD)) \to H^0(F_n) \to H^1(\mathcal{O}_X)$$

shows that it suffices to check that $\dim H^0(F_n) \to \infty$ as $n \to \infty$. Put $L = F_1$. Then L is an invertible sheaf on D, and from the exact sequence

$$0 \to F_{n-1} \to F_n \to L^n \to 0 \tag{7.11.2}$$

we see that the map $n \mapsto \dim H^0(F_n)$ is nondecreasing. By Riemann–Roch we have that $\chi(\mathcal{O}_X(nD)) = \chi(\mathcal{O}_X)$, and therefore, by additivity of the Euler–Poincaré characteristic, $\chi(F_n) = 0$ for all $n > 0$.

In the exact sequence

$$H^1(F_n) \to H^2(\mathcal{O}_X) \to H^2(\mathcal{O}_X(nD))$$

we have $H^2(\mathcal{O}_X(nD)) = 0$ for $n \gg 0$, because—using Serre duality—this is equivalent to showing that $|K - nD| = \varnothing$ for n sufficiently large, and this is true by the following general observation.

(7.11.3) *Remark.* If Δ is any divisor on a surface X and $D > 0$ is a strictly positive divisor, then $|\Delta - nD| = \varnothing$ for n sufficiently large.

Indeed, let H be an ample divisor on X. Then $(D \cdot H) > 0$, and therefore $(\Delta - nD \cdot H) < 0$ for all n sufficiently large. On the other hand, $(\Delta - nD \cdot H) \geq 0$ if $|\Delta - nD| \neq \varnothing$.

Returning to the proof of Theorem 7.11 (case $p_g > 0$), we see that $H^1(F_n) \neq 0$ for n sufficiently large (because $p_g = \dim H^2(\mathcal{O}_X) > 0$ and $H^2(\mathcal{O}_X(nD)) = 0$ for $n \gg 0$). Thus $\dim H^0(F_n) = \dim H^1(F_n) \neq 0$ for $n \gg 0$. By contradiction, assume that the sequence of integers $\{\dim H^0(F_n)\}$ is bounded above, and let n be the largest index for which $\dim H^0(F_{n-1}) < \dim H^0(F_n)$. From the exact sequence (7.11.2) we infer that L^n has a nonzero global section coming from a section $s \in H^0(F_n)$. Since D is an indecomposable curve of canonical type and L^n has degree zero on every component of D, Corollary 7.9 gives that $s|_D$ does not vanish at any point of D (i.e., $L^n \cong \mathcal{O}_D$); in other words, s does not vanish at any point of D. Since $\mathrm{Supp}(F_n) = D$, we get that s generates F_n as an $\mathcal{O}_X$-module at all points of X; that is, s defines a surjective homomorphism of $\mathcal{O}_X$-modules, $\mathcal{O}_X \to F_n$, whose kernel is exactly $\mathcal{O}_X(-nD)$. Thus we have proved that $\mathcal{O}_X/\mathcal{O}_X(-nD) \cong F_n \cong \mathcal{O}_X(nD)/\mathcal{O}_X$. Taking tensor powers of order $m \geq 1$ of F_n we get isomorphisms:

$$\mathcal{O}_X/\mathcal{O}_X(-nD) \cong F_n \cong \mathcal{O}_X(mnD)/\mathcal{O}_X(n(m-1)D) \cong F_{nm}/F_{n(m-1)}.$$

These isomorphisms induce the following diagram with exact rows and columns:

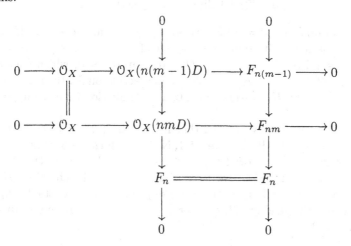

and from this we get a commutative diagram with exact rows and columns:

$$
\begin{array}{ccccccc}
H^1(F_{n(m-1)}) & \xrightarrow{\;\alpha\;} & H^1(F_{nm}) & \longrightarrow & H^1(F_n) & \longrightarrow & 0 \\
\downarrow & & \downarrow & & & & \\
H^2(\mathcal{O}_X) & = & H^2(\mathcal{O}_X) & & & & \\
\downarrow & & \downarrow & & & & \\
H^2(\mathcal{O}_X(n(m-1)D)) & = & H^2(\mathcal{O}_X(nmD)) & = & 0 & & \text{for } m \gg 0.
\end{array}
$$

Therefore the map α is nonzero (because $H^2(\mathcal{O}_X) \neq 0$), and consequently

$$\dim H^1(F_{nm}) = \dim \operatorname{Im}(\alpha) + \dim H^1(F_n) > \dim H^1(F_n).$$

This inequality contradicts the choice of n. Therefore the proof of the theorem is also complete in the case $p_g > 0$. □

Theorem 7.12. *Let X be a minimal surface with $(K^2) = 0$ and $(K \cdot C) \geq 0$ for every curve C on X. Then either $2K \sim 0$ or X contains at least one indecomposable curve of canonical type.*

Proof. If $|2K| \neq \varnothing$, let $D \in |2K|$. Then either $D \sim 0$, in which case $2K \sim 0$, or $D = \sum_{i=1}^p n_i E_i > 0$ with E_i the pairwise distinct integral components of D. In the second case we have $(D \cdot K) = 2(K^2) = 0$, so that $0 = (D \cdot K) = \sum_{i=1}^p n_i (K \cdot E_i)$. As $(K \cdot E_i) \geq 0$ (by hypothesis), we get $(K \cdot E_i) = 0$ for all i. Also, $(D \cdot E_i) = 2(K \cdot E_i) = 0$ for all i; in other words, D is of canonical type. In particular, X contains at least one indecomposable curve of canonical type.

Therefore there only remains to consider the case $|2K| = \varnothing$. Since $(K^2) = 0$, the Riemann–Roch theorem gives:

$$\dim |2K| + \dim |-K| \geq \chi(\mathcal{O}_X) - 2.$$

But $p_2 = \dim H^0(\omega_X^2) = 0$ implies $p_1 = p_g = 0$, and since $\dim |2K| = -1$ we have the inequality $\dim |-K| \geq -q$. If $q = 0$ we get $|-K| \neq \varnothing$, so that there exists $D \in |-K|$. If $D = 0$ then $2K \sim 0$, which contradicts $|2K| = \varnothing$. If $D > 0$, let H be a hyperplane section on X; then $(H \cdot D) > 0$, and therefore $(K \cdot H) = -(D \cdot H) < 0$, a contradiction with the hypotheses of the theorem.

In conclusion, we have proved that in the case $|2K| = \varnothing$ we have $q \geq 1$ (and clearly, $p_g = 0$). By Theorem 5.1, Noether's formula becomes $10 - 8q = b_2$; therefore $q = 1$. Now let $f : X \to B = \operatorname{Alb}(X)$ be the Albanese morphism of X. By (5.3) we have $\dim(B) = q = 1$. Therefore B is an elliptic curve, and f is surjective (or else f would be a constant map, and $f(X)$ would not generate B; see [Ser4]). Let $F_b = f^{-1}(b)$ be the fiber over a point $b \in B$.

If $p_a(F_b) = 0$, then since $(F_b^2) = 0$ we would have by the genus formula:

$$(K \cdot F_b) = -2,$$

and this would contradict the hypotheses of the theorem.

If $p_a(F_b) = 1$, then F_b would be a curve of canonical type, and the proof of the theorem would be complete.

Now assume that $p_a(F_b) \geq 2$. By the genus formula, we get $(K \cdot F_b) = 2p_a(F_b) - 2 \geq 2$. For each closed point $a \in B \setminus \{b\}$, where $b \in B$ is a fixed closed point of B, consider the exact sequence

$$0 \to \mathcal{O}_X(2K + F_a - F_b) \to \mathcal{O}_X(2K + F_a) \to \mathcal{O}_{F_b}(2K|_{F_b}) \to 0.$$

We have $|F_b - F_a - K| = \varnothing$; indeed, otherwise there would exist $D \sim F_b - F_a - K$, $D \geq 0$, and then $K \sim F_b - F_a - D$, so that $(K \cdot F_b) = (F_b^2) - (F_a \cdot F_b) - (D \cdot F_b) = -(D \cdot F_b) \leq 0$ (for the fibers of f are numerically equivalent to each other and $D \geq 0$), and this contradicts the inequality $(K \cdot F_b) \geq 2$, obtained earlier. Then, by Serre duality, we get that $H^2(2K + F_a - F_b) = 0$. On the other hand, the Riemann–Roch theorem gives $\chi(\mathcal{O}_X(2K + F_a - F_b)) = \chi(\mathcal{O}_X) = 1 - q + p_g = 0$. Therefore, for each $a \in B \setminus \{b\}$ we have one of the following two possibilities:

(a) $|2K + F_a - F_b| \neq \varnothing$, or

(b) $H^i(\mathcal{O}_X(2K + F_a - F_b)) = 0$ for all $i = 0, 1, 2$—and, in particular, the restriction homomorphism $H^0(\mathcal{O}_X(2K + F_a)) \to H^0(\mathcal{O}_{F_b}(2K|_{F_b}))$ is an isomorphism.

Now assume that $|2K + F_a - F_b| = \varnothing$ for all $a \in B$. Fix a nonzero section $s \in H^0(\mathcal{O}_{F_b}(2K|_{F_b}))$. Such a section exists because F_b is a curve of arithmetic genus ≥ 2 and $\deg(2K|_{F_b}) = 2(K \cdot F_b) = 4p_a(F_b) - 4$. Put $\Delta = \mathrm{div}_{F_b}(s)$. For every $a \in B \setminus \{b\}$, the section s can be lifted uniquely to a section $s_a \in H^0(\mathcal{O}_X(2K + F_a))$; let $D_a = \mathrm{div}_X(s_a)$. We get an algebraic family of effective divisors, $\{D_a\}_{a \in B \setminus \{b\}}$, parametrized by the curve $B \setminus \{b\}$, such that $D|_{F_b} = \Delta$ for all a. Moreover, if $a, a' \in B \setminus \{b\}, a \neq a'$, then D_a cannot be linearly equivalent to $D_{a'}$, or else we would have $F_a \sim F_{a'}$ (from $2K + F_a \sim 2K + F_{a'}$), or $a \sim a'$ on the elliptic curve B, which is an obvious contradiction. In particular, if $a \neq a'$ then $D_a \neq D_{a'}$. Then X is equal to the closure of the subset $\cup_{a \neq b} D_a$. Then the divisor D_b, the specialization of D_a as $a \rightsquigarrow b$, must contain F_b among its components, so that $D_b = F_b + D_b'$ for some $D_b' \geq 0$. But since $D_b \in |2K + F_b|$, we get that $D_b' \in |2K|$, contradicting the hypothesis that $|2K| = \varnothing$.

Therefore we have proved that $|2K + F_a - F_b| \neq \varnothing$ for at least one $a \in B$. If $D \in |2K + F_a - F_b|$, then D is of canonical type; the argument is similar to the one used earlier in this proof. $\square$

Corollary 7.13. Let X be a minimal surface with $(K^2) = 0$ and $(K \cdot C) \geq 0$ for all curves C on X. Then either $2K \sim 0$ or there exists an elliptic or quasielliptic fibration on X, $f : X \to B$.

Corollary 7.13 illustrates the central role played by elliptic and quasielliptic fibrations in the classification of algebraic surfaces. Our next objective
is to determine the canonical class of an elliptic or quasielliptic fibration
$f : X \to B$. From Corollary 7.3 we know that almost all the fibers of
f (except finitely many of them) are (geometrically) integral. Thus there
exists a finite number of closed points $b_1, \ldots, b_r \in B$ such that, for every
closed point $b \in B \setminus \{b_1, \ldots, b_r\}$, the fiber $F_b = f^{-1}(b)$ is an indecomposable curve of canonical type, and F_{b_i} $(i = 1, \ldots, r)$ is a curve of the form
$f^{-1}(b_i) = m_i P_i$, with P_i an indecomposable curve of canonical type and
$m_i \geq 2$ $(i = 1, \ldots, r, r \geq 0)$. This follows from the comments after Definition 7.6 and the fact that all the fibers of f are connected. The fibers
$F_{b_i} = m_i P_i$ with $m_i \geq 2$ and P_i indecomposable curve of canonical type,
$i = 1, \ldots, r$, are called the *multiple fibers* of the elliptic or quasielliptic
fibration $f : X \to B$.

From the base change theorems, cf. [Mum2, §5], we get that $R^1 f_* \mathcal{O}_X$ is
a coherent $\mathcal{O}_B$-module such that $(R^1 f_* \mathcal{O}_X) \otimes \Bbbk(b) \cong H^1(F_b, \mathcal{O}_{F_b})$ for all
$b \in B$. By Theorem 7.8 and Corollary 7.9 we have that $\dim H^1(F_b, \mathcal{O}_{F_b}) = 1$
for all $b \in B \setminus \{b_1, \ldots, b_r\}$; therefore the sheaf $R^1 f_* \mathcal{O}_X$ is invertible over
the open set $B \setminus \{b_1, \ldots, b_r\}$. On the other hand, since B is a nonsingular
curve, we have $R^1 f_* \mathcal{O}_X \cong L \oplus T$, where L is a locally free sheaf of finite
rank and T is the torsion part. Putting these observations together, we see
that L is an invertible sheaf and the support of T is contained in the set
$\{b_1, \ldots, b_r\}$; in particular, T is an $\mathcal{O}_B$-module of finite length.

If $b \in B$ is a closed point, then the Riemann–Roch theorem on X gives
$\chi(\mathcal{O}_X) = \chi(\mathcal{O}_X(-F_b))$, because F_b is a divisor of canonical type. From the
exact sequence

$$0 \to \mathcal{O}_X(-F_b) \to \mathcal{O}_X \to \mathcal{O}_{F_b} \to 0$$

and the additivity of the Euler–Poincaré characteristic we get $\chi(\mathcal{O}_{F_b}) = 0$,
so that $\dim H^1(\mathcal{O}_{F_b}) = \dim H^0(\mathcal{O}_{F_b})$. Using the same base change theorems
again, we get that

$$b \in \mathrm{Supp}(T) \iff \dim H^1(\mathcal{O}_{F_b}) \geq 2 \iff \dim H^0(\mathcal{O}_{F_b}) \geq 2.$$

This observation suggests the following definition.

Definition 7.14. The fibers F_b of the fibration f for which $\dim H^0(\mathcal{O}_{F_b}) \geq 2$ are called the *exceptional fibers* of f.

From what we said earlier it follows that every exceptional fiber of f is
a multiple fiber of f.

Theorem 7.15. *Let $f : X \to B$ be an elliptic or quasielliptic fibration,
and let $R^1 f_* \mathcal{O}_X = L \oplus T$ be the decomposition introduced earlier, with L
an invertible sheaf on B and T an $\mathcal{O}_B$-module of finite length. Then*

$$\omega_X \cong f^*(L^{-1} \otimes \omega_B) \otimes \mathcal{O}_X \left(\sum_{i=1}^{r} a_i P_i \right),$$

where

(a) $m_i P_i = F_{b_i}$ $(i = 1, \ldots, r)$ *are all the multiple fibers of* f,

(b) $0 \le a_i < m_i$,

(c) $a_i = m_i - 1$ *if* F_{b_i} *is not an exceptional fiber, and*

(d) $\deg(L^{-1} \otimes \omega_B) = 2p_a(B) - 2 + \chi(\mathcal{O}_X) + l(T)$, *where* $l(T)$ *is the length of* T.

Proof. If F_b is a nonmultiple fiber of f, then by Corollary 7.9 we have that $\omega_{F_b} \cong \mathcal{O}_{F_b}$. Since the conormal bundle of F_b in X is trivial, we have $\omega_X \otimes \mathcal{O}_{F_b} \cong \mathcal{O}_{F_b}$.

Thus, if $a_1, \ldots, a_m$ are m general points of B, we have an exact sequence:

$$0 \to \omega_X \to \omega_X \otimes \mathcal{O}_X \left(\sum_{i=1}^{m} F_{a_i} \right) \to \oplus_{i=1}^{m} \mathcal{O}_{F_{a_i}} \to 0,$$

whence a cohomology exact sequence:

$$0 \to H^0(\omega_X) \to H^0 \left(\omega_X \otimes \mathcal{O}_X \left(\sum_{i=1}^{m} F_{a_i} \right) \right) \to \oplus_{i=1}^{m} H^0(\mathcal{O}_{F_{a_i}}) \to H^1(\omega_X).$$

From this exact sequence we get the inequality

$$\dim H^0 \left(\omega_X \otimes \mathcal{O}_X \left(\sum_{i=1}^{m} F_{a_i} \right) \right) \ge \dim H^0(\omega_X) + \sum_{i=1}^{m} \dim H^0(\mathcal{O}_{F_{a_i}})$$
$$- \dim H^1(\omega_X)$$
$$= \dim H^0(\omega_X) + m - \dim H^1(\omega_X).$$

Therefore the complete linear system $|K + \sum_{i=1}^{m} F_{a_i}|$ is nonempty if $m > \dim H^1(\omega_X) - \dim H^0(\omega_X)$. Then let $D \in |K + \sum_{i=1}^{m} F_{a_i}|$. For any closed point $y \in B$ we have $(D \cdot F_y) = 0$, because F_y is a curve of canonical type. Therefore all the components of D are contained in fibers of f, and then K has the same property. Put

$$K \sim \sum_{j=1}^{s} b_j F_{y_j} + D, \quad b_j \in \mathbb{Z}, \ s \ge 0, \tag{7.15.1}$$

with $D \ge 0$ contained in a (finite) union of fibers of f, but not containing any fiber of f. Let $D_1, \ldots, D_t$ be the connected components of the divisor D, so that $\mathrm{Supp}(D_i) \cap \mathrm{Supp}(D_j) = \varnothing$ if $i \ne j$. Say D_i is contained in the fiber F_{z_i} $(i = 1, \ldots, t)$. From (2.6) we get $(D_i^2) \le 0$. If $(D_i^2) < 0$ for some i, then $(D_i \cdot E) < 0$ for at least one component E of D_i. Since the fiber F_{z_i}

is necessarily reducible (for $(D_i^2) < 0$), (2.6) shows that $(E^2) < 0$ as well. Using (7.15.1), we get:

$$(K \cdot E) = \sum_{j=1}^{s} b_j (F_{y_j} \cdot E) + \sum_{k=1}^{t} (D_k \cdot E) = (D_i \cdot E) < 0.$$

But $(E^2) < 0$ and $(K \cdot E) < 0$ imply that E is an exceptional curve of the first kind, contradicting the minimality of f.

Therefore we have proved that $(D_i^2) = 0$ for all $i = 1, \dots, t$. By (2.6) we get that D_i is a rational multiple of the fiber F_{z_i}. Therefore we can write

$$\omega_X \cong f^*(M) \otimes \mathcal{O}_X \left(\sum_{i=1}^{r} a_i P_i \right), \quad 0 \le a_i < m_i, \ a_i \in \mathbb{Z}, \qquad (7.15.2)$$

where M is an invertible sheaf on B.

(7.15.3) *Claim.* $f_* \mathcal{O}_X \left(\sum_{i=1}^{r} a_i P_i \right) \cong \mathcal{O}_B$ if $0 \le a_i < m_i$.

The inclusions $\mathcal{O}_X \subseteq \mathcal{O}_X \left(\sum_{i=1}^{r} a_i P_i \right) \subseteq \mathcal{O}_X \left(\sum_{i=1}^{r} (m_i - 1) P_i \right)$ imply inclusions

$$\mathcal{O}_B = f_* \mathcal{O}_X \subseteq f_* \mathcal{O}_X \left(\sum_{i=1}^{r} a_i P_i \right) \subseteq f_* \mathcal{O}_X \left(\sum_{i=1}^{r} (m_i - 1) P_i \right).$$

Therefore it suffices to show that $f_* \mathcal{O}_X \left(\sum_{i=1}^{r} (m_i - 1) P_i \right) \cong \mathcal{O}_B$. Since the problem is local on B, it is enough to show that $f_* \mathcal{O}_X((m_i - 1) P_i) \cong \mathcal{O}_B$ $(i = 1, \dots, r)$. Since $f^* \mathcal{O}_X((m_i - 1) P_i) \cong \mathcal{O}_B(b_i) \otimes f_* \mathcal{O}_X(-P_i)$ (projection formula), it suffices to show that $f_* \mathcal{O}_X(-P_i) \cong \mathcal{O}_B(-b_i)$. As $\mathcal{O}_B(-b_i) \subseteq f_* \mathcal{O}_X(-P_i) \subseteq \mathcal{O}_B$, it is enough to prove that the injective homomorphism $f_*(u) : f_* \mathcal{O}_X(-P_i) \to f_* \mathcal{O}_X = \mathcal{O}_B$ (induced by the inclusion homomorphism $u : \mathcal{O}_X(-P_i) \hookrightarrow \mathcal{O}_X$) is not an isomorphism. For this, consider the commutative diagram

$$
\begin{array}{ccc}
(f_* \mathcal{O}_X(-P_i))^{\otimes m_i} & \longrightarrow & f_* \mathcal{O}_X(-m_i P_i) \cong \mathcal{O}_B(-b_i) \\
{\scriptstyle (f_*(u))^{\otimes m_i}} \downarrow & & \downarrow {\scriptstyle f_*(u^{\otimes m_i})} \\
(f_* \mathcal{O}_X)^{\otimes m_i} & \xrightarrow{\ \cong\ } & f_*(\mathcal{O}_X^{\otimes m_i}) \cong \mathcal{O}_B
\end{array}
$$

If f were an isomorphism, then the left-hand vertical arrow would be an isomorphism. Since the right-hand arrow is always injective, it too would be an isomorphism. Therefore an invertible $\mathcal{O}_B$-module of degree -1 would be isomorphic to $\mathcal{O}_B$, and this is clearly a contradiction. Therefore we have finished the proof of (7.15.3).

From (7.15.2), (7.15.3) and the projection formula we get:

$$f_*(\omega_X) \cong M. \qquad (7.15.4)$$

To compute M, we need to recall Grothendieck's duality theorem for the morphism f. As X and B are projective and nonsingular and the fibers of f are curves on X (and therefore they are Gorenstein schemes), the morphism f is projective and Gorenstein. Therefore the dualizing complex of f is reduced to the invertible $\mathcal{O}_X$-module $\omega_{X/B} \cong \omega_X \otimes f^*(\omega_B^{-1})$.

If L' is an invertible $\mathcal{O}_X$-module, then the relative duality theorem for f says that there exists a canonical isomorphism

$$f_*(L') \cong \mathbf{Hom}_{\mathcal{O}_B}(R^1 f_*(L'^{-1} \otimes \omega_{X/B}), \mathcal{O}_B)$$
$$\cong \mathbf{Hom}_{\mathcal{O}_B}(R^1 f_*(L'^{-1} \otimes \omega_X) \otimes \omega_B^{-1}, \mathcal{O}_B) \quad \text{(projection formula)}$$
$$\cong \mathbf{Hom}_{\mathcal{O}_B}(R^1 f_*(L'^{-1} \otimes \omega_X), \omega_B).$$

Applying this theorem for $L' = \omega_X$, we get:

$$f_*(\omega_X) \cong \mathbf{Hom}_{\mathcal{O}_B}(R^1 f_* \mathcal{O}_X, \omega_B) \cong \mathbf{Hom}_{\mathcal{O}_B}(L, \omega_B) \cong L^{-1} \otimes \omega_B, \tag{7.15.5}$$

because $R^1 f_* \mathcal{O}_X = L \oplus T$ and the dual of the torsion part is zero.

Using (7.15.5) and (7.15.4), formula (7.15.2) becomes:

$$\omega_X \cong f^*(L^{-1} \otimes \omega_B) \otimes \mathcal{O}_X \left(\sum_{i=1}^r a_i P_i \right), \quad 0 \le a_i < m_i.$$

Now consider the Leray spectral sequence:

$$E_2^{pq} = H^p(B, R^q f_* \mathcal{O}_X) \implies H^{p+q}(X, \mathcal{O}_X).$$

The exact sequence in low terms associated to this spectral sequence [CE, Chap. XV, Theorem (5.11)] is:

$$0 \to H^1(\mathcal{O}_B) \to H^1(\mathcal{O}_X) \to H^0(R^1 f_* \mathcal{O}_X)$$
$$\to H^2(\mathcal{O}_B) = 0 \to H^2(\mathcal{O}_X) \to H^1(R^1 f_* \mathcal{O}_X) \to 0.$$

From this we get

$$\chi(\mathcal{O}_X) = \dim H^0(\mathcal{O}_X) - \dim H^1(\mathcal{O}_X) + \dim H^2(\mathcal{O}_X)$$
$$= \dim H^0(\mathcal{O}_B) - \dim H^1(\mathcal{O}_B) - \dim H^0(L \oplus T) + \dim H^1(L \oplus T)$$
$$= \chi(\mathcal{O}_B) - \chi(L) - \dim H^0(T)$$
$$= -\deg(L) - l(T),$$

and therefore

$$\deg(L) = -\chi(\mathcal{O}_X) - l(T).$$

Since $\deg(\omega_B) = 2p_a(B) - 2$, we have proved part (d) of the theorem.

Therefore there only remains to prove part (c). This follows from the following lemma.

(7.15.6) Lemma. *With the same notation as in Theorem 7.15, let α_i be the order of $\mathcal{O}_X(P_i) \otimes \mathcal{O}_{P_i}$ in $\mathrm{Pic}(P_i)$. Then:*

(a) *α_i divides m_i and $a_i + 1$.*

(b) *$\dim H^0(P_i, \mathcal{O}_{(\alpha_i+1)P_i}) \geq 2$ and $\dim H^0(P_i, \mathcal{O}_{\alpha_i P_i}) = 1$.*

(c) *$\dim H^0(P_i, \mathcal{O}_{nP_i})$ is a nondecreasing function of n.*

In particular, if $a_i < m_i - 1$, then $\alpha_i < m_i$, and therefore the fiber $m_i P_i$ is exceptional.

Proof. For simplicity, write m, P, a, α for m_i, P_i, a_i, α_i, respectively. Let $m > n \geq 1$. Then the canonical surjection $\mathcal{O}_{mP} \to \mathcal{O}_{nP} \to 0$ induces a surjection $H^1(P, \mathcal{O}_{mP}) \to H^1(P, \mathcal{O}_{nP}) \to 0$, so that the function $n \mapsto \dim H^1(P, \mathcal{O}_{nP})$ is nondecreasing. But the Riemann–Roch theorem and the fact that nP is a divisor of canonical type imply that $\chi(\mathcal{O}_X(-nP)) = \chi(\mathcal{O}_X)$, and from the exact sequence

$$0 \to \mathcal{O}_X(-nP) \to \mathcal{O}_X \to \mathcal{O}_{nP} \to 0$$

we get $\chi(\mathcal{O}_{nP}) = 0$. Therefore $\dim H^0(P, \mathcal{O}_{nP}) = \dim H^1(P, \mathcal{O}_{nP})$, and part (c) of the lemma is proved.

By the definition of α, we have an isomorphism $\mathcal{O}_X(\alpha P) \otimes \mathcal{O}_P \cong \mathcal{O}_P$, and thus an exact sequence:

$$0 \to \mathcal{O}_P \to \mathcal{O}_{(\alpha+1)P} \to \mathcal{O}_{\alpha P} \to 0,$$

which induces the cohomology exact sequence

$$0 \to H^0(\mathcal{O}_P) = \mathbb{k} \to H^0(\mathcal{O}_{(\alpha+1)P}) \overset{\text{restriction}}{\longrightarrow} H^0(\mathcal{O}_{\alpha P}),$$

which shows that $\dim H^0(\mathcal{O}_{(\alpha+1)P}) \geq 2$.

On the other hand, if $1 \leq j < \alpha$, then $L_j = \mathcal{O}_X(-jP) \otimes \mathcal{O}_P$ is an invertible $\mathcal{O}_P$-module whose degree on each component of P is equal to zero. Since L_j cannot be isomorphic to $\mathcal{O}_P$, Theorem (7.8 shows that $H^0(L_j) = 0$. Therefore the exact sequence

$$0 \to L_j \to \mathcal{O}_{(j+1)P} \to \mathcal{O}_{jP} \to 0$$

induces an injection $0 \to H^0(\mathcal{O}_{(j+1)P}) \to H^0(\mathcal{O}_{jP})$. Since $H^0(\mathcal{O}_P) \cong \mathbb{k}$ by Theorem 7.8, we have by induction that $H^0(\mathcal{O}_{\alpha P}) \cong \mathbb{k}$. Thus part (b) of the lemma is also proved.

Finally, we have isomorphisms

$$(\mathcal{O}_X(P) \otimes \mathcal{O}_P)^m \cong \mathcal{O}_P \qquad \text{and}$$
$$\mathcal{O}_X((a+1)P) \otimes \mathcal{O}_P \cong \omega_P,$$

where the first isomorphism is clear and the second follows from (7.15.2) and the adjunction formula. Recalling Corollary 7.9 and the definition of α, we see that α divides m and $a + 1$. This concludes the proof of Lemma 7.15.6 and Theorem 7.15.   □

Corollary 7.16. *In the hypotheses of Theorem 7.15, we have $(K^2) = 0$, where K is the canonical class of X.*

Corollary 7.17. *In the hypotheses and notation of Theorem 7.15 and Lemma 7.15.6, if $\dim H^1(\mathcal{O}_X) \leq 1$, then $a_i + 1 = m_i$ or $\alpha_i + a_i + 1 = m_i$.*

Proof. In the proof of (7.15.6) we have seen that $\chi(\mathcal{O}_{(\alpha+1)P}) = 0$ (where $\alpha = \alpha_i$, $P = P_i$, etc.), and (b) in the lemma gives $\dim H^0(P, \mathcal{O}_{(\alpha+1)P}) \geq 2$. By duality on the curve $(\alpha + 1)P$ we get

$$\dim H^0(P, \omega_{(\alpha+1)P}) = \dim H^1(P, \omega_{(\alpha+1)P}) = \dim H^0(P, \mathcal{O}_{(\alpha+1)P}) \geq 2.$$

The exact sequence

$$0 \to \omega_X \to \mathcal{O}_X((\alpha + 1)P) \otimes \omega_X \to \omega_{(\alpha+1)P} \to 0$$

induces the cohomology exact sequence

$$0 \to H^0(\omega_X) \to H^0(\mathcal{O}_X((\alpha + 1)P) \otimes \omega_X) \to H^0(\omega_{(\alpha+1)P}) \to H^1(\omega_X),$$

whence

$$\dim H^0(\mathcal{O}_X((\alpha + 1)P) \otimes \omega_X) + \dim H^1(\omega_X)$$
$$\geq \dim H^0(\omega_X) + \dim H^0(\omega_{(\alpha+1)P})$$
$$\geq 2 + \dim H^0(\omega_X).$$

Since $\dim H^1(\omega_X) = \dim H^1(\mathcal{O}_X) \leq 1$ by hypothesis, we get

$$\dim H^0(\mathcal{O}_X((\alpha + 1)P) \otimes \omega_X) > \dim H^0(\omega_X).$$

But $\omega_X = f^*(M) \otimes \mathcal{O}_X(aP) \otimes \cdots$, with $M = f_*(\omega_X)$; if $\alpha + 1 + a < m$, then (7.15.3) gives

$$H^0(\mathcal{O}_X((\alpha + 1)P) \otimes \omega_X) = H^0(M \otimes f_*(\mathcal{O}_X((\alpha + 1 + a)P) \otimes \cdots))$$
$$= H^0(M) = H^0(\omega_X),$$

contradicting the previous inequality.

Therefore $\alpha + 1 + a \geq m$, or $m - (a + 1) \leq \alpha$. Since $m - (a + 1) \geq 0$ and α divides both m and $(a + 1)$, $m - (a + 1)$ must be either 0 or α. $\square$

Theorem 7.18. *Let $f : X \to B$ be a quasielliptic fibration. Then the characteristic of the ground field $\Bbbk$ is 2 or 3. Moreover, the general fiber of f is a rational projective curve with exactly one singular point, which is an ordinary cusp.*

Proof. If $b \in B$ is a closed point such that the fiber F_b is integral, then $p_a(F_b) = 1$ and F_b singular imply that F_b is a rational curve with exactly one singular point, which can be

- either an ordinary cusp, or

- an ordinary double point (i.e., a double point with distinct tangents);

see [Ser3, Chap. IV].

We will show that the second alternative cannot occur for any fiber F_b of the quasielliptic fibration $f : X \to B$. Let Σ be the set of all points $x \in X$ where f is not smooth. Removing the points b_i for which the fiber F_{b_i} is not integral, put $\Sigma_0 = \Sigma \cap f^{-1}(B \setminus \{b_1, \ldots, b_n\})$. If t is a regular local parameter on B at a point $b = f(x)$ with $x \in \Sigma_0$, and u and v are regular local parameters at x on X, then f is described locally by a formal power series $f(u, v) \in k[[u, v]]$, corresponding to the completed local homomorphism $k[[t]] = \hat{O}_{B,b} \to k[[u, v]] = \hat{O}_{X,x}$. Since the fiber F_b is integral, the series $f(u, v)$ has the form:

- $f(u, v) = U(u, v) \cdot (u^2 + v^3)$ with $U(0, 0) \neq 0$, if x is an ordinary cuspidal point on F_b; or

- $f(u, v) = U(u, v) \cdot u \cdot v$ with $U(0, 0) \neq 0$, if x is an ordinary double point on F_b.

Since Σ_0 is given locally at x by the equations $\partial f / \partial u = \partial f / \partial v = 0$, in the latter case we have

$$\partial f / \partial u = \text{(unit)} \cdot v, \quad \partial f / \partial v = \text{(unit)} \cdot u,$$

and therefore in this case x is an isolated point of Σ_0. But then, by the definition of Σ_0 and the fact that f is a proper morphism, it follows that there exists a nonempty open set $V \subseteq B \setminus \{b_1, \ldots, b_n\}$ containing b such that the restriction of f to $f^{-1}(V)$ is a smooth morphism on $f^{-1}(V) \setminus \{x\}$, and this is impossible in a quasielliptic fibration.

We have thus proved that, for every $b \in B \setminus \{b_1, \ldots, b_n\}$, the fiber F_b is an integral curve with exactly one singular point, which is an ordinary cusp. Then Σ_0 is the locus of all such cuspidal points in $f^{-1}(B \setminus \{b_1, \ldots, b_n\})$. Also, $f(u, v)$ has the form

$$f(u, v) = U(u, v) \cdot (u^2 + v^3), \quad U(0, 0) \neq 0.$$

Now assume that $\text{char}(k) \neq 2$. Since $\partial f / \partial u = 2u \cdot U(u, v) + V(u, v)$ with $\text{ord}(V) \geq 2$, the equation $\partial f / \partial u$ defines, near x, a smooth curve that contains Σ_0. Therefore $\partial f / \partial u$ is a local equation of Σ_0 at x, and Σ_0 is nonsingular at each of its points. Furthermore, the restriction of f to Σ_0 is a bijection onto $B \setminus \{b_1, \ldots, b_n\}$. The intersection number $(\Sigma_0 \cdot F_b)$, $b \in B \setminus \{b_1, \ldots, b_n\}$, can be calculated easily as follows:

$$(\Sigma_0 \cdot F_b)_x = \dim(O_{\Sigma_0, x} / m_{B,b} O_{\Sigma_0, x}) = \dim k[[u, v]] / (f, \partial f / \partial u)$$
$$= \dim k[[u, v]] / (u^2 + v^3, u) = \dim k[[u]] / (v^3) = 3.$$

The field extension $\Bbbk(B) \hookrightarrow \Bbbk(\Sigma_0)$ is finite and purely inseparable, and therefore it has degree p^m for some integer $m \geq 0$, where $p = \operatorname{char}(\Bbbk)$. Moreover, $p^m = (\Sigma_0 \cdot F_b)$ for any $b \in B \setminus \{b_1, \ldots, b_n\}$, that is, $p^m = 3$. This implies that $p = 3$, that is, $\operatorname{char}(\Bbbk) = 3$.

$\square$

Remark 7.19. In characteristic 2 and 3 quasielliptic fibrations do exist. For examples, see [BM2] and Exercises 7.5 and 7.6 below.

EXERCISES

Exercise 7.1. Let C_1 and C_2 be two distinct nonsingular cubic curves in the complex projective plane $\mathbb{P}^2$, with equations $f_1, f_2 \in \mathbb{C}[T_0, T_1, T_2]_3$, and assume that the intersection $C_1 \cap C_2$ consists of nine different points $P_0, P_1, \ldots, P_8 \in \mathbb{P}^2$. Consider the pencil of cubics generated by C_1 and C_2, that is, the pencil corresponding to the two-dimensional vector space $V = \{\lambda_1 f_1 + \lambda_2 f_2 \mid \lambda_1, \lambda_2 \in \Bbbk\}$. Blowing up $\mathbb{P}^2$ at the points $P_0, P_1, \ldots, P_8$ we get a surface X together with a morphism $f : X \to \mathbb{P}^1$, whose fibers are the proper transforms of all the cubics belonging to the pencil. Show that the nine exceptional curves of the first kind $E_i \stackrel{\text{def}}{=} f^{-1}(P_i)$, $i = 0, 1, \ldots, 8$, are (pairwise distinct) sections of f. Show that one can choose the pencil such that the degenerate fibers of the elliptic fibration f are all irreducible and reduced (i.e., rational curves with an ordinary node or an ordinary cusp).

Exercise 7.2. Show that if in Exercise 7.1 the cubics C_1 and C_2 are general enough, then the surface X contains infinitely many exceptional curves of the first kind. [*Hint:* Let y_i be the intersection of the generic fiber F of f with the section E_i, $i = 0, 1, \ldots, 8$. Since F is an elliptic curve over the function field $\mathbb{C}(t)$ in one variable, the points y_i are $\mathbb{C}(t)$-rational points of F, whence F has a unique group structure over $\mathbb{C}(t)$ with y_0 as the origin. Let H be the subgroup of F generated by $\{y_1, \ldots, y_8\}$, which acts on X as a group of biregular automorphisms. Show that if C_1 and C_2 are general cubics, H is a free group of rank 8. Then $\{h(E_0)\}_{h \in H}$ is a family of pairwise distinct exceptional curves of the first kind.]

Exercise 7.3. Prove Pascal's theorem (which asserts that for a hexagon inscribed in a conic C, the three points of intersection of pairs of opposite sides are collinear), as an application of Bézout's theorem, by considering an appropriate pencil of cubics. (This idea is due to Plücker.) [*Hint:* If l_1 and l_1', l_2 and l_2', l_3 and l_3' are the equations of the opposite sides of the hexagon, consider the pencil of cubics $t l_1 l_2 l_3 + t' l_1' l_2' l_3'$, with $[t, t'] \in \mathbb{P}^1$. Using Bézout's theorem, show that one of the members of this pencil of plane curves of degree 3 is the union of C and a line L, the latter containing the three points of intersection.]

Exercise 7.4. Let X be a nonsingular surface of degree d in $\mathbb{P}^3$ that contains a straight line L of $\mathbb{P}^3$. Prove that the planes of $\mathbb{P}^3$ through L cut out on X a pencil $|F|$ of curves of degree $d - 1$ such that $(F^2) = 0$. Deduce that the projection $\mathbb{P}^3 \dashrightarrow \mathbb{P}^1$ of center L extends to a morphism $X \to \mathbb{P}^1$. [*Hint:* Use Exercise 1.6.]

Exercise 7.5. *An example of quasielliptic fibration in characteristic 3.* Assume that $\mathrm{char}(\mathbb{k}) = 3$. Then the versal deformation of the cusp $y^2 + x^3 = 0$ is $y^2 + x^3 + t_1 + t_2 x + t_3 x^2 = 0$, with t_1, t_2, and t_3 independent parameters. Inside the parameter space with coordinates (t_1, t_2, t_3), check that the set of points where the fiber still has a cusp is given by $t_2 = t_3 = 0$. Setting $x = x_1/x_0$, $y = x_2/x_0$, and $t = s_0/s_1$, it follows that in the pencil of cubic plane curves

$$s_0 x_0^3 + s_1(x_1^3 + x_0 x_2^2) = 0, \quad (s_0, s_1) \in \mathbb{P}^1,$$

the general member has a cusp as singularity. Eliminating the indeterminacies (by performing a number of quadratic transformations of $\mathbb{P}^2$), we get a quasielliptic fibration $f : X \to \mathbb{P}^1$ in characteristic 3.

Exercise 7.6. *An example of quasielliptic fibration in characteristic 2.* Along the lines of Exercise 7.4, produce an example of a quasielliptic fibration $f : X \to \mathbb{P}^1$ in characteristic 2. [*Hint:* Show that the versal deformation of the cusp $y^2 + x^3 = 0$ in characteristic 2 is $y^2 + x^3 + t_1 + t_2 x + t_3 y + t_4 xy = 0$, with t_1, t_2, t_3, and t_4 independent parameters, and the set of points where the fiber still has a cusp is given by $t_3 = t_4 = 0$.]

Exercise 7.7. Let C be a nonsingular irreducible curve of genus $g \geq 1$ on a surface X (over $\mathbb{k}$) such that $(C^2) = 0$. Let $N_C = \mathcal{O}_X(C) \otimes \mathcal{O}_C$, and $i \in \mathbb{Z}$, $i \geq 1$.

(i) If N_C is not a torsion element in $\mathrm{Pic}^0(C)$, show that $H^0(\mathcal{O}_{iC}) \cong \mathbb{k}$ for all $i \geq 1$.

(ii) If the order of N_C in $\mathrm{Pic}^0(C)$ is n, show that $H^0(\mathcal{O}_{iC}) \cong \mathbb{k}$ if and only if $i \leq n$.

Exercise 7.8. First let us make the following definition. An effective divisor $D = \sum_{i=1}^p n_i E_i$ on a surface X is called *of fiber type* if $(D \cdot E_i) = 0$, $\forall i = 1, \ldots, p$. D is called *indecomposable of fiber type* if D is connected and $\gcd(n_1, \ldots, n_p) = 1$. These concepts are useful for studying fibrations from a surface to a curve (cf. [Sak2*]). Fix an indecomposable divisor D of fiber type. By examining carefully the proof of Theorem 7.8 and its corollaries, prove the following statements:

(i) $\dim H^0(D, \mathcal{O}_D) = 1$.

(ii) If $L \in \mathrm{Pic}(D)$ is such that $\deg(L|E_i) = 0$, $\forall i = 1, \ldots, p$, then $H^0(D, L) \neq 0$ if and only if $L \cong \mathcal{O}_D$.

(iii) If $D' = \sum_{i=1}^{p} m'_i E_i$ is an effective divisor, then $(D'^2) = 0$ if and only if $D' = aD$ for some $a \in \mathbb{Q}$.

(iv) If D' is an effective divisor on X such that $(D' \cdot E_i) = 0$ for all i, then $D' = nD + D''$, with $n \geq 0$ and D'' effective with $\mathrm{Supp}(D'') \cap \mathrm{Supp}(D) = \varnothing$.

Exercise 7.9. Let C be an elliptic curve over $\Bbbk$ with $\mathrm{char}(\Bbbk) = 0$, and fix a point $c_0 \in C$. Then C has a unique structure of an Abelian variety with c_0 as the zero element. If $i : C \times C \to C \times C$ is the involution defined by $i(x, y) = (y, x)$, $\forall x, y \in C$, consider the quotient $X = C^{(2)} = (C \times C)/G$ (the two-fold symmetric product of C), where G is the group of order 2 generated by i. Denote by $\pi : C \times C \to X$ the canonical morphism.

(i) Prove that X is nonsingular. [*Hint:* Clearly, one has to prove that X is nonsingular only at points of the form $\pi(x, x) \in X$ with $x \in C$, and the problem is local on C. So we may assume that C is affine. Let t be a regular parameter on C at x, with t a regular function on C; then (t_1, t_2) is a regular system of parameters on the surface $C \times C$ at (x, x), where t_i is the pullback of t via the projection p_i of $C \times C$, $i = 1, 2$. Show that $(t_1 + t_2, t_1 t_2)$ is a regular system of parameters on X at the point $\pi(x, x)$.]

(ii) Let $g : C \times C \to C$ be the morphism defined by $g(x, y) = y - x$, and let $p : C \to C/G'$ be the canonical morphism onto the quotient of C by the subgroup G' of order 2 generated by the involution $i' : C \to C$, $i'(x) = -x$. Show that $C/G' \cong \mathbb{P}^1$ and that there exists a unique morphism $f : X \to C/G' \cong \mathbb{P}^1$ such that the following diagram commutes:

$$
\begin{array}{ccc}
C \times C & \xrightarrow{\ g\ } & C \\
{\scriptstyle \pi}\downarrow & & \downarrow{\scriptstyle p} \\
X = C^{(2)} & \xrightarrow[\ f\ ]{} & C/G' \cong \mathbb{P}^1
\end{array}
$$

(iii) Prove that f is an elliptic fibration with exactly three multiple fibers, each one with multiplicity 2 (corresponding to the three nonzero points of order 2 of C). Prove that $\chi(\mathcal{O}_X) = 0$ (cf. Exercise 11.8, X is birationally isomorphic to $C \times \mathbb{P}^1$), and use Theorem 7.15 to compute the canonical class of X (assume $\mathrm{char}(\Bbbk) = 0$). Show that $\mathcal{O}_X(\tilde{\Delta}) \cong \omega_X^{-2}$, where $\tilde{\Delta} = \pi(\Delta)$, with Δ the diagonal of $C \times C$.

Bibliographic References. Theorem 7.1, Lemma 7.2, and Corollary 7.3 are presented after [Sha2], and Proposition 7.4 after [Har1]. The concept of a curve of canonical type was introduced by Mumford in [Mum4], from which we took Theorem 7.8, Corollaries 7.9 and 7.10, and Theorems 7.11 and 7.12. The notion of an exceptional fiber of an elliptic or quasielliptic fibration appeared for the first time in [BM1], from which we took Theorem 7.15 and Corollary 7.17. Theorem 7.18 is presented after [Mum4] and [BM2].

8

Canonical Dimension of an Elliptic or Quasielliptic Fibration

Let $f : X \to B$ be an elliptic or quasielliptic fibration. Theorem 7.15 expresses the dualizing sheaf ω_X of X in the form

$$\omega_X = f^*(L^{-1} \otimes \omega_B) \otimes \mathcal{O}_X(\sum_{i=1}^{r} a_i P_i), \qquad 0 \leq a_i < m_i.$$

If $n \geq 1$ is a common multiple of the numbers $m_1, \ldots, m_r$, we get:

$$\omega_X^n = f^*(L^{-n} \otimes \omega_B^n) \otimes \mathcal{O}_X(\sum_{i=1}^{r} a_i n P_i)$$

$$= f^*(L^{-n} \otimes \omega_B^n \otimes \mathcal{O}_B(\sum_{i=1}^{r} a_i/m_i \cdot n b_i)).$$

Furthermore, from the same theorem we get the following.

Proposition 8.1. *If $n \geq 1$ is a common multiple of $m_1, \ldots, m_r$, then*

$$H^0(X, \omega_X^n) = H^0(B, L^{-n} \otimes \omega_B^n \otimes \mathcal{O}_B(\sum_{i=1}^{r} a_i/m_i \cdot n b_i)),$$

with

$$\deg(L^{-n} \otimes \omega_B^n \otimes \mathcal{O}_B(\sum_{i=1}^{r} a_i/m_i \cdot n b_i))$$

$$= n(2p_a(B) - 2) + n\chi(\mathcal{O}_X) + nl(T) + \sum_{i=1}^{r} na_i/m_i.$$

8.2. Now let X be any minimal surface. Then X belongs to exactly one of four classes of minimal surfaces, characterized by the following properties:

(a) There exists an integral curve C on X such that $(K \cdot C) < 0$.

(b) For every integral curve C on X we have $(K \cdot C) = 0$ (and in particular $(K^2) = 0$). In other words, $K \equiv 0$.

(c) $(K^2) = 0$, $(K \cdot C) \geq 0$ for every integral curve C on X, and there exists at least one integral curve C' such that $(K \cdot C') > 0$.

(d) $(K^2) > 0$ and $(K \cdot C) \geq 0$ for every integral curve C on X.

Indeed, the only remaining alternative would be $(K^2) < 0$ and $(K \cdot C) \geq 0$ for all integral curves C on X; but by Theorem 1.25 this is impossible. Note also that, in order to define these four classes of surfaces, it is essential to restrict one's attention to minimal models of surfaces; indeed, if X is not a minimal surface, then there exists at least one exceptional curve of the first kind, say, C, on X, and then $(K \cdot C) = -1 < 0$.

Remarks 8.3. Let X be a minimal surface.

(a) If X belongs to class (a), then $\kappa(X) = -\infty$, that is, $p_n = 0$ for all $n \geq 1$.

(b) If X belongs to class (b), then $\kappa(X) \leq 0$.

(c) If X has an elliptic or quasielliptic fibration structure given by $f : X \to B$, and if we put

$$\lambda(f) \overset{\text{def}}{=} 2p_a(B) - 2 + \chi(\mathcal{O}_X) + l(T) + \sum_{i=1}^{r} a_i/m_i$$

(using the notation from Proposition 8.1), then X cannot belong to class (d), and we have

   (i) X belongs to class (a) if and only if $\lambda(f) < 0$, and in this case $\kappa(X) = -\infty$.

   (ii) X belongs to class (b) if and only if $\lambda(f) = 0$, and in this case $\kappa(X) \leq 0$.

   (iii) X belongs to class (c) if and only if $\lambda(f) > 0$, and this happens if and only if $\kappa(X) = 1$.

Proof. (c) follows immediately from Proposition 8.1, (5.9), and (5.10).

(a) Let C be an integral curve on X with $(K \cdot C) < 0$. Then for every divisor $D \in \mathrm{Div}(X)$ there exists an integer n_D such that $|D + nK| = \varnothing$ for all $n \geq n_D$. Indeed, since $(K \cdot C) < 0$, there exists an integer n_D such that $n(K \cdot C) < -(D \cdot C)$ for all $n \geq n_D$. Assume that there exists $\Delta \in |D + nK|$, with $n \geq n_D$; then $(\Delta \cdot C) = (D \cdot C) + n(K \cdot C) < 0$. Since Δ is effective, this implies $(C^2) < 0$. But $(C^2) < 0$ and $(K \cdot C) < 0$ together imply that C is an exceptional curve of the first kind (using the genus formula), and this contradicts the hypothesis that X is minimal. We have thus proved that $|D + nK| = \varnothing$ for all $n \geq n_D$.

In particular, we can take $D = K$; thus there exists an integer n_0 ($= n_K + 1$) such that $|nK| = \varnothing$ for all $n \geq n_0$. As $R(X)$ is an integral domain (so that $p_{mn} = 0$ implies $p_m = 0$, for any positive integers m and n), we see that $p_n = 0$ for all $n \geq 1$.

(b) Assume that X belongs to class (b). By (5.10), we must show that $p_n \leq 1$ for all $n \geq 1$. By way of contradiction, assume that $p_n \geq 2$ for some $n \geq 1$. Then $\dim |nK| \geq 1$, that is, there exists a strictly positive divisor $0 < \Delta \in |nK|$. If H is a hyperplane section on X, we have $(\Delta \cdot H) > 0$, and therefore $(\Delta \cdot C) > 0$ for at least one irreducible component C of H. That is, $(nK \cdot C) > 0$. But then $(K \cdot C) > 0$, contradicting $K \equiv 0$. $\square$

8.4. Now let X be any minimal surface with $(K^2) = 0$ and $p_g \leq 1$; in particular, by Remark 8.3(b), every minimal surface belonging to class (b) satisfies these conditions. In this case Noether's formula becomes

$$10 - 8q + 12p_g = b_2 + 2\Delta.$$

Since $p_g \leq 1, 0 \leq \Delta \leq 2p_g$ and $\Delta = 2(q - s)$ is an even integer (cf. (5.1)), we see that Δ can only be either 0 or 2. Using all these observations, the preceding equation can be easily solved. The solutions are listed below.

1. $b_2 = 22,$ $b_1 = 0,$ $\chi(\mathcal{O}_X) = 2,$ $q = 0,$ $p_g = 1,$ and $\Delta = 0.$

2. $b_2 = 14,$ $b_1 = 2,$ $\chi(\mathcal{O}_X) = 1,$ $q = 1,$ $p_g = 1,$ and $\Delta = 0.$

3. $b_2 = 10,$ $b_1 = 0,$ $\chi(\mathcal{O}_X) = 1,$ $q = 0,$ $p_g = 0,$ and $\Delta = 0,$

     or $q = 1,$ $p_g = 1,$ and $\Delta = 2.$

4. $b_2 = 6,$ $b_1 = 4,$ $\chi(\mathcal{O}_X) = 0,$ $q = 2,$ $p_g = 1,$ and $\Delta = 0.$

5. $b_2 = 2,$ $b_1 = 2,$ $\chi(\mathcal{O}_X) = 0,$ $q = 1,$ $p_g = 0,$ and $\Delta = 0,$

     or $q = 2,$ $p_g = 1,$ and $\Delta = 2.$

Remark 8.5. If X belongs to class (b) and $p_g = 1$, then $K \sim 0$ (that is, $\omega_X \cong \mathcal{O}_X$), because $\omega_X \equiv 0$ and $H^0(\omega_X) \neq 0$ together imply that $\omega_X \cong \mathcal{O}_X$.

Theorem 8.6. If X is a minimal surface belonging to class (b) and having $b_2 = 2$, then $b_1 = 2$, and therefore $\mathrm{Alb}(X)$ is an elliptic curve, and the fibers of the Albanese morphism $f : X \to B = \text{Alb}(X)$ are either all smooth elliptic curves or all rational curves, each having exactly one singular point that is an ordinary cusp. Furthermore, the latter alternative is only possible in characteristic 2 or 3.

Proof. From the list of solutions in (8.4 we see that $b_2 = 2$ implies $b_1 = 2$, so that $s = 1$. From Theorem 5.3 we have that $\dim \text{Alb}(X) = s = 1$, and therefore $B = \text{Alb}(X)$ is an elliptic curve. Let $b \in B$ be a closed point, and let $F = F_b = f^{-1}(b)$. Since $(F^2) = (F \cdot K) = 0$, the genus formula gives $p_a(F) = 1$, so that $f : X \to B$ is an elliptic or quasielliptic fibration. The last assertion follows from Theorem 7.18. To show that all the fibers of f are irreducible we use the following lemma:

Lemma 8.7. *Let $f : X \to B$ be a fibration such that $f_* \mathcal{O}_X \cong \mathcal{O}_B$. Then $\rho = \text{rank} \, \text{NS}(X) \geq 2$, and if $\rho = 2$ then all the fibers of f are irreducible.*

Taking this lemma for granted (for the time being), we note that the Igusa–Severi inequality $\rho \leq b_2 = 2$ implies, by Lemma 8.7, that $\rho = 2$ (due to the existence of the fibration $f : X \to B = \text{Alb}(X)$) and that all the fibers of f are irreducible.

In the same list of solutions we find that $\chi(\mathcal{O}_X) = 0$, and since $p_a(B) = 1$ we have (with the notation from Proposition 8.1):

$$\deg(L^{-1} \otimes \omega_B) = 2p_a(B) - 2 + \chi(\mathcal{O}_X) + l(T) = l(T) \geq 0.$$

Since $\omega_X = f^*(L^{-1} \otimes \omega_B) \otimes \mathcal{O}_X \left(\sum_{i=1}^r a_i P_i \right) \equiv 0$, we see that

$$l(T) \cdot f^{-1}(y) + \sum_{i=1}^r a_i P_i \equiv 0$$

for any closed point $y \in B$. As the left-hand side is an effective divisor, it can be numerically equivalent to zero only if it is zero; therefore we have $l(T) = 0$ and $a_i = 0$ for all i. But $l(T) = 0$ means that $a_i = m_i - 1$ for all i, cf. (7.15), and therefore $m_i = 1$ for all i. In other words, the Albanese fibration $f : X \to B$ does not have any multiple fibers.

So far we have proved that all the fibers of f are integral curves. If the general fiber of f (or, equivalently, some closed fiber of f) is nonsingular, let $\omega \in \Gamma(\omega_B)$ be a nonzero regular differential form on B. Then $f^*(\omega)$ is a regular differential form on X, and it vanishes exactly at the points where f is not smooth. Thus $f^*(\omega)$ is a global section of the rank two vector bundle $\Omega^1_{X/k}$, and the zero locus $Z(f^*(\omega))$ is either empty or of pure codimension two. Then a general result of Grothendieck [Gro2] shows that $\deg(Z(f^*(\omega))) = c_2(\Omega^1_{X/k}) = c_2$, where c_2 is the second Chern class of X. Since $b_2 = b_1 = 2$, we have $c_2 = 2 - 2b_1 + b_2 = 0$, and therefore $f^*(\omega)$ is everywhere nonzero on X, so that f is smooth everywhere. Thus Theorem 8.6 is proved, modulo the proof of Lemma 8.7.

Proof of Lemma 8.7. Let F' be any closed fiber of f and let H be any hyperplane section on X. Since $(F'^2) = 0$ and $(H^2) > 0$, the classes of F' and H in $\mathrm{NS}(X) \otimes_{\mathbb{Z}} \mathbb{Q}$ are linearly independent, and therefore $\rho \geq 2$.

Now assume that $\rho = 2$. By way of contradiction, assume that there exists a reducible fiber $F = a_1 E_1 + \cdots + a_n E_n$, with $n \geq 2$, all $a_i \geq 1$, and $E_1, \ldots, E_n$ the mutually distinct irreducible components of F. By (2.6) we have $(E_1^2) < 0$. Therefore the classes of E_1 and F' in $\mathrm{NS}(X) \otimes_{\mathbb{Z}} \mathbb{Q}$ are linearly independent. Since $\rho = 2$, there exist $a, b \in \mathbb{Q}$ such that $H \equiv aE_1 + bF'$. But then, if F'' is a fiber of f distinct from F and F', we obviously have $(E_1 \cdot F'') = (F' \cdot F'') = 0$, and therefore $(H \cdot F'') = 0$; this is a contradiction, because H is a hyperplane section and F'' is a nonzero effective divisor on X. □

Definition 8.8. A minimal surface that belongs to class (b) and has $b_2 = 2$ is called *hyperelliptic* if all the fibers of the Albanese map are smooth elliptic curves.

Definition 8.9. A minimal surface that belongs to class (b) and has $b_2 = 2$ is called *quasihyperelliptic* if each fiber of the Albanese map is a rational curve with exactly one singular point, which is an ordinary cusp.

Theorem 8.10. *Let $f : X \to B = \mathrm{Alb}(X)$ be a hyperelliptic or quasihyperelliptic fibration. Then there exists another elliptic fibration, $g : X \to \mathbb{P}^1$, of X over the projective line $\mathbb{P}^1$.*

Proof. First we show that it suffices to find an indecomposable curve C of canonical type such that $(C \cdot F_t) > 0$ for all $t \in B$, where $F_t = f^{-1}(t)$.

Indeed, assume that such a curve C exists. Then by Theorem 7.11 there exists an elliptic or quasielliptic fibration $g : X \to B'$ such that $(F_t \cdot G_{t'}) > 0$ for all $t \in B$ and $t' \in B'$, where $G_{t'} = g^{-1}(t')$. If g were a quasielliptic fibration, then the general fiber $G_{t'}$ would be a rational curve, and therefore $f(G_t')$ would have to be a point on B, for B is an elliptic curve. Then $G_{t'} \subseteq F_t$ for some $t \in B$, and therefore $(F_t \cdot G_{t'}) = 0$, which is a contradiction. We have thus proved that g is, in fact, an elliptic fibration. As $\omega_X \equiv 0$, we have

$$\lambda(g) = 2p_a(B') - 2 + \chi(\mathcal{O}_X) + l(T') + \sum_{i=1}^{r} a_i'/m_i' = 0$$

by Remark 8.3(c), and therefore $p_a(B') \leq 1$. If $p_a(B') = 1$, then by the universal property of the Albanese variety we have that there exists a morphism $h : B \to B'$ such that $h \circ f = g$, and this again contradicts the inequality $(F_t \cdot G_{t'}) > 0$. Thus $p_a(B') = 0$, that is, $B' \cong \mathbb{P}^1$.

There only remains to prove the existence of a curve C of canonical type such that $(C \cdot F_t) > 0$.

Let H be a hyperplane section on X, and let D be a divisor of the form $D = aH + bF_0$, with a, b integers and F_0 the fiber of f over some fixed closed point $t_0 \in B$. Choose a and b such that

$$(D^2) = 0 \quad \text{and} \quad (D \cdot F_t) > 0. \tag{$*$}$$

(For example, $b = -(H^2)$ and $a = 2(H \cdot F_0)$ would work.)

Fix a divisor D with the properties of Equation $(*)$. For every closed point $t \in B$, put $D_t = D + F_t - F_0$. We claim that *there exists a closed point $t \in B$ such that $|D_t| \neq \varnothing$*.

To prove this claim, first we observe that $H^2(\mathcal{O}_X(D_t)) = 0$ for all $t \in B$. Indeed, by Serre duality we have to check that $|K - D_t| = \varnothing$ for all t. If there exists $\Delta \in |K - D_t|$ for some t, then $\Delta \equiv K - D_t \equiv -D_t \equiv -D$, so that $(\Delta \cdot F_0) = -(D \cdot F_0) < 0$, which is impossible, because $\Delta \geq 0$ and F_0 is an irreducible curve with $(F_0^2) = 0$.

Now, if the claim were false, we would have $H^0(\mathcal{O}_X(D_t)) = 0$ for all $t \in B$. By Riemann–Roch we have $\chi(\mathcal{O}_X(D_t)) = 0$ for all t; therefore we would also have $H^1(\mathcal{O}_X(D_t)) = 0$ for all t. But the exact sequence

$$0 \to \mathcal{O}_X(D_t) \to \mathcal{O}_X(D + F_t) \to \mathcal{O}_X(D) \otimes \mathcal{O}_{F_0} \to 0$$

induces the cohomology exact sequence

$$0 = H^0(\mathcal{O}_X(D_t)) \longrightarrow H^0(\mathcal{O}_X(D + F_t)) \xrightarrow{\epsilon_t} H^0(\mathcal{O}_X(D) \otimes \mathcal{O}_{F_0})$$
$$\longrightarrow H^1(\mathcal{O}_X(D_t)) = 0,$$

where ϵ_t is the canonical restriction homomorphism.

Since $(D \cdot F_0) > 0$, there exists a nonzero section $s \in H^0(F_0, \mathcal{O}_X(D) \otimes \mathcal{O}_{F_0})$; put $s_t = \epsilon_t^{-1}(s)$. As in the proof of (7.12), we have

$$X = \text{closure of } \cup_{t \neq t_0} \operatorname{div}(s_t),$$

with $\text{Supp}(\operatorname{div}(s_t) \cap F_0) \subseteq \operatorname{div}_{F_0}(s)$ for all $t \neq t_0$.

Taking $t \longrightarrow t_0$, we get $\operatorname{div}(s_{t_0}) = F_0 + \Delta \sim F_0 + D$ for some effective divisor $\Delta \in |D|$, which is a contradiction. Thus the claim is proved.

We have therefore proved that there exists an effective divisor $C \in |D_t|$ for some point $t \in B$. We have $(C^2) = ((D + F_t - F_0)^2) = (D^2) = 0$ and $(C \cdot F_0) = (D + F_t - F_0 \cdot F_0) = (D \cdot F_0) > 0$ (and in particular $C > 0$). We claim that C is of canonical type. Indeed, let $C = \sum_{i=1}^{n} m_i E_i$ with $E_1, \ldots, E_n$ the distinct irreducible components of C and $m_i \geq 1$. We have

$$(C \cdot E_j) = \sum_{i \neq j} m_i (E_i \cdot E_j) + m_j (E_j^2). \tag{$**$}$$

If $E = E_j$ is a curve with $(E^2) < 0$, then since $(K \cdot E) = 0$ we have $p_a(E) = 0$ and $(E^2) = -2$ by the genus formula. But in the hypotheses of Theorem 8.10 such a curve cannot exist. Indeed, E would have to be contained in a fiber of f, but this is impossible by Theorem 8.6.

Therefore in Equation $(**)$ we have $(E_j^2) \geq 0$ for all j. As $(E_i \cdot E_j) \geq 0$ for all $i \neq j$, we have $(C \cdot E_j) \geq 0$ for all j. But

$$(C^2) = \sum_{j=1}^{n} m_i (C \cdot E_i) = 0,$$

and therefore $(C \cdot E_j) = 0$ for all j. Since $K \equiv 0$ we have also that $(K \cdot E_j) = 0$ for all j; thus C is a curve of canonical type. $\qquad\square$

Theorem 8.11. *Let X be a minimal surface that belongs to class (b) or (c). Then $p_4 > 0$ or $p_6 > 0$. That is, if X belongs to class (b), then $4K \sim 0$ or $6K \sim 0$, and if X belongs to class (c) then either $|4K|$ or $|6K|$ contains a strictly positive divisor.*

Proof. By Corollary 7.13, if X belongs to class (b) or (c) then either $2K \sim 0$ or X admits an elliptic of quasielliptic fibration structure, $f : X \to B$. Of course, if $2K \sim 0$ then X belongs to class (b) and $p_2 \neq 0$ (so that $p_4 \neq 0$ and $p_6 \neq 0$ as well), and in this case there is nothing else to prove. We shall therefore assume that X admits an elliptic or quasielliptic fibration $f : X \to B$. We may also assume that $p_g = 0$ (if $p_g \overset{\text{def}}{=} p_1 > 0$ then $p_n > 0$ for all $n \geq 1$, and the conclusion of the theorem holds).

Since X is now a minimal surface with $(K^2) = 0$ and $p_g = 0$, we can use the list in (8.4). By Theorem 7.15 we have that

$$p_g = \dim H^0(B, L^{-1} \otimes \omega_B),$$

with $\deg(L^{-1} \otimes \omega_B) = 2p_a(B) - 2 + \chi(\mathcal{O}_X) + l(T)$. But we read in the list in (8.4) that $\chi(\mathcal{O}_X) \geq 0$, and therefore the Riemann–Roch theorem on the curve B implies that $[p_a(B) = 1, \chi(\mathcal{O}_X) = 0$ and $l(T) = 0]$ or $[p_a(B) = 0$ and $\chi(\mathcal{O}_X) + l(T) < 2]$.

Consider first the case with $p_a(B) = 1$, $\chi(\mathcal{O}_X) = 0$ and $l(T) = 0$. Since T is the set of points $b \in B$ such that F_b is an exceptional fiber, $l(T) = 0$ means that f doesn't have any exceptional fibers, and in particular, by Theorem 7.15 (c), $a_i = m_i - 1$ for all i. If f has a multiple fiber $m_1 P_1$, $m_1 \geq 2$, then the formula

$$\omega_X = f^*(L^{-1} \otimes \omega_B) \otimes \mathcal{O}_X \left(\sum_{i=1}^{r} (m_i - 1) P_i \right)$$

implies

$$\omega_X^2 = f^*(L^{-2} \otimes \omega_B^2) \otimes f^* \mathcal{O}_B(b_1 + \cdots) \otimes \mathcal{O}_X \left(\sum_{i=1}^{r} (m_i - 2) P_i \right);$$

but then, since $\deg(L^{-2} \otimes \omega_B^2 \otimes \mathcal{O}_B(b_1 + \cdots)) \geq 1$ and B is an elliptic curve, we have by the projection formula:

$$p_2 = \dim H^0(\omega_X) = \dim H^0(B, L^{-2} \otimes \omega_B^2 \otimes \mathcal{O}_B(b_1 + \cdots)) \geq 1,$$

and then again $p_4 > 0$ and $p_6 > 0$, so the theorem is proved in this case.

Thus, when $p_a(B) = 1$, $\chi(\mathcal{O}_X) = 0$ and $l(T) = 0$, there only remains to consider the case when f has no multiple fibers. Then by Theorem 7.15 we have

$$\omega_X = f^*(L^{-1} \otimes \omega_B),$$

and $\deg(L^{-1} \otimes \omega_B) = 2p_a(B) - 2 + \chi(\mathcal{O}_X) + l(T) = 0$. Therefore $\omega_X \equiv 0$, so that X belongs to class (b) in this case. Furthermore, we read in (8.4) that $p_g = 0$ and $\chi(\mathcal{O}_X) = 0$ occur (for the same surface X) only in case 5 in that list. That is, X is a hyperelliptic or quasihyperelliptic surface. Then, by Theorem 8.10, there exists another elliptic fibration $g : X \to \mathbb{P}^1$ of X over the projective line.

To complete the proof of Theorem 8.11 there only remains to analyze the case when X admits an elliptic or quasielliptic fibration $f : X \to B$ with $p_a(B) = 0$, that is, $B = \mathbb{P}^1$, and $p_g = 0$, that is, $\deg(L^{-1} \otimes \omega_B) = -2 + \chi(\mathcal{O}_X) + l(T) < 0$.

Since X belongs to class (b) or (c), from Remark 8.3(c) we have that

$$\lambda(f) = -2 + \chi(\mathcal{O}_X) + l(T) + \sum_{i=1}^{r} a_i/m_i \geq 0.$$

Put $\lambda' = -2 + \chi(\mathcal{O}_X) + l(T)$; thus $\lambda' < 0$ and $\lambda' + \sum_{i=1}^{r} a_i/m_i \geq 0$. Since $B = \mathbb{P}^1$, for every integer $n \geq 1$ we have

$$\dim |nK| = n\lambda' + \sum_{i=1}^{r} [na_i/m_i], \qquad (***)$$

if the right-hand side is nonnegative, where $[u]$ denotes the round-down (integral part) of a real number u.

We distinguish the following cases:

(A) $l(T) = 0$. In this case, by (7.15) and (7.15.6) we have $a_i = m_i - 1$ and $\alpha_i = m_i$ for all i. Moreover, since $\lambda' < 0$, we have $\chi(\mathcal{O}_X) \leq 1$.

If $\chi(\mathcal{O}_X) = 0$ then we must have at least three multiple fibers (for $\lambda' = -2$, so that $-2 + \sum_{i=1}^{r}(m_i - 1)/m_i \geq 0$). We have the following alternatives:

(a) There are at least four multiple fibers, and therefore $m_i \geq 2$ for $i = 1, 2, 3, 4, \ldots, r$, $r \geq 4$. Then from Equation $(***)$ we see that $|2K| \neq \varnothing$, that is, $p_2 > 0$.

(b) There are exactly three multiple fibers, all with $m_i \geq 3$. Then from Equation $(***)$ we get $|3K| \neq \varnothing$, that is, $p_3 > 0$.

(c) There are exactly three exceptional fibers, with $m_1 = 2, m_2, m_3 \geq 4$. Then $|4K| \neq \varnothing$.

(d) There are exactly three exceptional fibers, with $m_1 = 2, m_2 = 3, m_3 \geq 6$. Then $|6K| \neq \varnothing$.

If $\chi(\mathcal{O}_X) = 1$, then $\lambda' = -1$, and we have $-1 + \sum_{i=1}^{r}(m_i - 1)/m_i \geq 0$. Then we must have at least two multiple fibers, and we have $|2K| \neq \varnothing$.

(B) $l(T) = 1$. Then $\lambda' < 0$ implies $\chi(\mathcal{O}_X) = 0$. Since $p_g = 0$, this means that $q = 1$. Therefore we can use Corollary 7.17. Let $f^{-1}(b_1)$ be the exceptional fiber (there is exactly one exceptional fiber, because $l(T) = 1$). Then by Corollary 7.17 we have $a_1 = m_1 - 1$ or $a_1 = m_1 - 1 - \alpha_1$, with α_1 a common divisor of m_1 and $a_1 + 1$. Furthermore, there are at least two multiple fibers (for $\lambda' + \sum_{i=1}^{r} a_i/m_i \geq 0$), and $a_i = m_i - 1$ for all $i \geq 2$. We have the following alternatives:

(a') There are at least two multiple curves with $a_i = m_i - 1$. Then $|2K| \neq \varnothing$.

(b') The exceptional fiber has $m_1 = 3, a_1 = 1, \alpha_1 = 1$, and $m_2 \geq 3$. Then $|3K| \neq \varnothing$.

(c'_1) The exceptional fiber has $m_1 = 4, a_1 = 1, \alpha_1 = 2$, and $m_2 \geq 4$. Then $|4K| \neq \varnothing$.

(c'_2) The exceptional fiber has $m_1 = \beta_1 \alpha_1$ with $\beta_1 \geq 4$. Then $a_1/m_1 \geq 1/2$, and $|2K| \neq \varnothing$.

(d'_1) The exceptional fiber has $m_1 = 2\alpha_1, a_1 = \alpha_1 - 1, \alpha_1 \geq 3$, and $m_2 \geq 3$. Then $|3K| \neq \varnothing$.

(d'_2) The exceptional fiber has $m_1 = 3\alpha_1, a_1 = 2\alpha_1 - 1, \alpha_1 \geq 2$ (and $m_2 \geq 2$). Then $|2K| \neq \varnothing$.

This concludes the proof of Theorem 8.11.   □

As a corollary of the proof (more precisely, of the following fact: if X is a minimal surface with $p_g = 0$, if X belongs to class (b) or (c), and if X admits an elliptic or quasielliptic fibration, then $p_2 \neq 0$ or else there exists an elliptic or quasielliptic fibration $f : X \to \mathbb{P}^1$ over the projective line), we can easily prove the following refinement of Remark 8.3(c).

Corollary 8.12. Let $f : X \to B$ be an elliptic or quasielliptic fibration, with X a minimal surface. Then X cannot belong to class (d), and we have:

(i) [X belongs to class (a)] $\iff$ [$\lambda(f) < 0$] $\iff$ [$\kappa(X) = -\infty$] $\iff$ [$p_n = 0$ for all $n \geq 1$] $\iff$ [$p_4 = p_6 = 0$] $\iff$ [$p_{12} = 0$].

(ii) [X belongs to class (b)] $\iff$ [$\lambda(f) = 0$] $\iff$ [$\kappa(X) = 0$] $\iff$ [$nK \sim 0$ for some $n \geq 1$] $\iff$ [$4K \sim 0$ or $6K \sim 0$] $\iff$ [$12K \sim 0$].

(iii) [X *belongs to class* (c)] $\iff$ [$\lambda(f) > 0$] $\iff$ [$\kappa(X) = 1$] $\iff$ [*the linear system* $|nK|$ *contains a strictly positive divisor for some* $n \geq 1$] $\iff$ [$|4K|$ *or* $|6K|$ *contains a strictly positive divisor*] $\iff$ [$|12K|$ *contains a strictly positive divisor*].

Remark 8.13. By a general result of Raynaud [Ray], m_i/α_i is a power of the characteristic exponent p of $\Bbbk$ (i.e., $p = 1$ if char($\Bbbk$) $= 0$ and $p = $ char($\Bbbk$) if char($\Bbbk$) > 0). This shows, in particular, that the multiplicity of every exceptional fiber is divisible by p, and that in characteristic zero there are no exceptional fibers in any elliptic fibration (for the latter part we use Lemma 7.15.6).

Moreover, the same result of Raynaud shows that case (b') in the proof of Theorem 8.11 can occur only in characteristic 3, case (c'$_1$) can occur only in characteristic 2, and case (d'$_2$) can occur only in characteristic 3. Although this result is interesting in itself, we will not prove it here, because we will not use it in the sequel.

EXERCISES

Exercise 8.1. Prove that the elliptic fibration of Exercise 7.1 can have at most one multiple fiber. [*Hint:* Use Remarks 8.3 and 8.13.]

Exercise 8.2. Let $f : X \to B$ be an elliptic fibration, and let Alb(f) : Alb(X) $\to$ Alb(B) be the corresponding Albanese morphism. Assume that $\chi(X, \mathcal{O}_X) \geq 0$ (this hypothesis turns out to be automatically satisfied if X dominates a minimal surface of class (b) or (c) and $\Bbbk = \mathbb{C}$). Prove that either Alb(f) is an isomorphism or $q(X) = p_a(B) + 1$ and every nonsingular fiber F of f is isogenous to Ker(Alb(f)), that is, there exists a finite étale morphism $F \to$ Ker(Alb(f)). Analyze the special case of the hyperelliptic surfaces in view of Theorem 8.10. [*Hint:* Use Theorem 7.15 and the Leray spectral sequence $E_2^{pq} = H^p(X, R^q f_*(\mathcal{O}_X))$.]

Exercise 8.3. How many multiple fibers can the elliptic fibration $g : X \to \mathbb{P}^1$ from Theorem 8.10 have? What are the corresponding multiplicities?

Exercise 8.4. Let B be a nonsingular projective curve of genus g, and let L be a line bundle on B generated by its global sections. Let p_1 and p_2 be the two canonical projections of $B \times \mathbb{P}^2$ on B and $\mathbb{P}^2$, respectively. By Bertini's theorem, a general member $X \in |p_1^*(L) \otimes p_2^* \mathcal{O}_{\mathbb{P}^2}(3)|$ is a nonsingular projective surface. Show that the restriction $f : X \to B$ of p_1 is an elliptic fibration such that $\omega_X \cong f^*(\omega_B \otimes L)$. In particular, if $\deg(L) > 2 - 2g$ then X belongs to class (c) (or equivalently, in view of Theorem 9.4 below, $\kappa(X) = 1$).

Exercise 8.5. Let X be a minimal surface belonging to class (c), and let $D \in |nK|$ for some $n \geq 1$. Show that D is a curve of canonical type. Prove that $|nK|$ has no base points for $n \gg 0$. In particular, X admits a canonical elliptic fibration $f : X \to B$ induced (via the Stein factorization) by the morphism associated to the pluricanonical system $|nK|$ for $n \gg 0$.

Bibliographic References. Everything in this chapter is presented after [BM1].

9

The Classification Theorem According to Canonical Dimension

Theorem 9.1. *Let X be a minimal surface belonging to class (d), that is, $(K^2) > 0$ and $(K \cdot C) \geq 0$ for all curves C on X. Then $|2K| \neq \varnothing$, and for sufficiently large n the linear system $|nK|$ is free of base points and defines a morphism $\phi_n = \phi_{|nK|} : X \to \mathbb{P}(H^0(\mathcal{O}_X(nK))\check{})$ with the following properties: the image $X_n = \phi_n(X)$ is a normal surface having at most rational double points as singularities, and ϕ_n is an isomorphism of $X \setminus \phi_n^{-1}(\mathrm{Sing}(X_n))$ onto $X_n \setminus \mathrm{Sing}(X_n)$, where $\mathrm{Sing}(X_n)$ denotes the singular locus of X_n. In this case the canonical dimension of X is 2, and the canonical ring $R(X)$ is a finitely generated $\Bbbk$-algebra (of transcendence degree 3 over $\Bbbk$).*

Proof. First we prove that $|2K| \neq \varnothing$. By way of contradiction, if $|2K| = \varnothing$ (i.e., if $p_2 = 0$), then $\Delta = 0$ by (5.1), and therefore Noether's formula becomes:

$$10 = (K^2) + 8q + b_2,$$

whence $q \leq 1$. On the other hand, the Riemann–Roch theorem gives

$$\chi(\mathcal{O}_X(2K)) = (K^2) + \chi(\mathcal{O}_X). \tag{9.1.1}$$

By Serre duality we have

$$\chi(\mathcal{O}_X(2K)) = \dim H^0(\mathcal{O}_X(2K)) + \dim H^0(\mathcal{O}_X(-K))$$
$$- \dim H^1(\mathcal{O}_X(2K)) = - \dim H^1(\mathcal{O}_X(2K)) \leq 0,$$

for $H^0(\mathcal{O}_X(2K)) = 0$ by hypothesis, and $|-K| = \varnothing$ (indeed, otherwise there would exist an effective divisor $\Delta \in |-K|$, and therefore $(K^2) = (K \cdot -\Delta) = -(K \cdot \Delta) \leq 0$, contradicting the hypothesis).

On the other hand, $\chi(\mathcal{O}_X) = 1 - q + p_g = 1 - q \geq 0$; therefore from (9.1.1) and $(K^2) > 0$ we get $\chi(\mathcal{O}_X(2K)) > 0$, contradicting the inequality proved earlier. We have therefore proved that $|2K| \neq \varnothing$.

Now we analyze the complete linear system $|nK|$ with $n \gg 0$.

Let E be an integral curve on X such that $(K \cdot E) = 0$ (if such curves exist). Since $(K^2) > 0$, the Hodge index theorem (Theorem 2.4) implies that $(E^2) < 0$, because the divisor E cannot be numerically equivalent to zero. Since $0 \leq p_a(E) = 1/2(E^2) + 1/2(K \cdot E) + 1 = 1/2(E^2) + 1 \leq 0$, we must have $p_a(E) = 0$ and $(E^2) = -2$. Conversely, every integral curve E with $p_a(E) = 0$ and $(E^2) = -2$ has $(K \cdot E) = 0$.

If there are no such curves E on the surface X, then by the Nakai–Moishezon criterion the canonical divisor K is ample, and then everything is clear by (5.5)—except the finite generation of the canonical ring $R(X)$ as an algebra over $\Bbbk$.

Therefore there only remains to consider the case when such curves do exist on X. We will show that the set

$$\{E \subset X \mid E \text{ is an integral curve with } p_a(E) = 0 \text{ and } (E^2) = -2\}$$

is finite. This will follow from the Néron–Severi theorem on the finiteness of the base number $\rho = \operatorname{rank} \mathrm{NS}(X)$ and the following claim:

(9.1.2) Claim. *Let $E_1, \ldots, E_n$ be n pairwise distinct integral curves on X with $p_a(E_i) = 0$ and $(E_i^2) = -2$ for all $i = 1, \ldots, n$. Then the classes of $E_1, \ldots, E_n$ in $\mathrm{NS}(X) \otimes_{\mathbb{Z}} \mathbb{Q}$ are linearly independent over $\mathbb{Q}$, and the intersection matrix $\|(E_i \cdot E_j)\|_{i,j}$ is negative definite.*

Proof of Claim 9.1.2. First note that it is enough to show that $E_1, \ldots, E_n$ define linearly independent elements in the $\mathbb{Q}$-vector space $V = \mathrm{NS}(X) \otimes_{\mathbb{Z}} \mathbb{Q}$; indeed, the negative definiteness of the intersection matrix (which we will denote by M_n in the sequel) will then follow from the Hodge index theorem (Theorem 2.4), because $(K \cdot E_i) = 0$ for all i.

Now assume that there exists a linear dependence relation of the form

$$a_1 E_1 + \cdots + a_n E_n \equiv 0, \quad a_i \in \mathbb{Q}. \tag{9.1.3}$$

Put $Z \overset{\mathrm{def}}{=} \sum_{i=1}^n a_i E_i = Z_1 - Z_2$, with $Z_1 \geq 0$, $Z_2 \geq 0$, and Z_1, Z_2 without common components; that is,

$$Z_1 = \sum_{a_i \geq 0} a_i E_i, \quad Z_2 = -\sum_{a_i \leq 0} a_i E_i.$$

Since $Z \equiv 0$, we have $(Z \cdot Z_2) = 0$; that is, $(Z_1 \cdot Z_2) - (Z_2^2) = 0$. Since Z_1 and Z_2 have nonnegative coefficients and no common components, we have $(Z_1 \cdot Z_2) \geq 0$, and therefore $(Z_2^2) \geq 0$. But on the other hand, by the Hodge index theorem, $(K^2) > 0$ and $(K \cdot E_i) = 0$ for all $i = 1, \ldots, n$ imply that $(Z_2^2) \leq 0$, with equality only when $Z_2 \equiv 0$ (because Z_2 is a linear combination of the E_i). Therefore $Z_2 \equiv 0$, and then $Z_1 \equiv 0$ as well, because $Z = Z_1 - Z_2 \equiv 0$. However, since the E_i are integral curves on a projective surface and all the a_i are nonnegative, this is possible only if all the a_i are equal to zero; to see this, consider the intersection number against a hyperplane section of X. In other words, we proved the $\mathbb{Q}$-linear independence of $E_1, \ldots, E_n$ in V. $\qquad\square$

Returning to the proof of Theorem 9.1, let $E_1, \ldots, E_n$ be all the (pairwise distinct) integral curves on X with $(K \cdot E_i) = 0$. Since $p_a(E_i) = 0$ and $(E_i^2) = -2$ for all i and the intersection matrix $\|(E_i \cdot E_j)\|$ is negative definite, Theorem 3.15 guarantees the existence of a morphism $\phi : X \to Y$ that contracts the connected subconfigurations of the E_i to normal points, such that the restriction of ϕ is an isomorphism from $X \setminus \cup_{i=1}^n E_i$ to $Y \setminus \mathrm{Sing}(Y)$, ω_Y is invertible, and $\phi^*(\omega_Y) = \omega_X$.

The linear system $|\omega_X^n|$ is base-point free if and only if the linear system $|\omega_Y^n|$ is base-point-free. Therefore it will suffice to show that $|\omega_Y^n|$ is base-point-free for $n \gg 0$. But $(\omega_Y \cdot \omega_Y) = (\omega_X \cdot \omega_X) = (K^2) > 0$, and if F is any integral curve on Y and E is the proper transform of F by ϕ, then $(\omega_Y \cdot F) = (f^*(\omega_Y) \cdot E) = (\omega_X \cdot E) = (K \cdot E) > 0$, because E is not among the curves E_i with $(K \cdot E_i) = 0$. Therefore ω_Y is an ample invertible $\mathcal{O}_Y$-module, by the Nakai–Moishezon criterion. Thus $|\omega_Y^n|$ defines an embedding of Y in a projective space $\mathbb{P}$ for n sufficiently large. Consequently the complete linear system $|\omega_X^n|$ is base-point-free for such large n (for $\omega_X^n = \phi^*(\omega_Y^n)$), and the morphism ϕ_n is identified with ϕ.

As $\omega_Y^n \cong \phi_*\phi^*\omega_Y^n \cong \phi_*\omega_X^n$ for all $n \geq 1$, the canonical ring $R(X)$ is identified with $\Gamma_*(\omega_Y)$, and this ring has transcendence degree 3 over $\Bbbk$, by Lemma 5.5; therefore $\kappa(X) = 2$. (Note that the hypothesis that X is a nonsingular variety is not necessary in Lemma 5.5, which holds for any irreducible variety.)

Finally, the finite generation of the canonical ring $R(X)$ follows from the fact that the complete linear system $|\omega_X^n|$ is base-point-free for some sufficiently large n and the following result of Zariski: $\qquad\square$

Proposition 9.2 (Zariski). *Let X be a proper algebraic $\Bbbk$-scheme, and let L be an invertible $\mathcal{O}_X$-module. Put $S = \Gamma_*(L)$. If the complete linear system $|L^n|$ is base-point-free for some $n \geq 1$, then S is a finitely generated $\Bbbk$-algebra.*

We will prove this proposition at the end of the chapter (in an even more general form, which will allow us to prove yet another important result of Zariski).

9.3. We review the results about the classification of surfaces we have proved so far. Let X be a minimal surface. Then:

(i) [X is in class (a)] $\implies$ [$\kappa(X) = -\infty$] $\implies$ [$|4K| = \varnothing$ and $|6K| = \varnothing$].

(ii) [X is in class (b)] $\implies$ [$\kappa(X) = 0$] $\implies$ [$4K \sim 0$ or $6K \sim 0$].

(iii) [X is in class (c)] $\implies$ [$\kappa(X) = 1$] $\implies$ [$|4K|$ or $|6K|$ contains a strictly positive divisor, and $(K^2) = 0$].

(iv) [X is in class (d)] $\implies$ [$\kappa(X) = 2$], and in this case $|2K| \neq \varnothing$.

Since the classes (a), (b), (c), and (d) are mutually disjoint and exhaust all the minimal models of (smooth and projective) surfaces, and similarly for the minimal surfaces with canonical dimension $-\infty, 0, 1$, and 2, we have the following classification theorem.

Theorem 9.4. *Let X be a minimal surface. Then:*

(i) *[There exists an integral curve C on X such that $(K \cdot C) < 0$]* $\iff$ *[$\kappa(X) = -\infty$]* $\iff$ *[$p_4 = p_6 = 0$]* $\iff$ *[$p_{12} = 0$].*

(ii) *[$(K \cdot C) = 0$ for every integral curve C on X]* $\iff$ *[$\kappa(X) = 0$]* $\iff$ *[$4K \sim 0$ or $6K \sim 0$]* $\iff$ *[$12K \sim 0$].*

(iii) *[$(K^2) = 0, (K \cdot C) \geq 0$ for all integral curves C on X, and there exists an integral curve C' with $(K \cdot C') > 0$]* $\iff$ *[$\kappa(X) = 1$]* $\iff$ *[$(K^2) = 0$ and $|4K|$ or $|6K|$ is represented by a strictly positive divisor]* $\iff$ *[$(K^2) = 0$ and $|12K|$ is represented by a strictly positive divisor].*

(iv) *[$(K^2) > 0$ and $(K \cdot C) \geq 0$ for every integral curve C on X]* $\iff$ *[$\kappa(X) = 2$]. In this case we have $|2K| \neq \varnothing$.*

Definition 9.5. A surface X is *of general type* if $\kappa(X) = 2$, and *of special type* if $\kappa(X) \leq 1$. If X is a minimal surface of general type, then $X_n = \phi_n(X)$ (with the notation of Theorem 9.1) is called *the canonical model of order n of X, or the n-canonical model of X.*

The surfaces of general type are the two-dimensional analogue of the nonsingular projective curves of genus $g \geq 2$. If X is a minimal surface of general type, then for n sufficiently large the n-canonical model X_n is a normal projective surface with at most rational double points as singularities, the morphism ϕ_n is birational (ϕ_n is indeed a morphism for $n \gg 0$, because the linear system $|nK|$ is base-point-free), and the pair (X_n, ϕ_n) is independent of n. If X_n is, in fact, nonsingular for some $n \gg 0$, then ϕ_n is an isomorphism, and therefore ω_X is an ample invertible $\mathcal{O}_X$-module. This happens if and only if X does not contain any integral curve E with $p_a(E) = 0$ and $(E^2) = -2$.

9.6. Examples of Surfaces of General Type. In (5.8)(ii) we saw that the product of two nonsingular projective curves C_1 and C_2 of genera $g_1, g_2 \geq 2$ is a surface of general type. We give a few more examples here.

Example 9.6.1 Let X be a nonsingular projective surface embedded in $\mathbb{P}^n$, such that X is a complete intersection in $\mathbb{P}^n$. That is, the homogeneous ideal $\mathfrak{I}(X)$ of X in the polynomial ring $\Bbbk[T_0, \ldots, T_n]$ (the homogeneous coordinate ring of $\mathbb{P}^n$) is generated by $n - 2$ homogeneous polynomials $f_1, \ldots, f_{n-2}$ of respective degrees, say, $d_1, \ldots, d_{n-2} > 0$. In geometric language we say that X is the (complete) intersection of the hypersurfaces $H_1, \ldots, H_{n-2}$, where H_i is the hypersurface in $\mathbb{P}^n$ with equation $f_i = 0$, $i = 1, \ldots, n - 2$. A general result of Serre [Ser1, §78] shows that X is projectively normal, that is, the Serre homomorphism (see the proof of (5.5), [Ser1], or [EGA II]) is an isomorphism, and $H^1(X, \mathcal{O}_X(s)) = 0$ for all $s \in \mathbb{Z}$, where $\mathcal{O}_X(s)$ denotes the restriction of $\mathcal{O}_{\mathbb{P}^n}(s)$ to X: $\mathcal{O}_X(s) = \mathcal{O}_{\mathbb{P}^n}(s) \otimes \mathcal{O}_X$. Furthermore, $\omega_X = \mathcal{O}_X((\sum_{i=1}^{n-2} d_i) - n - 1)$.

If $\sum_{i=1}^{n-2} d_i > n + 1$, then ω_X is an ample invertible $\mathcal{O}_X$-module. In particular, if E is any integral curve on X, then $(\omega_X \cdot E) > 0$, so that E cannot be an exceptional curve of the first kind (if it were, then we would have $(\omega_X \cdot E) = -1$). In conclusion, if $\sum_{i=1}^{n-2} d_i > n + 1$, then X is a minimal surface of general type, with irregularity $q = \dim H^1(\mathcal{O}_X) = 0$. For example, if X is a nonsingular surface of degree $d \geq 5$ in $\mathbb{P}^3$, then X is a minimal surface of general type with $q = 0$.

Example 9.6.2 Let X' be the Fermat surface in $\mathbb{P}^3$ given by the equation

$$T_0^5 + T_1^5 + T_2^5 + T_3^5 = 0.$$

If the characteristic of $\Bbbk$ is not 5, then, as we have seen above, X' is a nonsingular minimal surface of general type with $q(X') = 0$. Let ξ be a primitive root of order 5 of unity, and let $u : X' \to X'$ be the automorphism of X' defined by $u(t_0, t_1, t_2, t_3) = (t_0, \xi t_1, \xi^2 t_2, \xi^3 t_3)$. Then u is an automorphism of order 5 of X' ($u^5 = \mathrm{id}_{X'}$). It is easy to see that u has no fixed points on X'. Let G be the subgroup of order 5 of $\mathrm{Aut}(X')$ generated by u, where $\mathrm{Aut}(X')$ denotes the group of algebraic automorphisms of X'. Since u has no fixed points on X', the quotient surface $X \overset{\text{def}}{=} X'/G$ is a nonsingular projective surface.

Let $f : X' \to X$ be the canonical morphism. Then the canonical $\mathcal{O}_X$-module homomorphism $\alpha : \mathcal{O}_X \to f_*(\mathcal{O}_{X'})$ splits, that is, there exists an $\mathcal{O}_X$-module homomorphism $\beta : f_*(\mathcal{O}_{X'}) \to \mathcal{O}_X$ such that $\beta \circ \alpha = \mathrm{id}$. Indeed, since f is a finite morphism, the trace homomorphism $T : f_*(\mathcal{O}_{X'}) \to \mathcal{O}_X$ is defined, and we can take $\beta = \frac{1}{5}T$. Consequently the homomorphism induced in cohomology, $H^1(\mathcal{O}_X) \to H^1(f_*(\mathcal{O}_{X'})) = H^1(\mathcal{O}_{X'})$, is (split) injective. As $H^1(\mathcal{O}_{X'}) = 0$ (for X' is a complete intersection), we have $H^1(\mathcal{O}_X) = 0$.

On the other hand, $\omega_{X'} = \mathcal{O}_{X'}(1)$ (see Example 9.6.1), and since X' is projectively normal we have:

$$p_g(X') = \dim H^0(X', \omega_{X'}) = \dim H^0(X', \mathcal{O}_{X'}(1)) = 4,$$

whence $\chi(\mathcal{O}_{X'}) = 1 - q(X') + p_g(X') = 5$. Now we use the following general result.

Proposition 9.7. *Let* $f : X' \to X$ *be an étale morphism of degree* n *between two surfaces* X' *and* X. *Then* $\chi(\mathcal{O}_{X'}) = n\chi(\mathcal{O}_X)$. $\qquad\square$

Taking this proposition for granted for the time being, in our situation we have $\chi(\mathcal{O}_X) = 1$, that is, $1 - q(X) + p_g(X) = 1$. Since, as we have seen, $q(X) = 0$, we get $p_g(X) = 0$. Also, since f is an étale morphism, we have $f^*(\omega_X) \cong \omega_{X'}$, and therefore $\omega_{X'}$ ample on X' implies ω_X ample on X (see [EGA III] or [Har3, p. 25]). In particular, X is a minimal surface.

Thus X is an example of a minimal surface of general type with $q = p_g = 0$. This surface X is known as *Godeaux' surface*.

Proof of Proposition 9.7. Proposition 9.7 is true for nonsingular projective varieties X' and X of any dimension; it can be proved easily, in this generality, as a consequence of the Riemann–Roch–Grothendieck theorem [BS]. Here we will give a direct proof for surfaces, based on Noether's formula.

Since f is an étale morphism, we have $f^*(\Omega^1_{X/k}) = \Omega^1_{X'/k}$; indeed, f is a smooth morphism of relative dimension zero, and the identification above follows from the canonical exact sequence

$$0 \to f^*(\Omega^1_{X/k}) \to \Omega^1_{X'/k} \to \Omega^1_{X'/X} = 0,$$

cf. [AK, p. 147] or [Har1, p. 269]. Therefore $f^*(T_X) = T_{X'}$, where T_X and $T_{X'}$ are the tangent bundles of X and X', and we have:

$$\mathbf{c_2}(X') = \mathbf{c_2}(T_{X'}) = \mathbf{c_2}(f^*(T_X)) = f^*\mathbf{c_2}(T_{X'}) = f^*\mathbf{c_2}(X),$$

where we use boldface $\mathbf{c_2}$ to indicate classes of zero-cycles in the Chow rings $A(X')$ and $A(X)$.

Taking degrees in this formula, we get:

$$c_2(X') = (\deg(f)) \cdot c_2(X) = nc_2(X).$$

On the other hand, as $\omega_{X'} = f^*(\omega_X)$, (1.18) gives

$$(\omega_{X'} \cdot \omega_{X'}) = n \cdot (\omega_X \cdot \omega_X).$$

Using Noether's formula, we finally get

$$\chi(\mathcal{O}_{X'}) = \tfrac{1}{12}[(\omega_{X'} \cdot \omega_{X'}) + c_2(X')]$$
$$= \tfrac{1}{12}[n(\omega_X \cdot \omega_X) + nc_2(X)]$$
$$= n\chi(\mathcal{O}_X). \qquad\square$$

Using the explicit computation of the canonical class of an elliptic or quasielliptic surface (i.e., a surface admitting an elliptic or a quasielliptic fibration over a curve), Corollary 7.13 and Proposition 9.2, we can complete parts (ii) and (iii) of Theorem 9.4 and make them more precise, as follows.

Theorem 9.8. *Let X be a minimal surface. The following are equivalent:*

(i) $\kappa(X) = 0$.

(ii) X *belongs to class* (b).

(iii) $12K \sim 0$.

Moreover, if these conditions are satisfied, then the canonical ring $R(X)$ is a finitely generated $\Bbbk$-algebra of transcendence degree one.

Theorem 9.9. *Let X be a minimal surface. The following are equivalent:*

(i) $\kappa(X) = 1$.

(ii) X *belongs to class* (c).

(iii) *The complete linear system $|12K|$ contains a strictly positive divisor and $(K^2) = 0$.*

(iv) *There exists an elliptic or quasielliptic fibration $f : X \to B$ with $\lambda(f) > 0$.*

(v) *There exist an elliptic or quasielliptic fibration $f : X \to B$, a strictly positive divisor D on the curve B, and an $n > 0$, such that $nK \sim f^*(D)$.*

Moreover, if these conditions are satisfied, then the complete linear system $|nK|$ is base-point-free for some $n > 0$, and the canonical ring $R(X)$ is a finitely generated $\Bbbk$-algebra of transcendence degree two.

Corollary 9.10 (Zariski–Mumford). *The canonical ring $R(X)$ of any nonsingular projective surface X is a finitely generated $\Bbbk$-algebra.*

Proof. By (6.3) there exists a birational morphism $f : X \to Y$ to a minimal model Y, and by (5.7) f induces an isomorphism $R(X) \cong R(Y)$. Thus, without loss of generality, we may assume that X is a minimal surface. If $\kappa(X) = -\infty$, then $R(X) = \Bbbk$. If $\kappa(X) = 0$ then $R(X)$ is a finitely generated algebra by Theorem 9.8; the same is true if $\kappa(X) = 1$, by Theorem 9.9; and if $\kappa(X) = 2$, by Theorem 9.1. $\qquad\square$

There only remains to prove Proposition 9.2. As we have already said, we will in fact prove a much more general result of Zariski. But first we need a few preparations.

9.11. Let $m \geq 1$ be a fixed natural number, and let $\mathbb{Z}^m$ be the additive group $\mathbb{Z} \times \cdots \times \mathbb{Z}$ (with m factors). On $\mathbb{Z}^m$ consider the natural partial ordering $(a_1, \dots, a_m) \leq (b_1, \dots, b_m)$ if and only if $a_i \leq b_i$ for all $i = 1, \dots, m$. We write $(a_1, \dots, a_m) \ll (b_1, \dots, b_m)$ if $a_i \ll b_i$ for all i.

An *m-graded ring* S is a commutative ring S together with a direct sum decomposition $S = \oplus_{a \in \mathbb{Z}^m} S_a$ with S_a an Abelian group for every $a \in \mathbb{Z}^m$, such that $[x \in S_a, y \in S_b] \implies [xy \in S_{a+b}]$. In particular, S_0 is a subring of S, and S_a is an S_0-module for every $a \in \mathbb{Z}^m$. For $m = 1$ we get the usual definition of a graded ring. The definition of m-graded S-modules should now be clear. If $E = \oplus_{a \in \mathbb{Z}} E_a$ is an m-graded S-module, then the elements of $\oplus_{a_1 + \cdots + a_m = n} E_a$ are called *of total degree* n.

In what follows, we shall consider only m-graded rings S with $S_a = 0$ for all $a = (a_1, \ldots, a_m)$ with at least one $a_i < 0$ (because this is all we need). On the other hand, modules may have elements of negative total degree.

For each i between 1 and m we put $e_i = (0, \ldots, 0, 1, 0, \ldots, 0)$, with 1 in the ith position. Let S be an m-graded ring, and let E be an m-graded S-module. We say that E is *i-finite* if there exists a natural number n such that the multiplication maps $S_{e_i} \times E_a \to E_{e_i + a}$ are surjective for all $a = (a_1, \ldots, a_m) \in \mathbb{Z}^m$ with $a_i \geq n$. If E is i-finite for all $i = 1, \ldots, m$, then we say that E is *polyfinite*.

Lemma 9.12. *Let S be an m-graded ring and E an m-graded S-module. Let S' be the S_0-subalgebra of S generated by its elements of total degree one. Then:*

(i) *If E is finitely generated as an S'-module, then E is polyfinite.*

(ii) *Conversely, assume in addition that every E_a is finitely generated as an S_0-module, and that $E_a = 0$ for all $a \ll 0$. Then E being polyfinite implies that E is finitely generated as an S'-module.*

Proof. We will make use of the following obvious properties:

(a) If E is i-finite then $E(a)$ is also i-finite for every $a \in \mathbb{Z}^m$, where $E(a)$ is the m-graded S-module with $E(a)_b \stackrel{\text{def}}{=} E_{a+b}$.

(b) If E and E' are i-finite, then so is $E \oplus E'$.

(c) Every quotient of an i-finite m-graded S-module is again i-finite.

(d) Every m-graded S-module E is also an m-graded S'-module, and E is polyfinite as an S-module if and only if it is polyfinite as an S'-module.

S' is clearly polyfinite as a module over itself. If E is finitely generated as an S'-module, then E is a quotient of a direct sum of S'-modules of the form $S'(a)$, and the properties mentioned earlier show that E is polyfinite as an S'-module, or equivalently, as an S-module.

Conversely, if E is polyfinite, then there exists $b \in \mathbb{Z}^m$ such that the multiplication maps $S_{e_i} \times E_a \to E_{e_i + a}$ are surjective for all $i = 1, \ldots, m$ and all $a \in \mathbb{Z}^m$ such that $a_i \geq b_i$. This immediately implies that E is generated by $\oplus_{a \leq b} E_a$ as a module over S'. But the hypotheses imply that $\oplus_{a \leq b} E_a$ is finitely generated as a module over S_0, and therefore E is finitely generated as a module over S'. $\qquad\square$

Proposition 9.13. *Let* $\Bbbk$ *be a field, and let* $V_1, \ldots, V_m$ *($m \geq 1$) be* m *finite-dimensional vector spaces over* $\Bbbk$. *Put*

$$Y \stackrel{\text{def}}{=} \mathbb{P}(V_1) \times \cdots \times \mathbb{P}(V_m),$$

where $\mathbb{P}(V_i)$ *is the projective space associated to the* $\Bbbk$-*vector space* V_i.
Let F *be a coherent* $\mathcal{O}_Y$-*module. For every* $a \in \mathbb{Z}^m$, *put*

$$F(a) \stackrel{\text{def}}{=} F \otimes p_1^*(\mathcal{O}(a_1)) \otimes \cdots \otimes p_m^*(\mathcal{O}(a_m)),$$

where p_i *is the canonical projection of* Y *onto* $\mathbb{P}(V_i)$. *Then:*

(i) *The natural Serre homomorphism*

$$S \stackrel{\text{def}}{=} S(V_1) \otimes \cdots \otimes S(V_m) \rightarrow \oplus_{a \in \mathbb{Z}^m} H^0(Y, \mathcal{O}_Y(a))$$

is an isomorphism of m-*graded rings, where* $S(V_i)$ *is the symmetric* $\Bbbk$-*algebra of* V_i.

(ii) $H^q(Y, \mathcal{O}_Y(a)) = 0$ *for all* $q \geq 1$ *and all* $a \geq 0$.

(iii) $\oplus_{a \geq 0} H^q(Y, F(a))$ *is a finitely generated* m-*graded* S-*module,* $\forall q \geq 0$.

(iv) *For* $q \geq 1$, *we have* $H^q(Y, F(a)) = 0$ *for all* $a \gg 0$.

Proof. The case $m = 1$ is a well-known theorem of Serre; see, for example, [EGA III] or [FAC].

Next we reduce the proof of the proposition to a special case. Namely, assume that parts (i), (ii), and (iii) are true for F of the form $F = \mathcal{O}_Y(b)$ (arbitrary m). We show that this implies (iii) and (iv) for arbitrary F. Since Y can be embedded into a projective space by the Segre immersion, via the very ample invertible $\mathcal{O}_Y$-module $L = \mathcal{O}_Y(1, \ldots, 1)$, every coherent $\mathcal{O}_Y$-module is a quotient of a finite direct sum F' of modules of the form L^j for various integers j. Thus we have an exact sequence

$$0 \rightarrow G \rightarrow F' \rightarrow F \rightarrow 0,$$

which induces a cohomology exact sequence

$$H^q(Y, F') \rightarrow H^q(Y, F) \rightarrow H^{q+1}(Y, G).$$

Since (iii) and (iv) are true for F', and also for arbitrary F if $q > \dim(X)$, it follows by decreasing induction on q that (iii) and (iv) are, in fact, true for arbitrary F.

To complete the proof of the proposition, there only remains to show that (i), (ii), and (iii) hold for $F = \mathcal{O}_Y(b)$. We will prove this by induction on $m \geq 1$, the case $m = 1$ being covered by Serre's theorem. Assume that (i), (ii), and (iii) are true for a given $m \geq 1$, $Y = \mathbb{P}(V_1) \times \cdots \times \mathbb{P}(V_m)$, and

$F = \mathcal{O}(b)$, $\forall b \in \mathbb{Z}^m$. Let V be another finite-dimensional $\Bbbk$-vector space, and consider the Cartesian diagram

$$
\begin{array}{ccc}
Y \times \mathbb{P}(V) & \xrightarrow{\quad g \quad} & \mathbb{P}(V) = \mathbb{P} \\
{\scriptstyle f}\downarrow & & \downarrow{\scriptstyle h} \\
Y & \xrightarrow{\quad p \quad} & \mathrm{Spec}(\Bbbk)
\end{array}
$$

If $a \in \mathbb{Z}^m$ and $n \in \mathbb{Z}$ then $\mathcal{O}_{Y \times \mathbb{P}}(a, n) = f^* \mathcal{O}_Y(a) \otimes g^* \mathcal{O}_{\mathbb{P}}(n)$. The natural map $\mathcal{O}_Y(a) \otimes R^q f_*(g^* \mathcal{O}_{\mathbb{P}}(n)) \to R^q f_*(\mathcal{O}_{Y \times \mathbb{P}}(a, n))$ is an isomorphism (projection formula). Also, since p is a flat morphism, the natural map $p^* R^q h_* \mathcal{O}_{\mathbb{P}}(n) \to R^q f_*(g^* \mathcal{O}_{\mathbb{P}}(n))$ is an isomorphism. From these two isomorphisms we get that $R^q f_* \mathcal{O}_{Y \times \mathbb{P}}(a, n) = \mathcal{O}_Y(a) \otimes H^q(\mathbb{P}, \mathcal{O}_{\mathbb{P}}(n))$, whence

$$
H^p(Y, \mathcal{O}_Y(a)) \otimes_{\Bbbk} H^q(\mathbb{P}, \mathcal{O}_{\mathbb{P}}(n)) = H^p(Y, R^q f_* \mathcal{O}_{Y \times \mathbb{P}}(a, n)). \qquad (*)
$$

By the inductive hypothesis, the natural map

$$
S_a \otimes S^n(V) \to H^0(Y, \mathcal{O}_Y(a)) \otimes_{\Bbbk} H^0(\mathbb{P}, \mathcal{O}_{\mathbb{P}}(n)) \cong H^0(Y \times \mathbb{P}, \mathcal{O}_{Y \times \mathbb{P}}(a, n))
$$

is an isomorphism; thus (i) is proved for $Y \times \mathbb{P}(V)$.

The inductive hypothesis for (ii) and the isomorphism $(*)$ imply that $H^p(Y, R^q f_* \mathcal{O}_{Y \times \mathbb{P}}(a, n)) = 0$ if $(a, n) \geq 0$ and either $p > 0$ or $q > 0$. Therefore the Leray spectral sequence

$$
E_2^{pq} = H^p(Y, R^q f_* \mathcal{O}_{Y \times \mathbb{P}}(a, n)) \implies H^{p+q}(Y \times \mathbb{P}, \mathcal{O}_{Y \times \mathbb{P}}(a, n))
$$

degenerates, and then

$$
H^p(Y, f_* \mathcal{O}_{Y \times \mathbb{P}}(a, n)) \cong H^p(Y \times \mathbb{P}, \mathcal{O}_{Y \times \mathbb{P}}(a, n)).
$$

Since for $p > 0$ we have isomorphisms

$$
H^p(Y, f_* \mathcal{O}_{Y \times \mathbb{P}}(a, n)) \cong H^p(Y, \mathcal{O}_Y(a)) \otimes_{\Bbbk} H^0(\mathbb{P}, \mathcal{O}_{\mathbb{P}}(n)),
$$

we see that (ii) holds for $Y \times \mathbb{P}$ if it holds for Y.

Finally, by the inductive hypothesis we have that $\oplus_{a \geq 0} H^p(Y, \mathcal{O}_Y(b + a))$ is a finitely generated S-module for all $b \in \mathbb{Z}^m$, and $\oplus_{n \geq 0} H^q(\mathbb{P}, \mathcal{O}_{\mathbb{P}}(n' + n))$ is a finitely generated $S(V)$-module for all $n' \in \mathbb{Z}$. By the isomorphism $(*)$ we see that their tensor product, $\oplus_{(a,n) \geq 0} H^p(Y, R^q f_* \mathcal{O}_{Y \times \mathbb{P}}(b + a, n' + n))$, is a finitely generated $S \otimes_{\Bbbk} S(V)$-module. Therefore the limit of the corresponding Leray spectral sequence is also finitely generated. Thus the proof of Proposition 9.13 is complete. $\qquad\square$

Theorem 9.14 (Zariski). *Let X be a proper scheme over $\Bbbk$, and let F be a coherent $\mathcal{O}_X$-module. Let $L_1, \dots, L_m$ be m invertible $\mathcal{O}_X$-modules, $m \geq 1$. Let T be an m-graded subring of the m-graded ring $\oplus_{a \geq 0} H^0(X, L_1^{a_1} \otimes \cdots \otimes$*

$L_m^{a_m}$) *(where multiplication is induced by the tensor product), and let E be a graded T-submodule of $\oplus_{a\geq 0}H^n(X, F \otimes L_1^{a_1} \otimes \cdots \otimes L_m^{a_m})$ ($n \geq 0$). If the linear systems $|T_{e_i}|$ are base-point-free for all $i = 1,\ldots, m$, then E is polyfinite.*

Proof. Let $V_i = T_{e_i} \subset H^0(X, L_i)$. Since the linear systems $|V_i|$ are base-point-free, they define morphisms $f_i : X \to \mathbb{P}(V_i)$ such that $f_i^*(\mathcal{O}(1)) \cong L_i$. Let $f : X \to Y = \mathbb{P}(V_1) \times \cdots \times \mathbb{P}(V_m)$ be the morphism with components $f_1, \ldots, f_m$. Then $f^*(\mathcal{O}(a)) \cong L_1^{a_1} \otimes \cdots \otimes L_m^{a_m}$ for all $a \in \mathbb{Z}^m$. Since f is proper, the sheaves $R^q f_*(F)$ are coherent on Y. By Proposition 9.13 the m-graded S-module

$$E_2^{pq} = \oplus_{a\geq 0}H^p(Y, R^q f_*(F(a)))$$

is finitely generated, and therefore the limit $\oplus_{a\geq 0}H^n(X, F(a))$ is also a finitely generated S-module. Since S is Noetherian, the S-submodule E is also finitely generated. (Note that the ring T is a quotient of S.) Finally, since $T_{e_i} = S_{e_i}$, Lemma 9.12 shows that E is polyfinite as a T-module. $\square$

9.15. *Proof of Proposition 9.2.* With the same notation as in Proposition 9.2, let $L_1 = L^n$, $F = L^i$ ($0 \leq i < n$), $T = \oplus_{a\geq 0}H^0(X, L_1^a) = \Gamma_*(L_1) = \Gamma_*(L^n) = S^{(n)}$. Then $T_1 = H^0(X, L^n)$, and by hypothesis the linear system $|T_1|$ is base-point-free. Let $E = \oplus_{a\geq 0}H^0(X, F \otimes L_1^a) = \oplus_{a\geq 0}H^0(X, L^{an+i})$. By Theorem 9.14 we have that E is a polyfinite T-module (i.e., a finitely generated T-module, since $m = 1$ in this case). That is, for each i between 0 and $n - 1$ the natural maps

$$H^0(X, L^n) \otimes H^0(X, L^{an+i}) \to H^0(X, L^{(a+1)n+i})$$

are surjective for all $a \gg 0$. Since the $\Bbbk$-vector spaces $H^0(X, L^s)$ are finite-dimensional for all $s \geq 0$, this implies that S is finitely generated as a $\Bbbk$-algebra.

Remark 9.16. The proof of Proposition 9.2 uses only the special case of Theorem 9.14 with $m = 1$. In this case Theorem 9.14 has a much shorter proof (although the proof uses the same ideas). The proof of Theorem 9.14 given here is due to A. Ogus [Ogu]. Zariski's original proof is more complicated and doesn't use sheaf theory. To illustrate the usefulness of considering several linear systems simultaneously (i.e., the case $m > 1$), we will prove the following result, also due to Zariski.

Theorem 9.17 (Zariski [Zar1]). *Let X be a nonsingular projective variety of dimension at least two, and let $|D|$ be a complete linear system on X having only a finite number of base points. Then the linear system $|nD|$ is base-point-free for all sufficiently large integers n.*

Proof. Let H be a generic hyperplane section on X; in particular, H should not contain any base point of $|D|$. Put $L_{ij} = \mathcal{O}_X(iH + jD)$. Let

$$\sigma_{ij} : H^0(X, L_{ij}) \to H^0(H, L_{ij} \otimes \mathcal{O}_H)$$

be the canonical restriction homomorphism, $\sigma = (\sigma_{ij})_{(i,j)\geq 0}$, and $T = \operatorname{Im}(\sigma) = \oplus_{(i,j)\geq 0} \operatorname{Im}(\sigma_{ij})$. T is a 2-graded subring of $\oplus_{(i,j)\geq 0} H^0(H, L_{ij} \otimes \mathcal{O}_H)$.

Since the linear systems $|T_{10}|$ and $|T_{01}|$ are base-point-free on H, T is polyfinite by (9.14); thus there exists $N > 0$ such that for every $i \geq 0$ and every $j \geq N$, the canonical homomorphism

$$T_{i,j-1} \otimes_k T_{01} = (\operatorname{Im}(\sigma_{i,j-1})) \otimes_k (\operatorname{Im}(\sigma_{01})) \xrightarrow{\tau} \operatorname{Im}(\sigma_{ij})$$

is surjective. Consider the commutative diagram with exact rows:

$$
\begin{array}{ccccc}
H^0(X, L_{i,j-1}) \otimes H^0(X, L_{01}) & \to & \operatorname{Im}(\sigma_{i,j-1}) \otimes \operatorname{Im}(\sigma_{01}) & \to & 0 \\
\downarrow{\scriptstyle v} & & \downarrow{\scriptstyle \tau} & & \\
0 \to H^0(X, L_{i-1,j}) \xrightarrow{\;u\;} H^0(X, L_{ij}) & & \longrightarrow \operatorname{Im}(\sigma_{ij}) & \longrightarrow & 0
\end{array}
$$

in which u is the homomorphism defined by $u(s) = s \otimes h$, where $h \in H^0(X, L_{10})$ is a section with $\operatorname{div}_X(h) = H$. The surjectivity of τ implies that $H^0(X, L_{ij}) = \operatorname{Im}(v) + \operatorname{Im}(u)$.

On the other hand, for a fixed $j \geq N$ and for $i \gg 0$, L_{ij} is generated by global sections. Let x be a base point of $|D|$. Then x is a base point for every divisor of the form $\operatorname{div}_X(s)$ with $s \in \operatorname{Im}(v)$. Since $H^0(X, L_{ij}) = \operatorname{Im}(v) + \operatorname{Im}(u)$ and $|L_{ij}|$ is base-point-free, we see that x cannot be a base point for the linear system $|L_{i-1,j}| + H$, and therefore x is not a base point for the linear system $|L_{i-1,j}|$. By decreasing induction on i we see that x is not a base point for $|L_{ij}|$ for any $i \geq 0$. In particular, x is not a base point for $|L_{0j}| = |jD|$. $\qquad\square$

Remark. It is clear from the proof that the condition that X be nonsingular is not needed: if X is any irreducible projective variety and D is a Cartier divisor on X, then the theorem is still true.

EXERCISES

Exercise 9.1. Let Y be an irreducible curve on a nonsingular projective surface X such that $(Y^2) > 0$. Show that the k-algebra $\oplus_{i=0}^{\infty} H^0(X, \mathcal{O}_X(iY))$ is finitely generated. [*Hint:* Use Exercise 1.8 and Proposition 9.2.]

Exercise 9.2. In Exercise 1.8 assume that X is a $K3$ surface. Prove that for large n the singularities of the surface S_n are all rational double points, $\omega_{S_n} \cong \mathcal{O}_{S_n}$, and $H^1(S_n, \mathcal{O}_{S_n}) = 0$.

Exercise 9.3. Generalize Exercises 1.8 and 9.1 to the case when Y is an effective divisor on a nonsingular projective variety X such that $\mathcal{O}_X(Y) \otimes \mathcal{O}_Y$ is ample.

Exercise 9.4. Produce examples of minimal surfaces of general type X containing nonsingular rational curves (resp. nonsingular rational curves C with $(K \cdot C) = 0$).

Exercise 9.5. Prove that the Fermat surface X with equation $x_0^d + x_1^d + x_2^d + x_3^d = 0$ in $\mathbb{P}^3$ ($d \geq 2$, d not a multiple of $\mathrm{char}(\mathbb{k})$ if $\mathrm{char}(\mathbb{k}) > 0$) admits a fibration $f : X \to \mathbb{P}^1$ whose fibers are plane curves of degree $d - 1$.

Exercise 9.6. Let D be an effective divisor on a surface X such that $|D|$ has no fixed components, and let $U = X \setminus \mathrm{Supp}(D)$. Prove that there is a proper morphism $f : U \to V$, with V an affine variety. [*Hint:* Apply Theorem 9.17.]

Exercise 9.7. Let D be an effective divisor on a surface X such that $|D|$ has no fixed components, and let $U = X \setminus \mathrm{Supp}(D)$. Prove that $H^0(U, \mathcal{O}_U)$ is a finitely generated $\mathbb{k}$-algebra. [*Hint:* Use Theorem 9.17 and Proposition 9.2.]

Exercise 9.8. Let $f : X' \to X$ be a surjective morphism between two surfaces over $\mathbb{k}$, with $\mathrm{char}(\mathbb{k}) = 0$. Prove that $\kappa(X) \leq \kappa(X')$, with equality if f is étale.

Bibliographic References. Theorem 9.1 is due to Mumford [Mum3]; it motivates the study of rational double points and the contractability criterion 3.15. The proof given here is a slight modification of that in [Mum3]. Theorems 9.4, 9.8, and 9.9 follow from the results in Chapter 8 and Theorem 9.1. Proposition 9.2 was proved in [Zar1], together with its generalization (9.13). Zariski's original proofs are complicated; we chose to present Ogus's simpler, cohomological proofs. Theorem 9.17 is also due to Zariski (loc. cit.) Example (9.6.2) is due to Godeaux; the presentation given here follows [Bea].

10

Surfaces with Canonical Dimension Zero (char($\Bbbk$) $\neq 2, 3$)

10.1. Recall the list of minimal surfaces X with $(K^2) = 0$ and $p_g \leq 1$ from Chapter 8:

1. $b_2 = 22,$ $b_1 = 0,$ $\chi(\mathcal{O}_X) = 2,$ $q = 0,$ $p_g = 1,$ and $\Delta = 0.$

2. $b_2 = 14,$ $b_1 = 2,$ $\chi(\mathcal{O}_X) = 1,$ $q = 1,$ $p_g = 1,$ and $\Delta = 0.$

3. $b_2 = 10,$ $b_1 = 0,$ $\chi(\mathcal{O}_X) = 1,$ $q = 0,$ $p_g = 0,$ and $\Delta = 0,$
         or $q = 1,$ $p_g = 1,$ and $\Delta = 2.$

4. $b_2 = 6,$ $b_1 = 4,$ $\chi(\mathcal{O}_X) = 0,$ $q = 2,$ $p_g = 1,$ and $\Delta = 0.$

5. $b_2 = 2,$ $b_1 = 2,$ $\chi(\mathcal{O}_X) = 0,$ $q = 1,$ $p_g = 0,$ and $\Delta = 0,$
         or $q = 2,$ $p_g = 1,$ and $\Delta = 2.$

We have already seen that every minimal surface X with $\kappa(X) = 0$ has $(K^2) = 0$ and $p_g \leq 1$, and therefore it is in one of the five classes of surfaces listed above.

Theorem 10.2. *There are no minimal surfaces X with $\kappa(X) = 0$ and $b_2 = 14$.*

Proof. By way of contradiction, assume that such a surface X exists. From the list above we see that $b_1 = 2 = 2s$, so that $s = \dim \underline{\text{Pic}}^0(X) = 1$.

In particular, the Picard variety $P(X)$ is nontrivial. That is, there exists an invertible $\mathcal{O}_X$-module L that is algebraically equivalent to zero but not isomorphic to $\mathcal{O}_X$. By the Riemann–Roch theorem we have

$$\chi(L) = 1/2(L \cdot L \otimes \omega_X^{-1}) + \chi(\mathcal{O}_X) = \chi(\mathcal{O}_X) = 1$$

(we see that $\chi(\mathcal{O}_X) = 1$ in the list above), and therefore $H^0(L) \neq 0$ or $H^2(L) \neq 0$. Since $\omega_X \equiv 0$ and $p_g = 1$, we have $\omega_X \cong \mathcal{O}_X$, and therefore $H^2(L) \cong H^0(L^{-1})$ by Serre duality. Thus $H^0(L) \neq 0$ or $H^0(L^{-1}) \neq 0$, and therefore $L \cong \mathcal{O}_X$ or $L^{-1} \cong \mathcal{O}_X$ (because $L \equiv 0$), which is a contradiction. $\qquad\square$

Theorem 10.2 and the list in 10.1 show that there are four classes of minimal surfaces X with $\kappa(X) = 0$, according to the second Betti number $b_2 = b_2(X)$: $b_2 = 22$, $b_2 = 10$, $b_2 = 6$, and $b_2 = 2$.

Theorem 10.3. *Let X be a minimal surface. Then the following two sets of conditions are equivalent:*

(a) $\kappa(X) = 0$ *and* $b_2 = 22$;

(b) $K \sim 0$ *and* $q = 0$.

Surfaces X that satisfy these conditions are called K3 surfaces, and they have the following additional properties:

(i) $\mathrm{Pic}^\tau(X) = 0$; *that is, the Picard group $\mathrm{Pic}(X)$ of X is torsionfree, and therefore, by the Néron–Severi theorem, $\mathrm{Pic}(X)$ is a free Abelian group of finite rank.*

(ii) *If $f : X' \to X$ is a connected étale covering of X, then f is an isomorphism. In other words, X is an algebraically simply connected surface.*

Proof. First we prove the equivalence of (a) and (b). If (a) holds, then $q = 0$ follows from the list in 10.1, and $K \sim 0$ follows from $\omega_X \equiv 0$ and $p_g = 1$ (which we read in the same list). Conversely, if $K \sim 0$, then clearly $\kappa(X) = 0$. From the list in 10.1 we see that the only possibilities for $q = 0$ are $b_2 = 22$ and the first subcase of $b_2 = 10$—and the first subcase of $b_2 = 10$ is excluded because then we would have $p_g = 0$, contradicting $K \sim 0$.

Now we prove property (i). Let L be any invertible $\mathcal{O}_X$-module such that $L \equiv 0$. By the Riemann–Roch theorem we have that $\chi(L) = \chi(\mathcal{O}_X) = 2$, and therefore $H^0(L) \neq 0$ or $H^2(L) \neq 0$. Since $\omega_X \cong \mathcal{O}_X$ we have that $H^0(L) \neq 0$ or $H^0(L^{-1}) \neq 0$ (by Serre duality), and since $L \equiv 0$ we get $L \cong \mathcal{O}_X$ or $L^{-1} \cong \mathcal{O}_X$. Thus $\mathrm{Pic}^\tau(X) = 0$.

Finally we prove (ii). Let $f : X' \to X$ be an étale covering of degree n of X. Since $\omega_{X'} = f^*(\omega_X) \cong \mathcal{O}_{X'}$, we have $\kappa(X') = 0$, X' is a minimal surface, and $p_g(X') = 1$. Therefore X' is in the list in 10.1. Now Proposition 9.7 gives that $\chi(\mathcal{O}_{X'}) = n\chi(\mathcal{O}_X) = 2n$. But we see in the list in 10.1 that there are no surfaces X' with $\chi(\mathcal{O}_{X'}) > 2$. Therefore n must be equal to 1, that is, f must be an isomorphism. $\square$

10.4. Examples of K3 Surfaces. Let X be a smooth complete intersection surface in $\mathbb{P}^n$, $n \geq 3$. Assume that X is the intersection of hypersurfaces $H_1, \ldots, H_{n-2}$ of degrees $d_1, \ldots, d_{n-2}$. For X to be a K3 surface we must have $\sum_{i=1}^{n-2} d_i = n + 1$. Indeed, if X is a K3 surface, then $\omega_X = \mathcal{O}_X(\sum_{i=1}^{n-2} d_i - n - 1) \cong \mathcal{O}_X$. Conversely, if X is a (smooth) complete intersection of hypersurfaces $H_1, \ldots, H_{n-2}$ of degrees $d_1, \ldots, d_{n-2}$ with $\sum_{i=1}^{n-2} d_i = n + 1$, then X is a K3 surface, because $\omega_X \cong \mathcal{O}_X$ and $H^1(\mathcal{O}_X) = 0$—see Example 9.6.1. In particular, every smooth surface of degree 4 in $\mathbb{P}^3$ is a K3 surface. Also, every smooth complete intersection of three quadrics in $\mathbb{P}^5$ is a K3 surface.

Another important class of K3 surfaces is the Kummer surfaces associated to Abelian varieties of dimension 2.

10.5. Kummer Surfaces. Let k be an algebraically closed field of characteristic different form 2, and let X be an Abelian variety of dimension 2 over k. Consider the automorphism $\tau : X \to X$ given by $\tau(x) = -x$ (the negative of x for the group structure of X). Clearly $\tau^2 = \mathrm{id}_X$. Thus we have an action of the group $G = \{\mathrm{id}_X, \tau\}$ on X. A (closed) point $x \in X$ is a fixed point for τ if and only if the order of x in the Abelian group $(X, +)$ is 1 or 2. As the characteristic of k is different from 2, an elementary result about Abelian varieties [Mum2, § 6] shows that there are exactly 16 such points on X, say, $x_1 = 0, x_2, \ldots, x_{16}$. Let $\pi : \tilde{X} \to X$ be the blow-up of X with center the (reduced) discrete closed subscheme $\{x_1, \ldots, x_{16}\}$. Since $x_1, \ldots, x_{16}$ are fixed points for τ, it follows immediately that there exists a unique automorphism $\sigma : \tilde{X} \to \tilde{X}$ such that $\tau \circ \pi = \pi \circ \sigma$. Then $\sigma^2 = \mathrm{id}_{\tilde{X}}$, so that we can consider the two-dimensional quotient variety $Y = \tilde{X}/\sigma$. Y is called the *Kummer surface* associated to the Abelian surface X. Some of the properties of the Kummer surface Y are given in the following theorem.

Theorem 10.6. *With the hypotheses and notation from (10.5), Y is a nonsingular projective surface. Moreover, Y is a (minimal) K3 surface that contains 16 pairwise disjoint curves $E_1, \ldots, E_{16}$ with $p_a(E_i) = 0$ and $(E_i^2) = -2$, $i = 1, \ldots, 16$, and if H is a hyperplane section on Y, then the classes of $H, E_1, \ldots, E_{16}$ are linearly independent elements of $\mathrm{Pic}(Y)$.*

Proof. Let $\phi : \tilde{X} \to Y$ be the canonical morphism; ϕ is a finite morphism of degree 2. Since $\tilde{X}$ is a projective surface and ϕ is surjective, Y is a complete two-dimensional variety. If we can show that Y is nonsingular, then Theorem 1.28 will show that Y is projective.

Let's show therefore that Y is nonsingular. Let $F_i = \pi^{-1}(x_i)$, $i = 1, \ldots, 16$, be the exceptional curves of the first kind on $\tilde{X}$ with respect to the blow-up morphism π. As ϕ is an étale morphism outside of $\cup_{i=1}^{16} F_i$, there only remains to prove that Y is nonsingular at points of the form $\phi(x)$ with $x \in \cup_{i=1}^{16} F_i$. X being an Abelian variety, the translation t_{x_i} induces an isomorphism between $\mathcal{O}_{X,x_i}$ and $\mathcal{O}_{X,x_1} = \mathcal{O}_{X,0}$. Therefore it will suffice to consider the points of Y of the form $\phi(x)$ with $x \in F_1$. If $T_{X,0}$ and $\mathfrak{m}_0$ are, respectively, the tangent space to X at 0 and the maximal ideal of $\mathcal{O}_{X,0}$, then a general result about Abelian varieties [Bor, 3.2] shows that the tangent map $(d\tau)_0 : T_{X,0} \to T_{X,0}$ is the map $a \mapsto -a$ of $T_{X,0}$, and therefore its dual, the map $\bar{\tau}^* : \mathfrak{m}_0/\mathfrak{m}_0^2 \to \mathfrak{m}_0/\mathfrak{m}_0^2$ is the map $b \mapsto -b$ of $\mathfrak{m}_0/\mathfrak{m}_0^2 = T'_{X,0}$. Thus if (u, v) is a regular system of parameters on X at 0 (that is, a minimal system of generators of $\mathfrak{m}_0$), we have that $(\tau^*(u), \tau^*(v))$ is also a regular system of parameters on X at 0, which coincides with $(-u, -v)$ modulo $\mathfrak{m}_0^2$. Now put $u' = u - \tau^*(u), v' = v - \tau^*(v)$; thus $u' \equiv 2u$ (mod $\mathfrak{m}_0^2$), $v' \equiv 2v$ (mod $\mathfrak{m}_0^2$). Because the characteristic of k is different from 2, we have that (u', v') is a regular system of parameters on X at 0, which satisfies the additional conditions:

$$\tau^*(u') = -u', \quad \tau^*(v') = -v'. \tag{$*$}$$

Let x be an arbitrary point on F_1. We may assume that $u'' = \pi^*(u'), v'' = \pi^*(v')/\pi^*(u')$ is a regular system of parameters on $\tilde{X}$ at x. From Equation $(*)$ we can see immediately, using the definition of σ, that

$$\sigma^*(u'') = -u'', \quad \sigma^* v'' = v''.$$

This implies that (u''^2, v'') is a regular system of parameters on Y at the point $\phi(x)$. We have thus proved that Y is nonsingular.

Now let $E_i = \phi(F_i)$ (with the reduced structure), $i = 1, \ldots, 16$. If $i \neq j$, then clearly $E_i \cap E_j = \varnothing$. On the other hand, since u'' is a local equation of F_1 at x, we have that u''^2 is a local equation of E_1 at $\phi(x)$. Therefore E_1 is a nonsingular curve and $\phi^{-1}(E_1) = 2F_1$ (scheme-theoretically). From (1.18) we have that

$$(\phi^{-1}(E_1) \cdot \phi^{-1}(E_1)) = \deg(\phi) \cdot (E_1^2),$$

whence $(E_1^2) = -2$. Since $p_a(F_1) = 0$, we also have $p_a(E_1) = 0$. The homogeneity argument we used earlier shows that $(E_i^2) = -2$ and $p_a(E_i) = 0$ for all $i = 1, \ldots, 16$.

Now let $h, e_1, \ldots, e_{16}$ be the classes of $H, E_1, \ldots, E_{16}$ in $\text{Num}(Y) = \text{Pic}(Y)/\text{Pic}^\tau(Y)$. We show that these elements are linearly independent. Let $a_0 h + \sum_{i=1}^{16} a_i e_i = 0$ with $a_0, a_1, \ldots, a_{16}$ integers. As $(e_i \cdot e_j) = (E_i \cdot E_j) = 0$ for all $i \neq j$, we have

$$a_0^2 (h \cdot h) = \sum_{i=1}^{16} a_i^2 (e_i \cdot e_i) = -2 \sum_{i=1}^{16} a_i^2 \leq 0,$$

and therefore $a_0 = 0$, because $(h \cdot h) = (H^2) > 0$. Now from $\sum_{i=1}^{16} a_i e_i = 0$ we get

$$0 = \left(e_j \cdot \sum_{i=1}^{16} a_i e_i \right) = -2a_j,$$

and therefore $a_j = 0$, for all $j = 1, \ldots, 16$.

As a corollary of what we have just proved, we get that $\rho \geq 17$, and since $\rho \leq b_2$ (the Igusa–Severi inequality), we get $b_2 \geq 17$.

To complete the proof of Theorem 10.6 it will suffice to show that $\omega_Y \cong \mathcal{O}_Y$. Indeed, in that case Y is a minimal surface with $\kappa(Y) = 0$, and the list in 10.1 shows that the only such surfaces with $b_2 \geq 17$ have $b_2 = 22$, so that Y is a K3 surface. Furthermore, if Y is a K3 surface, then Theorem 10.3 shows that $\mathrm{Pic}(Y) = \mathrm{Num}(Y)$.

Now we show that $\omega_Y \cong \mathcal{O}_Y$. Since X is an Abelian variety, we may consider the translation-invariant regular differential form ω of degree 2 on X that induces the form $du' \wedge dv'$ at the origin. Using Equation $(*)$ it follows that $\pi^*(\omega)$ is a nonzero regular differential form on $\tilde{X}$ that is invariant under the automorphism σ. Therefore $\pi^*(\omega)$ has the form $\pi^*(\omega) = \sigma^*(\omega')$ for a nonzero rational differential form ω' of degree 2 on Y. Clearly the divisor of ω' is supported on $\cup_{i=1}^{16} E_i$. If $y = \phi(x) \in E_1$, we have (considering the definitions and notation introduced earlier):

$$\pi^*(\omega) = d(\pi^*(u')) \wedge d(\pi^*(v')) = du'' \wedge d(u'' \cdot v'')$$
$$= u'' \cdot du'' \wedge dv'' = \tfrac{1}{2} d(u''^2) \wedge dv''.$$

Thus ω' is a regular differential form on Y with $\mathrm{div}_Y(\omega') = 0$, and therefore $\omega_Y \cong \mathcal{O}_Y$. $\qquad\square$

Remarks 10.7.

(a) If Y is the Kummer surface associated with the Abelian surface X, then $H^0(Y, \Omega_Y) = 0$, where $\Omega_Y = \Omega^1_{Y/k}$.

To see this we use the following elementary remark: if $f : X' \to X$ is a birational map of surfaces, then f induces an isomorphism $f^* : H^0(X, \Omega_X) \to H^0(X', \Omega_{X'})$.

Taking this for granted for a moment, assume by contradiction that there exists a nonzero 1-form $\omega' \in H^0(Y, \Omega_Y)$. Then $\phi^*(\omega')$ is a σ-invariant 1-form in $H^0(\tilde{X}, \Omega_{\tilde{X}})$, and since π induces an isomorphism $\pi^* : H^0(X, \Omega_X) \to H^0(\tilde{X}, \Omega_{\tilde{X}})$ we get a nonzero, τ-invariant 1-form $\omega \in H^0(X, \Omega_X)$. We reached a contradiction, because $H^0(X, \Omega_X)$ has a basis du', dv' where u' and v' satisfy the equations in $(*)$ in the proof of Theorem 10.6.

There only remains to show that a birational map f induces an isomorphism $f^* : H^0(X, \Omega_X) \to H^0(X', \Omega_{X'})$. Since f is defined on the complement of a finite set of X' and induces an isomorphism of a dense open subset of X' onto a dense open subset of X, we have that f induces an injective homomorphism $f^* : H^0(X, \Omega_X) \to H^0(X' \setminus \{z_1, \ldots, z_m\}, \Omega_{X'})$. The latter linear space is canonically identified with $H^0(X', \Omega_{X'})$, because a rational 1-form on X' is either regular on all of X' or the poles form a whole divisor. Finally, everything we said so far applies equally to the birational map $f^{-1} : X \to X'$; this shows that f^* is an isomorphism.

(b) The Kummer surface Y associated with the Abelian surface X can also be obtained as follows: first consider the quotient surface X/τ; this surface has exactly 16 singular points, corresponding to the 16 fixed points of τ. Using the regular system of parameters (u', v') on X at 0 that satisfies the equations in $(*)$ in the proof of Theorem 10.6, we see easily that all these singularities are rational double points that can each be desingularized by just one blow-up. It is easy to see that Y coincides with the surface obtained from X/τ by blowing up the 16 rational double points.

10.8. From the list in 10.1 and Theorem 10.2 we see that the next class of minimal surfaces with $\kappa(X) = 0$ (in terms of the second Betti number) are the surfaces with $\omega_X \equiv 0$ and $b_2 = 10$. Then $b_1 = 0$ and $\chi(\mathcal{O}_X) = 1$. In this case there are two possible subcases:

(a) $p_g = 0, q = 0$ and $\Delta = 0$, or

(b) $p_g = 1, q = 1$ and $\Delta = 2$.

Regarding alternative (b) we have the following result.

Theorem 10.9. *If the characteristic of* k *is not* 2, *then there are no minimal surfaces* X *with* $\kappa(X) = 0$, $b_2 = 10$, *and* $p_g = 1$.

Proof. By way of contradiction, assume that such a surface exists. From (10.1) we have that $q = \dim H^1(\mathcal{O}_X) = 1$ and $\Delta = 2$. From Theorem 5.1 we have that $\Delta = 2 \implies p = \text{char}(k) > 0$. On the other hand, since $q > 0$, there exists a 1-cocycle $a = \{a_{ij}\} \in Z^1(\mathcal{U}, \mathcal{O}_X)$ that is not cohomologous to zero, with $\mathcal{U} = \{U_i\}$ a finite covering of X with affine open sets. By a well-known procedure we can associate to the cocycle a a G_a-bundle $\pi : V \to X$ (where G_a is the additive group $(k, +)$), as follows: $\pi^{-1}(U_i) = \mathbb{A}^1 \times U_i$ for each i, and the patching is given by the following identification: if $x \in U_i \cap U_j$, then the pairs $(z_i, x) \in \mathbb{A}^1 \times U_i$ and $(z_j, x) \in \mathbb{A}^1 \times U_j$ represent the same point in V if and only if $z_i = a_{ij}(x) + z_j$.

Since $p_g = 1$ and $\omega_X \equiv 0$, there exists a regular form ω of degree 2 on X that does not vanish anywhere. Then the family $\{dz_i \wedge \omega\}_i$ of forms of degree 3 defines a unique regular form of degree 3 on V that doesn't vanish anywhere. Thus V is a nonsingular variety of dimension 3 with $\omega_V \cong \mathcal{O}_V$.

Since $\dim H^1(\mathcal{O}_X) = 1$ and $a^p = \{a_{ij}^p\}$ is a 1-cocycle of $\mathcal{O}_X$, there exists a $\lambda \in \Bbbk$ such that

$$a_{ij}^p = \lambda a_{ij} + b_i - b_j, \quad \text{with } b_i \in \Gamma(U_i, \mathcal{O}_X).$$

It is easy to see that the family of regular functions $\{z_i^p - \lambda z_i - b_i\}_i$, with $z_i^p - \lambda z_i - b_i \in \Gamma(\mathbb{A}^1 \times U_i, \mathcal{O}_V)$, defines a unique nonzero function $f \in \Gamma(V, \mathcal{O}_V)$ (i.e., $f|_{\mathbb{A}^1 \times U_i} = z_i^p - \lambda z_i - b_i$). Let X' be the surface in V defined by the equation $f = 0$. Therefore X' is a Gorenstein surface. If $\lambda \neq 0$, then the restriction $\pi' = \pi|_{X'} : X' \to X$ of π to X' is an étale morphism, and therefore in this case X' is nonsingular. If $\lambda = 0$ then X' is given in $\mathbb{A}^1 \times U_i$ by the equation $z_i^p - b_i = 0$. Since $b_i^{1/p} \notin \Gamma(U_i, \mathcal{O}_X)$—or else the cocycle a would be cohomologous to zero—we have that in this case X' is reduced and connected. Either way we have $\dim H^0(X', \mathcal{O}_{X'}) = 1$. Furthermore, since the conormal bundle of X' in V is trivial and $\omega_V \cong \mathcal{O}_V$, we infer that $\omega_{X'} \cong \mathcal{O}_{X'}$.

We have $\chi(\mathcal{O}_{X'}) \leq \dim H^0(\mathcal{O}_{X'}) + \dim H^2(\mathcal{O}_{X'}) = \dim H^0(\mathcal{O}_{X'}) + \dim H^0(\mathcal{O}_{X'}) = 2$.

Finally, it is easy to check that we have the following global filtration of $\pi_*' \mathcal{O}_{X'}$:

$$0 \subset \mathcal{O}_X \subset [\mathcal{O}_X \oplus z_i \mathcal{O}_X] \subset \cdots \subset [\mathcal{O}_X \oplus z_i \mathcal{O}_X \oplus \cdots \oplus z_i^{p-1} \mathcal{O}_X] = \pi_*' \mathcal{O}_{X'},$$

with $[\mathcal{O}_X \oplus z_i \mathcal{O}_X \oplus \cdots \oplus z_i^n \mathcal{O}_X]/[\mathcal{O}_X \oplus z_i \mathcal{O}_X \oplus \cdots \oplus z_i^{n-1} \mathcal{O}_X] \cong \mathcal{O}_X$ for all $n = 1, \ldots, p - 1$. Since the Euler–Poincaré characteristic is additive, we get $\chi(X, \pi_*' \mathcal{O}_{X'}) = p\chi(\mathcal{O}_X)$. Since π' is a finite morphism, we have $\chi(X, \pi_*' \mathcal{O}_{X'}) = \chi(\mathcal{O}_{X'})$. Therefore the inequality obtained earlier becomes $p\chi(\mathcal{O}_X) \leq 2$, and therefore $p \leq 2$. $\square$

10.10. Theorem 10.9 shows that if char($\Bbbk$) $\neq 2$ then the only minimal surfaces X with $\kappa(X) = 0$ and $b_2 = 10$ are those in subcase (a) above; that is, they have $\kappa(X) = 0$, $b_2 = 10$, $b_1 = 0$, $\chi(\mathcal{O}_X) = 1$, $p_g = 0$, $q = 0$, and $\Delta = 0$. These surfaces are called *Enriques surfaces*, or sometimes *classical Enriques surfaces* to distinguish them from the nonclassical Enriques surfaces; the nonclassical Enriques surfaces are the minimal surfaces X with $\kappa(X) = 0$, $b_2 = 10$ and $p_g = 1$, which do exist in characteristic 2, cf. [BM2]. In this chapter we set out to study the minimal surfaces X with $\kappa(X) = 0$ only in characteristic different from 2 and 3; therefore we will consider only classical Enriques surfaces.

Proposition 10.11. *If X is an Enriques surface, then $\omega_X \not\cong \mathcal{O}_X$ and $\omega_X^2 \cong \mathcal{O}_X$, so that $p_2 = 1$.*

Proof. Since $p_g = 0$, we have $\omega_X \not\cong \mathcal{O}_X$. On the other hand, we have

$$\chi(\mathcal{O}_X(-K)) = \chi(\mathcal{O}_X) = 1$$

by the Riemann–Roch Theorem. Therefore

$$\dim H^0(\mathcal{O}_X(-K)) + \dim H^2(\mathcal{O}_X(-K)) \geq 1.$$

By Serre duality we have $\dim H^2(\mathcal{O}_X(-K)) = \dim H^0(\mathcal{O}_X(2K))$. On the other hand, since $-K \not\sim 0$ and $-K \equiv 0$, we have $H^0(\mathcal{O}_X(-K)) = 0$. Therefore $\dim H^0(\mathcal{O}_X(2K)) \geq 1$, that is, $p_2 \geq 1$. As $\kappa(X) = 0$, we have also $p_2 \leq 1$, so that in fact $p_2 = 1$. Finally, since $\omega_X^2 = \mathcal{O}_X(2K) \equiv 0$, we get $\omega_X^2 \cong \mathcal{O}_X$. □

Corollary 10.12. *If X is an Enriques surface, then the order of ω_X in* $\mathrm{Pic}(X)$ *is 2.*

Proposition 10.13. *If X is an Enriques surface, then* $\mathrm{Pic}^\tau(X) \cong \mathbb{Z}/2\mathbb{Z}$.

Proof. Since $p_g = 0$, the scheme $\underline{\mathrm{Pic}}^\tau(X)$ is reduced (cf. Theorem 5.1). As $H^1(\mathcal{O}_X)$ is the tangent space to the algebraic group $\underline{\mathrm{Pic}}^\tau(X)$ at the origin, we have that $\underline{\mathrm{Pic}}^\tau(X)$ is a finite group. Now let L be any invertible $\mathcal{O}_X$-module that is numerically equivalent to zero. By the Riemann–Roch theorem we have that $\chi(L) = \chi(\mathcal{O}_X) = 1$, and therefore either $H^0(L) \neq 0$ or $H^2(L) \neq 0$; or equivalently, by Serre duality, that either $H^0(L) \neq 0$ or $H^0(\omega_X \otimes L^{-1}) \neq 0$. Since both L and $\omega_X \otimes L^{-1}$ are numerically equivalent to zero, we get that either $L \cong \mathcal{O}_X$ or $L \cong \omega_X$. □

Proposition 10.14. *Let X be an Enriques surface. If* char(k) $\neq 2$*, then there exists an étale covering of degree 2, $\pi : X' \to X$, with X' a K3 surface and the structural group of π isomorphic to $\mathbb{Z}/2\mathbb{Z}$.*

Proof. Let $(f_{ij}) \in Z^1(\{U_i\}), \mathcal{O}_X^*$ be a 1-cocycle that represents the invertible $\mathcal{O}_X$-module ω_X in $\mathrm{Pic}(X) = H^1(X, \mathcal{O}_X^*)$. Since $\omega_X^2 \cong \mathcal{O}_X$, the 1-cocycle (f_{ij}^2) is a coboundary, and therefore we can write

$$f_{ij}^2 = g_i/g_j \text{ on } U_i \cap U_j, \ g_i \in \Gamma(U_i, \mathcal{O}_X^*).$$

We define the covering $\pi : X' \to X$ locally by the equation $z_i^2 = g_i$ on $U_i \times \mathbb{A}^1$ and gluing via $z_i/z_j = f_{ij}$ along $(U_i \times \mathbb{A}^1) \cap (U_j \times \mathbb{A}^1)$. π is a connected étale covering, because the characteristic of k is different from 2. Therefore $\omega_{X'} \cong \pi^*(\omega_X)$, so that X' is a minimal surface with $\kappa(X') = 0$. By Proposition 9.7 we have $\chi(\mathcal{O}_{X'}) = 2\chi(\mathcal{O}_X) = 2$, and then the list in 10.1 shows that X' is a K3 surface. □

Proposition 10.15. *Let X be a surface that admits an étale and connected covering of degree 2, $\pi : X' \to X$, with X' a K3 surface. If the characteristic of k is different from 2, then X is an Enriques surface.*

Proof. Since π is an étale morphism, we have $\omega_{X'} \cong \pi^*(\omega_X)$, and since $\omega_{X'} \cong \mathcal{O}_{X'}$ we get $\omega_X \equiv 0$, so that X is a minimal surface with $\kappa(X) = 0$. Proposition 9.7 shows that $\chi(\mathcal{O}_X) = \frac{1}{2}\chi(\mathcal{O}_{X'}) = 1$, because X' is a K3 surface. By Theorems 10.2 and 10.9 and the list in 10.1, X is an Enriques surface.   □

10.16. Examples of Enriques Surfaces. Consider in $\mathbb{P}^5_{\mathbb{k}}$ (char(k) $\neq 2$) the surface X' given by three equations of the form

$$f_i(x_0, \ldots, x_5) = Q_i(x_0, x_1, x_2) + Q'_i(x_3, x_4, x_5) \quad (i = 1, 2, 3),$$

where Q_i and Q'_i are quadratic forms in three variables.

By Bertini's theorem, a generic surface of this kind is nonsingular. Consider the involution

$$\sigma : \mathbb{P}^5 \to \mathbb{P}^5, \quad \sigma(x_0, x_1, x_2, x_3, x_4, x_5) = (x_0, x_1, x_2, -x_3, -x_4, -x_5).$$

Then $\sigma(X') = X'$. On the other hand, the set of fixed points of σ is the union of the two planes $x_0 = x_1 = x_2 = 0$ and $x_3 = x_4 = x_5 = 0$. Generically, the three conics $Q_i = 0$ (resp. the three conics $Q'_i = 0$) in $\mathbb{P}^2$ have no points in common. Then the restriction $\sigma' = \sigma|_{X'}$ is an involution without fixed points on X'. Therefore we may consider the (nonsingular projective) surface $X = X'/\sigma'$. The canonical morphism $\pi : X' \to X$ is then an étale covering of X degree 2. By (10.4) X' is a K3 surface, and then by (10.15) X is an Enriques surface.

Theorem 10.17. *Every Enriques surface X is elliptic or quasielliptic; in characteristic different from 2 or 3 it is always an elliptic surface.*

Proof. By (7.11) it would suffice to find an integral curve C on X with $p_a(C) = 1$. Indeed, since $K \equiv 0$, we would then have that C is an indecomposable curve of canonical type.

By way of contradiction, assume that X does not contain any integral curve C with $p_a(C) = 1$.

(C'^2) is an even integer for every integral curve on X, because $(C'^2) = 2p_a(C') - 2$. If C' is a generic hyperplane section on X, then $(C'^2) > 0$, and therefore $p_a(C') > 1$. In particular, X contains integral (and even smooth) curves C' of arithmetic genus strictly greater than 1.

Henceforth fix an integral curve C on X with $p_a(C) > 1$ and $p_a(C)$ minimal with these conditions.

(10.17.1) Lemma. *With the preceding hypotheses and notation, let $|\Delta| \neq \varnothing$ be a complete linear system with $(\Delta^2) > 0$. Then the moving part $|D|$ of $|\Delta|$ is irreducible (i.e., it contains at least one irreducible member) and $(D^2) > 0$.*

Proof of Lemma 10.17.1. We have $\dim |\Delta| \geq 1$ by the Riemann–Roch theorem.

First assume that $|\Delta|$ has no fixed components; under this assumption, we will show that $|\Delta|$ is irreducible. If $|\Delta|$ has no fixed components, then it has a finite number of base points. Let $f : X \dashrightarrow B$ be the rational map obtained by the Stein factorization of the rational map $\phi_{|\Delta|} : X \dashrightarrow B' = \phi_{|\Delta|}(X) \subset \mathbb{P}(H^0(\mathcal{O}_X(\Delta))^{\check{}})$, where B is the normalization of B' in the rational function field $\Bbbk(X)$. As $\mathcal{O}_X(\Delta) \cong \phi_{|\Delta|}^*(\mathcal{O}_{B'}(1))$, we have $\mathcal{O}_X(\Delta) \cong f^*(L)$, where $L = \pi^*(\mathcal{O}_{B'}(1))$ is an ample line bundle on B ($\pi : B \to B'$ is the normalization morphism, which is finite).

If $|\Delta|$ were reducible, then $\dim(B) = 1$ (that is, $|\Delta|$ would be composite with a pencil of curves). Furthermore, since the general fiber of f is irreducible, cf. (7.3), we would have $n = \deg(L) > 1$.

If $g : X' \to X$ is a composite of quadratic transformations such that $f \circ g$ is a morphism, then $H^1(\mathcal{O}_{X'}) = H^1(\mathcal{O}_X) = 0$. From the exact sequence

$$0 \to H^1(\mathcal{O}_B) \to H^1(\mathcal{O}_{X'}) = 0$$

we then infer that B is isomorphic to $\mathbb{P}^1$. Therefore we have

$$\dim |\Delta| = \dim |L| = \deg(L) = n \quad \text{(because } B \cong \mathbb{P}^1).$$

On the other hand, we have by the Riemann–Roch theorem that

$$\dim |\Delta| \geq \frac{n^2}{2}(F^2) \geq n^2,$$

where F is an integral fiber of f. (Indeed, we have $(F^2) \geq 2$, because (F^2) is an even integer and $(\Delta^2) > 0$.)

From these computations we get $n \geq n^2$, and this contradicts $n > 1$. Thus we have shown that $|\Delta|$ is irreducible if it doesn't have fixed components.

Now consider the general case, that is, with $|\Delta|$ arbitrary. If we can show that $(D^2) > 0$, then the linear system $|D|$ is irreducible by what we have already proved. Therefore there only remains to show that $(D^2) > 0$.

Let $D' = \sum_i n_i E_i$ be a member of $|D|$, with $n_i \geq 1$ and E_i pairwise distinct irreducible curves. Since $|D|$ doesn't have fixed components, we have $(D^2) \geq 0$. Indeed, if $D_i \in |D|$ is an effective divisor whose support does not contain E_i, then $(D \cdot E_i) = (D_i \cdot E_i) \geq 0$, and therefore $(D^2) = \sum_i n_i(D \cdot E_i) \geq 0$.

If $(D^2) = 0$, then $(D \cdot E_i) = 0$ for all i, because in any case $(D \cdot E_i) \geq 0$ for all i, and $\sum_i n_i(D \cdot E_i) = (D^2) = 0$. Since $K \equiv 0$, we would have that D' is a divisor of canonical type, and therefore the conclusion of Theorem 10.17 would be true (by (7.11)), which is a contradiction. We have thus proved that $(D^2) > 0$. $\qquad \square$

(10.17.2) Lemma. *Let C be an integral curve on X with $p_a(C) > 1$ and $p_a(C)$ minimal. Then*

$$\dim |C| = \dim |K + C| = p_a(C) - 1 = \frac{1}{2}(C^2)$$

and the linear system $|K + C|$ is irreducible and reduced.

Proof of Lemma 10.17.2. The exact sequence

$$0 \to \mathcal{O}_X(-C) \to \mathcal{O}_X \to \mathcal{O}_C \to 0$$

induces the long exact sequence of cohomology

$$0 \to H^0(\mathcal{O}_X) \to H^0(\mathcal{O}_C) \to H^1(\mathcal{O}_X(-C)) \to H^1(\mathcal{O}_X) = 0.$$

As $H^0(\mathcal{O}_X) = H^0(\mathcal{O}_C) = \Bbbk$, we get $H^1(\mathcal{O}_X(-C)) = 0$, and therefore, by Serre duality, $H^1(\mathcal{O}_X(K+C)) = 0$. We also have $\dim H^2(\mathcal{O}_X(K+C)) = \dim H^0(\mathcal{O}_X(-C)) = 0$, and therefore

$$\dim |K+C| = \frac{1}{2}(C^2) = p_a(C) - 1$$

by the Riemann–Roch theorem.

Let $|D|$ be the moving part of the linear system $|K+C|$, and let Γ be the fixed part. From (10.17.1) we know that $|D|$ is irreducible and that $p_a(D) > 1$. If the general member of $|D|$ were not reduced, then by Bertini's theorem we would have that the general member of $|D|$ has the form $p^e F$, with F some integral curve on X, where p is the characteristic exponent of $\Bbbk$ and $e \geq 1$. Then we would have

$$p_a(D) - 1 \leq \dim |D| = \dim |K+C| = p_a(C) - 1$$

by the Riemann–Roch theorem, and therefore $1 < p_a(D) \leq p_a(C)$.

On the other hand,

$$p_a(F) = \frac{1}{2}(F^2) + 1, \quad p_a(D) = \frac{1}{2}p^{2e}(F^2) + 1.$$

Then $1 < p_a(F) < p_a(D) \leq p_a(C)$, contradicting the minimality of $p_a(C)$. Therefore the general member of $|D|$ is integral, and $p_a(D) = p_a(C)$, that is, $(D^2) = (C^2)$. But

$$(C^2) = ((D+\Gamma)^2) = (D^2) + (D \cdot \Gamma) + (\Gamma^2),$$

whence

$$2(D \cdot \Gamma) + (\Gamma^2) = 0. \tag{$*$}$$

Since $|D|$ has no fixed components and Γ is effective, we have that $(D \cdot \Gamma) \geq 0$, because $(D \cdot E) \geq 0$ for every integral curve E on X (to see that, it suffices to take a divisor $D_0 \in |D|$ whose support does not contain E). Therefore Equation $(*)$ implies that $(\Gamma^2) \leq 0$.

If $(\Gamma^2) < 0$, then $(D \cdot \Gamma) > 0$ by Equation $(*)$; using Equation $(*)$ again, we have:

$$(C \cdot \Gamma) = (D+\Gamma \cdot \Gamma) = (D \cdot \Gamma) + (\Gamma^2) = -(D \cdot \Gamma) < 0;$$

but this is impossible for an integral curve C with $(C^2) > 0$.

If $(\Gamma^2) = 0$, then Equation ($*$) implies that $(D \cdot \Gamma) = 0$ as well. Since $(D^2) > 0$, the Hodge index theorem implies that $\Gamma \equiv 0$, and therefore that $\Gamma = 0$, because in any case Γ is an effective divisor on X.

We have therefore proved that $\Gamma = 0$, that is, the complete linear system $|C + K|$ is free of fixed components, and its general member is irreducible and reduced. Now the same argument as in the beginning of the proof shows that

$$\dim H^1(\mathcal{O}_X(C)) = \dim H^1(\mathcal{O}_X(K + C + K)) = 0.$$

The Riemann–Roch theorem then gives

$$\dim |C| = \frac{1}{2}(C^2) + \dim H^1(\mathcal{O}_X(C)) = \frac{1}{2}(C^2). \quad \square$$

(10.17.3) Corollary. *With the same hypotheses and notation, we have*

$$\dim |2C| = 2(C^2) = 4p_a(C) - 4.$$

Proof of Corollary 10.17.3. By the Riemann–Roch theorem we have

$$\dim |2C| = 2(C^2) + \dim H^1(\mathcal{O}_X(2C)).$$

There only remains to prove that $H^1(\mathcal{O}_X(2C)) = 0$, or equivalently (by Serre duality) that $H^1(\mathcal{O}_X(K - 2C)) = 0$. The exact sequence

$$0 \to \mathcal{O}_X(K - 2C) \to \mathcal{O}_X(K - C) \to F \to 0,$$

with F an invertible sheaf on C with $\deg(F) = -(C^2) < 0$, induces the cohomology exact sequence

$$0 = H^0(F) \to H^1(\mathcal{O}_X(K - 2C)) \to H^1(\mathcal{O}_X(K - C)).$$

But in the proof of Lemma 10.17.2 we have shown that $H^1(\mathcal{O}_X(C)) = 0$, which by Serre duality gives $H^1(\mathcal{O}_X(K - C)) = 0$. Therefore we have $H^1(\mathcal{O}_X(K - 2C)) = 0$, as required. $\quad \square$

(10.17.4) Now let C be an integral curve on X with minimal value of $g = p_a(C) > 1$. From Lemma 10.17.2 and Corollary 10.17.3 we have

$$\dim |C| = g - 1,$$
$$\dim |2C| = 4g - 4.$$

Put

$$V_1 = \{\Delta \in |2C| \mid \exists C_1, C_2 \in |C| \text{ such that } \Delta = C_1 + C_2\}$$

and

$$V_2 = \{\Delta \in |2C| = |2C + 2K| \mid \exists C_1', C_2' \in |C + K| \text{ such that } \Delta = C_1' + C_2'\}.$$

(10.17.5) Lemma. *The sets V_1 and V_2 are closed algebraic subsets of $|2C| \cong \mathbb{P}^{4g-4}$, and each of these two sets has dimension $2g - 2$.*

Proof of Lemma 10.17.5. Consider the map $h : |C| \times |C| \to |2C|$ given by $h(C_1, C_2) = C_1 + C_2$. Here $|C| \cong \mathbb{P}^{g-1}, |2C| \cong \mathbb{P}^{4g-4}$, and V_1 is just $h(|C| \times |C|) \subset |2C|$. It suffices to show that h is an algebraic morphism and that at least one fiber of h is zero-dimensional; for then V_1 will be an algebraic subset of $|2C|$, because $|C| \times |C|$ is a projective variety, and the dimension of the image V_1 will be equal to that of $|C| \times |C|$, that is, $2g - 2$. Let $C_1 \in |C|$ be an integral curve. Let

$$L(C_1) = \{f \in \Bbbk(X)^* \mid \mathrm{div}_X(f) + C_1 \geq 0\} \cup \{0\}.$$

Then $H^0(\mathcal{O}_X(C)) \cong L(C_1)$, and $|C| = \mathbb{P}(H^0(\mathcal{O}_X(C))) \cong \mathbb{P}(L(C_1))$. This isomorphism is given by $\hat{f} \in \mathbb{P}(L(C_1)) \mapsto \mathrm{div}_X(f) + C_1 \in |C|$.

Under these identifications, h is induced by the bilinear map

$$L(C_1) \times L(C_1) \to L(2C_1) \cong H^0(\mathcal{O}_X(2C)), \quad (f, g) \mapsto fg.$$

Therefore h is an algebraic map. Indeed, if we introduce affine coordinates on $L(C_1)$ and $L(2C_1)$, and if we use the same coordinates as homogeneous coordinates on $|C|$ and $|2C|$, respectively, then h is given, in these homogeneous coordinates, by a collection of bilinear forms in the homogeneous coordinates of $|C|$; and a bilinear form is, indeed, bihomogeneous, as is required to define an algebraic map from a product of projective space to a projective space. Also, the fiber above the point $2C_1 \in |2C|$ consists of one point only, namely, $(C_1, C_1) \in |C| \times |C|$. This is because there is only one way of writing $C_1 + C_1$ as a sum of two effective divisors, namely, as the sum of C_1 and C_1. Therefore at least this fiber is zero-dimensional.

Thus we have established that V_1 is an algebraic subset of $|2C|$, and that its dimension is $2g - 2$. The proof for V_2 is entirely analogous. $\square$

(10.17.6) *Proof of Theorem 10.17: Conclusion.* By Lemma 10.17.5 we have that $V_1 \cap V_2 \neq \varnothing$, and therefore there exists a divisor $\Delta \in |2C| = |2C + 2K|$ such that

$$\Delta = C_1 + C_2, \ C_1, C_2 \in |C|,$$
$$\Delta = C_1' + C_2', \ C_1', C_2' \in |C + K|.$$

Let Γ be the common part of the effective divisors C_1, C_2, C_1', and C_2'. Consider the linear system of equations for D_1, D_2, D_3, and D_4:

$$\begin{array}{ll} D_1 + D_2 = C_1 - \Gamma & D_1 + D_3 = C_1' - \Gamma \\ D_3 + D_4 = C_2 - \Gamma & D_2 + D_4 = C_2' - \Gamma \end{array} \qquad (*)$$

If we put

$$C_1 - \Gamma = \sum_i a_i E_i \qquad C_1' - \Gamma = \sum_i c_i E_i$$

$$C_2 - \Gamma = \sum_i b_i E_i \qquad C_2' = \Gamma = \sum_i d_i E_i$$

with E_i pairwise distinct integral curves on X and a_i, b_i, c_i, d_i nonnegative integers with $a_i + b_i + c_i + d_i > 0$ and $a_i + b_i = c_i + d_i$ for all i, and if we put

$$D_1 = \sum_i x_i E_i \qquad D_3 = \sum_i z_i E_i$$

$$D_2 = \sum_i y_i E_i \qquad D_4 = \sum_i u_i E_i$$

then the previous linear system becomes

$$x_i + y_i = a_i \qquad x_i + z_i = c_i$$
$$z_i + u_i = b_i \qquad y_i + u_i = d_i. \qquad (**)$$

As the right-hand sides of the equations in $(*)$ have no common components, we have that $a_i b_i c_i d_i = 0$ for all i. It is then easy to see that the system $(**)$ has a solution in nonnegative integers x_i, y_i, z_i, u_i that satisfies the supplementary condition that $x_i u_i = y_i z_i = 0$ for all i.

Thus we obtain a solution (D_1, D_2, D_3, D_4) of the linear system $(*)$ that satisfies the following conditions:

(a) $D_i \geq 0, i = 1, 2, 3, 4$.

(b) The divisors D_1 and D_4 (resp. D_2 and D_3) have no common components.

Furthermore, we have $D_1 + D_3 \sim D_1 + D_2 + K$, $D_2 + D_4 \sim D_2 + D_1 + K$; that is, $D_3 \sim D_2 + K$ and $D_4 \sim D_1 + K$.

This implies $D_i > 0$ for $i = 1, 2, 3, 4$ (because K is not linearly equivalent to any effective divisor, and $-K \sim K$), and $(D_i^2) \geq 0$ for $i = 1, 2, 3, 4$. For example: $(D_2^2) = (D_2 \cdot D_3) \geq 0$, because D_2 and D_3 have no common components.

Put $D_i = \sum_j n_{ij} F_{ij}$, with F_{ij} mutually distinct integral curves. Then $(D_i \cdot F_{ij}) \geq 0$ for all i and j. For example, for $i = 2$: if F_{2j} is a component of D_2, then F_{2j} is not a component of D_3; hence $(D_2 \cdot F_{2j}) = (D_3 \cdot F_{2j}) \geq 0$. If $(D_i^2) = 0$ for some i, then the equation

$$\sum_j n_{ij}(D_i \cdot F_{ij}) = (D_i^2) = 0$$

implies $(D_i \cdot F_{ij}) = 0$ for all j. Since $K \equiv 0$, this means that D_i is a divisor of canonical type, and then, by (7.11), that X is elliptic or quasielliptic, which is a contradiction.

Therefore $(D_i^2) > 0$ for $i = 1, 2, 3, 4$. Let then $|C'|$ be the moving part of the complete linear system $|D_1|$. By Lemma 10.17.1 we know that $|C'|$ is irreducible and $(C'^2) > 0$. By the Riemann–Roch theorem we have $\dim |D_2| \geq \frac{1}{2}(D_2^2) \geq 1$, so that D_2 cannot be contained in the fixed part of the complete linear system $|D_1 + D_2|$ (see the final observation below). Therefore $\dim |D_1| < \dim |D_1 + D_2|$.

By Lemma 10.17.2 we have $\dim |C| = \frac{1}{2}(C^2)$, and $\dim |C'| \geq \frac{1}{2}(C'^2)$. The earlier inequalities imply that $(C'^2) < (C^2)$. Put $C'' = C'_{\mathrm{red}}$. Then $0 < (C''^2) \leq (C'^2) < (C^2)$, and therefore $1 < p_a(C'') < p_a(C)$, which contradicts the choice of C. This completes the proof of Theorem 10.17.

We have used the following trivial observation: if $|E| \neq \varnothing$ is any complete linear system and F is its fixed part, then $\dim |F| = 0$ (and then, of course, $\dim |F_1| = 0$ for all $F_1 \leq F$). Indeed, denote the moving part of $|E|$ by $|E'|$, and assume, by way of contradiction, that $\dim |F| \geq 1$. Then there exists $F' \in |F|$, $F' \neq F$. $F' - F$ cannot be effective (because it is not zero but it is numerically equivalent to zero). Therefore there is at least one integral curve on X, say, C, such that the coefficient of C in F is strictly greater than the coefficient of C in F'. Since E' is free of fixed components, there exists $G \in |E'|$ such that C is not contained in the support of G. Then $F' + G \in |E|$, but $F \not\leq F' + G$ (because the coefficient of C in $F' + G$ is equal to the coefficient of C in F', and that is strictly less than the coefficient of C in F). This contradicts the fact that F is the fixed part of $|E|$. $\qquad\square$

10.18. Next we study minimal surfaces X with $\kappa(X) = 0$ and $b_2 = 6$. From the list in 10.1 we have $b_1 = 4, \chi(\mathcal{O}_X) = 0, q = 2, p_g = 1$, and $\Delta = 0$.

First we remark that every two-dimensional Abelian variety X has these invariants, that is, $\kappa(X) = 0, b_2 = 6$, etc. Indeed, if X is an Abelian variety of arbitrary dimension, say, d, then

$$\dim H^p(X, \mathcal{O}_X) = \binom{d}{p}$$

by [Mum2, §13]. Therefore if X is an Abelian surface then $K \sim 0$, $q = 2$, and $p_g = 1$. On the other hand, we have $\dim \underline{\mathrm{Pic}}^0(X) = \dim(X)$ for every Abelian variety X; hence $\Delta = 0$ by Theorem 5.1. From the list in 10.1 we see that $b_2 = 6$.

Now we show that, conversely, the only minimal surfaces X with $\kappa(X) = 0$ and $b_2 = 6$ are the Abelian surfaces.

Theorem 10.19. *Every minimal surface X with $\kappa(X) = 0$ and $b_2 = 6$ is an Abelian surface.*

Proof. To avoid technical complications, we give the proof only in characteristic zero. However, the result holds in arbitrary characteristic; see [BM1]. (A large part of the argument below works in arbitrary characteristic; we indicate explicitly the points where characteristic zero is used.)

Let $\phi : X \to \text{Alb}(X)$ be the Albanese morphism of X. Since $q = 2$ and $\Delta = 0$, we have $\dim \text{Alb}(X) = 2$ by (5.2). First we show that ϕ is surjective. By way of contradiction, assume that ϕ is not surjective. Then $B = \phi(X)$ must be a curve ($\phi(X)$ generates $\text{Alb}(X)$, so $\phi(X)$ cannot be a point). We will prove that B is a nonsingular curve of genus 2. For the time being, let $\pi : \tilde{B} \to B$ be the normalization of B; π is a finite birational morphism. By the universal property of $\text{Alb}(X)$, there exists a unique morphism $f : \text{Alb}(X) \to \text{Alb}(\tilde{B})$ that closes the following commutative diagram:

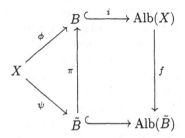

Let $\pi' = f|_B : B \to \tilde{B}$. Then one sees immediately that $\pi' \circ \pi = \text{id}_{\tilde{B}}$ and $\pi \circ \pi' = \text{id}_B$, and therefore that π is an isomorphism. Thus B is a nonsingular curve. Since $B = \tilde{B}$, the previous diagram reduces to the following commutative diagram:

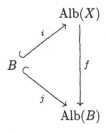

By the universal property of the Albanese variety $\text{Alb}(B)$ we get a unique morphism $g : \text{Alb}(B) \to \text{Alb}(X)$ of Abelian varieties that closes the commutative diagram

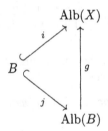

Since $i(B)$ generates $\mathrm{Alb}(X)$ and $j(B)$ generates $\mathrm{Alb}(B)$, we see easily that f and g are isomorphisms. In particular, $\dim \mathrm{Alb}(B) = 2$, so that the genus of B is 2.

Next, using the Nakai–Moishezon criterion, we show that B is an ample divisor on $A \stackrel{\mathrm{def}}{=} \mathrm{Alb}(X)$. The genus formula and the isomorphism $\omega_A \cong \mathcal{O}_A$ give $(B^2) = 2$. To prove that B is an ample divisor on A there only remains to show that $(B \cdot C) > 0$ for every integral curve C on A. Fix such a curve C, and pick a point $z \in C$ (any point will do). As B generates A, there are two points x and y on B such that $z = x - y$ in the group operation of A. Thus $x \in B \cap (y + C) = B \cap t^*_{-y}(C)$, where t_{-y} is the translation $v \mapsto -y + v$ on A. But the square formula [Mum2, §6] implies that $t^*_{-y}(C)$ is algebraically equivalent to C, and therefore

$$(B \cdot C) = (B \cdot t^*_{-y}(C)) > 0,$$

for $B \cap t_{-y}(C) \neq \varnothing$ and $(B^2) > 0$.

Thus B is an ample divisor on A. Now let n be an integer that is not divisible by the characteristic of $\Bbbk$, and consider the Cartesian diagram:

$$
\begin{array}{ccc}
X \times_{\mathrm{Alb}(X)} \mathrm{Alb}(X) = X' & \xrightarrow{\ \psi\ } & \mathrm{Alb}(X) \\
{\scriptstyle f}\downarrow & & \downarrow{\scriptstyle n} \\
X & \longrightarrow & \mathrm{Alb}(X)
\end{array}
$$

Since (multiplication by) n is an étale morphism of degree n^4, cf. [Mum2, §6], we have that f is also an étale morphism, and in particular that X' is also a nonsingular surface. We have $\psi(X') = n^{-1}(\psi(X)) = n^{-1}(B)$. Therefore the restriction of n to $\psi(X')$ is an étale morphism of degree n^4. Since n is a finite morphism, B is ample on A and $\psi(X') = n^{-1}(B)$, we have that $\psi(X')$ is an ample divisor on A. Since A is a nonsingular projective surface, the Enriques–Severi–Zariski–Serre theorem implies that the curve $\psi(X')$ is connected. Thus we have an étale and connected covering of degree n^4, $\psi(X') \to B$. Since the genus of B is 2, the genus of the nonsingular curve $\psi(X')$ is strictly greater than 2. Let X'_0 be a connected component of X' such that $\psi(X'_0) = \psi(X')$. Then we have the commutative diagram:

$$
\begin{array}{ccc}
X'_0 & \xrightarrow{\ u\ } & \mathrm{Alb}(X'_0) \\
{\scriptstyle \psi}\downarrow & & \downarrow \\
\psi(X') & \xrightarrow{\ v\ } & \mathrm{Alb}(\psi(X')) = \mathrm{Jac}(\psi(X'))
\end{array}
$$

Since $v(\psi(X'_0))$ generates $\mathrm{Alb}(X')$, we have that the right-hand vertical arrow is surjective. Therefore $q(X'_0) \geq \dim \mathrm{Alb}(X'_0) \geq \dim \mathrm{Alb}(\psi(X')) = $ genus of $\psi(X') > 2$. Thus we have constructed a (nonsingular projective) surface X_0' with $\omega_{X_0'} \cong \mathcal{O}_{X_0'}$ and $q(X_0') > 2$. But we have seen in (10.1) that there are no such surfaces. We have thus reached a contradiction, which we derived from the assumption that the morphism ϕ was not surjective.

Hence $\phi : X \to A = \mathrm{Alb}(X)$ is a surjective morphism of surfaces. In particular, ϕ is a morphism of finite degree. Our next objective is to prove that ϕ is, in fact, a finite morphism. To that end, consider the Stein factorization of ϕ:

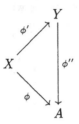

where ϕ'' is finite and ϕ' is birational. By way of contradiction, assume that ϕ is not finite. Then ϕ' is not an isomorphism. Hence there exists an integral curve E on X such that $\phi'(E)$ is a point on Y. Then we have $(E^2) < 0$ by (2.7). $\omega_X \cong \mathcal{O}_X$ and the genus formula show that $p_a(E) = 0$ and $(E^2) = -2$. Therefore Y is a normal surface with at most a finite number of rational double points as singularities. By Theorem 3.15 we have that the dualizing sheaf ω_Y is invertible and $\omega_X \cong \phi'^*(\omega_Y)$. Since $\omega_X \cong \mathcal{O}_X$, we have also that $\omega_Y \cong \mathcal{O}_Y$.

Now consider the finite morphism ϕ''. After removing a finite number of points in A (namely, the images by ϕ'' of all the rational double points of Y), we get a flat finite morphism $\tilde{\phi} : Y' \to A'$ of nonsingular surfaces, with $\omega_{A'} \cong \mathcal{O}_{A'}$ and $\omega_{Y'} \cong \mathcal{O}_{Y'}$. Since we are in characteristic zero, the morphism $\tilde{\phi}$ is separable. Let ω be a nonzero regular differential form of degree 2 on A'. Then $\tilde{\phi}^*(\omega)$ is a nonzero regular differential form on Y'. Since $\tilde{\phi}^*(\omega)$ vanishes exactly at the points of Y' where $\tilde{\phi}$ is not an étale morphism, and since $\omega_{Y'} \cong \mathcal{O}_{Y'}$, it follows that $\tilde{\phi}$ is étale. Finally, the theorem on the purity of the branch locus, cf. [AK, Gro6], shows that ϕ'' is étale everywhere on Y, and therefore that Y is nonsingular, because A is nonsingular. This implies that ϕ' is an isomorphism, and therefore we have proved that the Albanese morphism $\phi : X \to \mathrm{Alb}(X)$ is finite.

In fact, we have also proved that ϕ is étale. Now a result of Lang and Serre [Mum2, §18] shows that X has a structure of Abelian variety such that ϕ is an isogeny. $\square$

10.20. The last class of minimal surfaces X with $\kappa(X) = 0$ consists of the surfaces with $b_2 = 2$. Then the list in 10.1 gives $b_1 = 2$ and $\chi(\mathcal{O}_X) = 0$. In this case there are two possibilities:

(a) $p_g = 0$, $q = 1$, and $\Delta = 0$; or

(b) $p_g = 1$, $q = 2$, and $\Delta = 2$.

This class of surfaces was partially studied in §8, cf. (8.6) and (8.9), where we saw that there are two types of such surfaces: hyperelliptic and quasihyperelliptic. Quasihyperelliptic surfaces exist only in characteristic 2 or 3. In the remainder of this chapter we shall analyze the hyperelliptic surfaces in characteristic other than 2 and 3.

Let X be a hyperelliptic surface (char(k) $\neq$ 2, 3). By Theorem 8.6, the Albanese morphism $f : X \to \mathrm{Alb}(X) = B$ has the following properties: B is a nonsingular elliptic curve, and each fiber of f is a nonsingular elliptic curve. Also, Theorem 8.9 shows that there exists a second elliptic fibration, $g : X \to \mathbb{P}^1$, which is transversal to f. Comparing these two elliptic fibrations, we will derive the structure of hyperelliptic surfaces.

First we need to prove the following general result.

Theorem 10.21. *Let X and X' be two minimal surfaces with $\kappa(X) \geq 0$ and $\kappa(X') \geq 0$, and let $\phi : X \to X'$ be a birational map. Then ϕ is an isomorphism.*

Proof. We must show that ϕ and ϕ^{-1} are morphisms. For example, let's show that ϕ is a morphism. Let

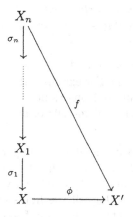

be a sequence of n quadratic transformations such that $f = \phi \circ \sigma_1 \circ \cdots \circ \sigma_n$ is a morphism, with n minimal. We must show that $n = 0$. By way of contradiction, assume that $n > 0$. Let $E \subset X_n$ be the exceptional curve of the last blow-up, σ_n. If $f(E)$ were a point on X', then f would factor through a morphism $f' : X_{n-1} \to X'$,

contradicting the minimality of n. Therefore $f(E) = F$ is a curve on X'.

We claim that $(\omega_{X'} \cdot F) < 0$. If we prove this claim, then X' belongs to class (a) in the classification of (8.2), and therefore $\kappa(X') = -\infty$ by Remark 8.3(a). This is a contradiction, derived from the assumption that ϕ is not a morphism.

Therefore there only remains to show that $(\omega_{X'} \cdot F) < 0$. The proof is by induction on the number of quadratic transformations in the factorization of f as a composite of such transformations.

First consider a blow-up $\sigma : U \to V$ of a surface V at a point $v \in V$. Let $D \subset U$ be the exceptional curve of σ. If $E' \subset U$ is an integral curve such that $E'' = \sigma(E')$ is a curve on V, then we have

$$(\omega_U \cdot E') = (\sigma^*(\omega_V) \otimes \mathcal{O}_U(D) \cdot \sigma^*(E'')) - (\sigma^*(\omega_V) \otimes \mathcal{O}_U(D) \cdot mD),$$

where $m = (E'' \cdot D)$. Thus

$$(\omega_U \cdot E') = (\sigma^*(\omega_V) \cdot \sigma^*(E'')) - m(D^2) = (\omega_V \cdot E'') + m \geq (\omega_V \cdot E'').$$

Returning to the proof of the theorem, we have $(\omega_{X_n} \cdot E) = -1$, because E is an exceptional curve of the first kind on X_n, and therefore by this observation and induction we have $(\omega_{X'} \cdot F) < 0$. $\qquad\square$

Corollary 10.22. *If X is a surface with $\kappa(X) \geq 0$, then the birational isomorphism class of X contains a unique minimal model (up to biregular isomorphism).*

Remark 10.23. If $\kappa(X) = -\infty$, then the corresponding statement in Corollary 10.22 becomes false. For example, the minimal surface $X = \mathbb{P}^2$ is birationally isomorphic, but not isomorphic, to the minimal surface $\mathbb{P}^1 \times \mathbb{P}^1$. We will classify all the minimal models of surfaces with $\kappa(X) = -\infty$ in Chapter 12.

10.24. We return now to the analysis of hyperelliptic surfaces (in characteristic different from 2 and 3). We use the notation from (10.20). Let $F_b = f^{-1}(b)$, with $b \in B$, and let $F'_c = g^{-1}(c)$, with $c \in \mathbb{P}^1$. As $b_2 = 2$ and $\rho \leq b_2$, Lemma 8.7 implies that all the fibers of g are irreducible. (This fact can also be proved directly, as follows: if some fiber F'_c were reducible, with irreducible components $E_1, \ldots, E_n$, $n \geq 2$, then $(E_1^2) < 0$ by (2.6). Since $\omega_X \equiv 0$, the genus formula gives $p_a(E_1) = 0$ and $(E_1^2) = -2$. Since E_1 cannot be contained in any fiber of f, it follows that the rational curve E_1 dominates the elliptic curve B, which is a contradiction.)

If F'_c is a multiple fiber of g, then $F''_c = (F'_c)_{\text{red}}$ is in any case an elliptic curve: $p_a(F''_c) = 1$, and F''_c cannot be a rational curve (with an ordinary cusp, cf. Theorem 7.18 and its proof), because F''_c dominates B. Therefore in this case we have $F'_c = mF''_c$ with $m \geq 2$ and F''_c a nonsingular elliptic curve. Put $S = \{c \in \mathbb{P}^1 \mid F'_c \text{ is a multiple fiber of } g\}$. Then S is a finite subset of $\mathbb{P}^1$.

For each point $c \in \mathbb{P}^1$, put $f_c = f|_{F'_c} : F'_c \to B$, the restriction of f to F'_c. If $c \in \mathbb{P}^1 \setminus S$, then f_c is an étale morphism by Hurwitz's formula, because

F'_c and B are elliptic curves, and f_c induces a homomorphism of algebraic groups,

$$f_c : \underline{\mathrm{Pic}}^0(B) \to \underline{\mathrm{Pic}}^0(F'_c).$$

But if we fix a base point $b_0 \in B$ (viewed as the zero element) on an elliptic curve B, then the correspondence

$$b \mapsto \text{the class of } \mathcal{O}_B(b - b_0) \text{ in } \underline{\mathrm{Pic}}^0(B)$$

defines an isomorphism of algebraic groups between B and $\underline{\mathrm{Pic}}^0(B)$. Next we note that $\underline{\mathrm{Pic}}^0(F'_c)$ acts canonically on F'_c (with $c \in \mathbb{P}^1 \setminus S$) by translations: if L is an invertible $\mathcal{O}_{F'_c}$-module of degree 0, then L maps the point $x \in F'_c$ to the unique point $y \in F'_c$ such that $L \otimes \mathcal{O}_{F'_c}(x) = \mathcal{O}_{F'_c}(y)$. Using the homomorphism f_c, we get an action of B on F'_c,

$$B \times F'_c \to F'_c,$$

for each $c \in \mathbb{P}^1 \setminus S$.

Since $\{f_c\}_{c \in \mathbb{P}^1 \setminus S}$ is an algebraic family of homomorphisms of algebraic groups, we get an action of B on $g^{-1}(\mathbb{P}^1 \setminus S) \subset X$,

$$\sigma_0 : B \times g^{-1}(\mathbb{P}^1 \setminus S) \to g^{-1}(\mathbb{P}^1 \setminus S).$$

In particular, every element $b \in B$ defines an automorphism of $g^{-1}(\mathbb{P}^1 \setminus S)$, which by Theorem 10.21 extends (uniquely) to an automorphism of X, because X is a minimal model with $\kappa(X) = 0$. Therefore the action σ_0 extends uniquely to an action

$$\sigma : B \times X \to X.$$

The action of B on X can be described explicitly as follows: if $b \in B$ and $x \in F'_c \subset X$ (with $c \in \mathbb{P}^1 \setminus S$), then $b \cdot x = y$, where y is the unique point of F'_c such that

$$f^* \mathcal{O}_B(b - b_0) \otimes \mathcal{O}_{F'_c}(x) = \mathcal{O}_{F'_c}(y).$$

Applying the norm $N_{F'_c/B}$ to both sides of the equation, we get

$$\mathcal{O}_B(nb - nb_0 + f(x)) \cong \mathcal{O}_B(f(y)),$$

where $n = \deg(f_c) = (F_b \cdot F'_c)$. Hence we get the commutative diagrams

$$
\begin{array}{ccc}
X & \overset{b}{\longrightarrow} & X \\
{\scriptstyle f}\downarrow & & \downarrow{\scriptstyle f} \\
B & \underset{t_{nb}}{\longrightarrow} & B
\end{array}
\qquad\qquad
\begin{array}{ccc}
B \times X & \overset{\sigma}{\longrightarrow} & X \\
{\scriptstyle \mathrm{id}_B \times f}\downarrow & & \downarrow{\scriptstyle f} \\
B \times B & \longrightarrow & B
\end{array}
$$

$$(b, b') \longmapsto nb + b'$$

Let $B_0 = F_{b_0}$ and $A_n = \text{Ker}(n_B : B \to B)$, where n_B is multiplication by n on B; A_n is considered with the natural structure of group subscheme of B (if char($\Bbbk$) $= 0$, then A_n is simply a finite subgroup of B). The commutative diagrams above show that the fibers of f are invariant under the action of A_n on X; in particular, A_n acts on B_0. Denote this action by

$$\alpha : A_n \to \text{Aut}(B_0) = \text{the group scheme of automorphisms of } B_0.$$

The action σ of B on X induces a morphism

$$\tau : B \times B_0 \to X,$$

which (due to the commutative diagrams above) completes the commutative diagram

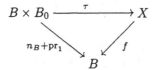

Note that $\tau(b, x) = \tau(b', x') \iff \sigma(b - b', x) = x' \iff b - b' \in A_n$ and $\alpha(b - b')(x) = x'$. The first equivalence is proved as follows: $\boxed{\Rightarrow}$ $\sigma(b - b', x) = \sigma(-b', \sigma(b, x)) = \sigma(-b, \tau(b, x)) = \sigma(-b', \tau(b', x')) = \sigma(-b', \sigma(b', x')) = \sigma(b' - b', x') = \sigma(0, x') = x'$. $\boxed{\Leftarrow}$ $\tau(b', x') = \sigma(b', x') = \sigma(b', \sigma(b - b', x)) = \sigma(b' + b - b', x) = \sigma(b, x) = \tau(b, x)$.

Thus X is isomorphic to the quotient of $B \times B_0$ by the action of A_n given by

$$a \cdot (b, b_0) = (b + a, \alpha(a)(b_0)), \quad \forall a \in A_n, \forall b \in B, \forall b_0 \in B_0.$$

Substituting the elliptic curve $B/\text{Ker}(\alpha)$ for B, we have proved the following.

Theorem 10.25. *Every hyperelliptic surface X has the form $X = B_1 \times B_0/A$, where B_0 and B_1 are elliptic curves, A is a finite group subscheme of B_1, and A acts on the product $B_1 \times B_0$ by $a \cdot (b_1, b_0) = (b_1 + a, \alpha(a)(b_0))$, with $a \in A, b_1 \in B_1, b_0 \in B_0$, and $\alpha : A \to \text{Aut}(B_0)$ a suitable injective homomorphism. Moreover, the two elliptic fibrations of X are given by: $f : B_1 \times B_0/A \to B_1/A = B$ (an elliptic curve) and $g : B_1 \times B_0 \to B_0/\alpha(A) \cong \mathbb{P}^1$.*

10.26. Using Theorem 10.25 and the relatively simple structure of the automorphism group $G = \text{Aut}(B_0)$ of an elliptic curve B_0 over a field $\Bbbk$ of characteristic different from 2 and 3, the next objective is to enumerate all the possible cases for such hyperelliptic surfaces.

Let $H \subset G$ be the subgroup consisting of all the translations of B_0 (H is isomorphic to the underlying group of the elliptic curve B_0 by the map $t_b \in H \leftrightarrow b \in B_0$), and let $F = \mathrm{Aut}(B_0, 0)$ be the subgroup of G consisting of the automorphisms of B_0 that fix the origin $0 \in B_0$. An elementary rigidity result for Abelian varieties, cf. [Mum2, §4] or [Sha1, p. 221], implies immediately the fact that every element $\sigma \in F$ is a group homomorphism, that is, $\sigma(b + b') = \sigma(b) + \sigma(b')$ for all $b, b' \in B_0$. Thus $\sigma \circ t_b = t_{\sigma(b)} \circ \sigma$ for all $b \in B_0$ and all $\sigma \in F$. Also, for every $u \in G$ we have obviously $u = t_{u(0)} \circ \sigma$ with $\sigma \in F$. Therefore the subgroup generated by $H \cup F$ is equal to G and H is a normal subgroup in G; furthermore, we clearly have $H \cap F = \{\mathrm{id}_{B_0}\}$. Hence G is the semidirect product of the normal subgroup H and the subgroup F, $G = H \rtimes F$; that is, $\mathrm{Aut}(B_0)$ is isomorphic to the semidirect product of the group B_0 (which is isomorphic to H) and the group $\mathrm{Aut}(B_0, 0)$.

For an elliptic curve B_0, let $j(B_0)$ be the j-invariant of B_0; see [Har1] or [Tat]. Then (still for char(k) $\neq 2, 3$) we have

$$\mathrm{Aut}(B_0, 0) = \{\mathrm{id}_{B_0}, -\mathrm{id}_{B_0}\} \cong \mathbb{Z}/2\mathbb{Z} \text{ if } j(B_0) \neq 0, 12^3;$$
$$\mathrm{Aut}(B_0, 0) \cong \mathbb{Z}/4\mathbb{Z} \text{ if } j(B_0) = 12^3; \quad \text{and}$$
$$\mathrm{Aut}(B_0, 0) \cong \mathbb{Z}/6\mathbb{Z} \text{ if } j(B_0) = 0.$$

Since the subgroup H is isomorphic to B_0, we have $\mathrm{Aut}(B_0) \cong B_0 \rtimes \mathrm{Aut}(B_0, 0)$ (semidirect product). Then, in the notation of Theorem 10.25, we cannot have $\alpha(A) \subset B_0$. Indeed, if $\alpha(A)$ were a subgroup of B_0, then the quotient $B_0/\alpha(A)$ would be an elliptic curve, while we know that $B_0/\alpha(A) \cong \mathbb{P}^1$. Let then $a \in A$ be an element of A such that the image of $\alpha(a)$ in the quotient group $\mathrm{Aut}(B_0)/B_0$ generates the image of the subgroup $\alpha(A)$ in $\mathrm{Aut}(B_0)/B_0 \cong \mathrm{Aut}(B_0, 0)$ (from what we said earlier, this last group is cyclic of order 2, 4, or 6). Then $\alpha(a)$ is an automorphism of B_0 that has a fixed point. Indeed, we have $\alpha(a) = \sigma \circ t_b$ for suitable $\sigma \in \mathrm{Aut}(B_0, 0) \setminus \{\mathrm{id}_{B_0}\}$ and $b \in B_0$. To find a fixed point z of $\alpha(a)$ means to find $z \in B_0$ such that $\sigma(z) + \sigma(b) = z$. But since $\sigma \neq \mathrm{id}_{B_0}$, the endomorphism $\sigma - \mathrm{id}_{B_0}$ of B_0 is surjective, and therefore $-\sigma(b)$ has a preimage $z \in B_0$ under $\sigma - \mathrm{id}_{B_0}$.

Choosing the origin of B_0 so that it coincides with a fixed point of $\alpha(a)$, we see immediately that $\alpha(A)$ is a direct product of the form

$$\alpha(A) \cong A_0 \times \mathbb{Z}/n\mathbb{Z},$$

with A_0 a (finite) closed group subscheme of B_0 and n one of the integers 2, 3, 4, or 6. Since A_0 and $\alpha(a)$ commute and $\alpha(a) \in \mathrm{Aut}(B_0, 0)$, we have $A_0 \subset U = \{b \in B_0 \mid \alpha(a)(b) = b\}$. In view of [Tat], the set U can be calculated explicitly; still in characteristic different from 2 and 3, we have the following possibilities:

(a) $n = 2$, and then $U = \mathrm{Ker}(2_{B_0}) = \mathbb{Z}/2\mathbb{Z} \times \mathbb{Z}/2\mathbb{Z}$;

(b) $n = 3$, and then $U = \mathbb{Z}/3\mathbb{Z}$;

(c) $n = 4$, and then $U = \mathbb{Z}/2\mathbb{Z}$;

(d) $n = 6$, and then $U = \{0\}$.

Putting all these observations together, we find the following possibilities for hyperelliptic surfaces in characteristic different from 2 and 3:

List 10.27 (Bagnera–DeFranchis).

(a_1) $(B_1 \times B_0)/(\mathbb{Z}/2\mathbb{Z})$, with the generator a of $\mathbb{Z}/2\mathbb{Z}$ acting on $B_1 \times B_0$ by

$$a \cdot (b_1, b_0) = (b_1 + a, -b_0).$$

(a_2) $(B_1 \times B_0)/(\mathbb{Z}/2\mathbb{Z})^2$, with generators a and g of $(\mathbb{Z}/2\mathbb{Z})^2$ such that $\alpha(a) \in \mathrm{Aut}(B_0, 0)$ being the automorphism $b_0 \mapsto -b_0$ of B_0, and $\alpha(g) \in B_0$ corresponding to a translation t_c on B_0, with $c \in B_0$ a point of order 2. In this case the action is given by

$$a \cdot (b_1, b_0) = (b_1 + a, -b_0) \quad \text{and}$$
$$g \cdot (b_1, b_0) = (b_1 + g, b_0 + c).$$

(b_1) $(B_1 \times B_0)/(\mathbb{Z}/3\mathbb{Z})$, with a generator a of $\mathbb{Z}/3\mathbb{Z}$ such that $\alpha(a) = \omega \in \mathrm{Aut}(B_0, 0)$ an automorphism of order 3. This case appears only when $j(B_0) = 0$; the action is given by

$$a \cdot (b_1, b_0) = (b_1 + a, \omega(b_0)).$$

(b_2) $(B_1 \times B_0)/(\mathbb{Z}/3\mathbb{Z})^2$, with generators a and g of $(\mathbb{Z}/3\mathbb{Z})^2$ such that $\alpha(a) = \omega \in \mathrm{Aut}(B_0, 0)$ an automorphism of order 3 and $\alpha(g) \in B_0$ corresponding to a translation t_c of B_0. This case is possible only when $j(B_0) = 0$; the action is given by

$$a \cdot (b_1, b_0) = (b_1 + a, \omega(b_0)) \quad \text{and}$$
$$g \cdot (b_1, b_0) = (b_1 + g, b_0 + c).$$

Furthermore, in this case we have $\omega(c) = c$ and c is a point of order 3 on B_0.

(c_1) $(B_1 \times B_0)/(\mathbb{Z}/4\mathbb{Z})$, with $\alpha(a) = i \in \mathrm{Aut}(B_0, 0)$ an automorphism of order 4, where a is a generator of $\mathbb{Z}/4\mathbb{Z}$. This case occurs only when $j(B_0) = 12^3$; the action is

$$a \cdot (b_1, b_0) = (b_1 + a, i(b_0)).$$

(c_2) $(B_1 \times B_0)/[(\mathbb{Z}/2\mathbb{Z}) \times (\mathbb{Z}/4\mathbb{Z})]$, with a a generator of $0 \times (\mathbb{Z}/4\mathbb{Z})$ and g the generator of $(\mathbb{Z}/2\mathbb{Z}) \times 0$ such that $\alpha(a) = i \in \mathrm{Aut}(B_0, 0)$ an automorphism of order 4 and $\alpha(g) \in B_0$ corresponding to a translation t_c of B_0. This case is possible only if $j(B_0) = 12^3$; the action is given by

$$a \cdot (b_1, b_0) = (b_1 + a, i(b_0)) \quad \text{and}$$
$$g \cdot (b_1, b_0) = (b_1 + g, b_0 + c).$$

In this case we have $i(c) = c$, and c is a point of order 2.

(d) $(B_1 \times B_0)/(\mathbb{Z}/6\mathbb{Z})$, with a a generator of $\mathbb{Z}/6\mathbb{Z}$, $\alpha(a) = -\omega$ with $\omega \in \mathrm{Aut}(B_0, 0)$ an automorphism of order 3, and the action given by

$$a \cdot (b_1, b_0) = (b_1 + a, -\omega(b_0)).$$

This case occurs only when $j(B_0) = 0$.

It is easy to see that these are indeed the only possibilities for existence of hyperelliptic surfaces. For example, the case $n = 2$ and $A_0 = \mathrm{Ker}(2_{B_0})$ is impossible, because in that case we would have $A \cong (\mathbb{Z}/2\mathbb{Z})^3$, but $A \subset B_1$, and B_1 has only four points of order 2.

EXERCISES

Exercise 10.1. In Exercise 9.2 assume also that Y is a nonhyperelliptic curve of genus $g \geq 3$. Then by a classical theorem of M. Noether [ACGH*], the canonical ring $R(Y, \omega_Y)$ is generated by its homogeneous part of degree one. Using this result, prove that Y is very ample on S_n and $\dim(|Y|) = g$.

Exercise 10.2. Prove that there are nonsingular surfaces of degree 4 in $\mathbb{P}^3$ that contain nonsingular rational curves.

Exercise 10.3. Let C be an irreducible reduced curve of arithmetic genus g on a $K3$ surface over $\mathbb{C}$.

(i) Prove that the normal bundle of C in X is isomorphic to ω_C and $(C^2) = 2g - 2$.

(ii) Prove that $H^1(X, \mathcal{O}_X(C)) = 0$ and $\dim(|C|) = g$.

(iii) If C is nonsingular and $g \geq 1$, show that $|C|$ has no base points, and the restriction to C of the morphism $\phi : X \to \mathbb{P}^g$ associated to $|C|$ coincides with the morphism defined by the canonical system of C. [*Hint*: Use the fact that the canonical system of C has no base points.]

(iv) If $g \geq 2$, show that the morphism ϕ is finite if and only if there are no curves D on X such that $(C \cdot D) = 0$.

(v) If $g = 2$, then the morphism $\phi : X \to \mathbb{P}^2$ is generically finite of degree 2, and is ramified along a curve of degree 6. [*Hint:* If L is a general line in $\mathbb{P}^2$ then the degree of $\phi^{-1}(L) \in |C|$ is 2, whence by Hurwitz's formula the degree of the ramification locus of $\phi^{-1}(L) \to L$ is 6.]

(vi) If $g \geq 3$, there are two possibilities:

(a) The morphism ϕ is birational onto its image and ϕ blows down to points exactly those curves D with $(C \cdot D) = 0$ (then necessarily $p_a(D) = 0$ and $(D^2) = -2$); in this case the general curve of $|C|$ is nonhyperelliptic; or

(b) The morphism ϕ is of degree 2 onto a rational (possibly singular) surface of degree $g - 1$ in $\mathbb{P}^g$; in this case a general curve of $|C|$ is hyperelliptic.

Exercise 10.4. Let C be a connected nonsingular curve of genus $g \geq 2$ on a $K3$ surface X.

(i) If $g = 2$, show that the morphism $\phi_{|3C|}$ is birational.

(ii) If $g \geq 3$, show that the morphism $\phi_{|2C|}$ is birational.

Exercise 10.5. Prove that if $f : X \to \mathbb{P}^2$ is a finite morphism of degree 2, ramified along a nonsingular curve of degree 6 in $\mathbb{P}^2$, then X is a $K3$ surface (assume that char(k) $\neq$ 2). Prove that the same holds for a finite morphism $f : X \to \mathbb{P}^1 \times \mathbb{P}^1$ of degree 2 ramified along a nonsingular curve of bidegree $(4, 4)$. [*Hint:* Use the following general result on n-cyclic coverings. If Y is a nonsingular projective variety, L a line bundle on Y and $s \in H^0(Y, L^n)$ such that the divisor $B = \mathrm{div}_Y(s)$ is nonsingular, then the n-cyclic covering $\pi : X \to Y$ ramified along B has the following properties: X is nonsingular, projective, $\mathcal{O}_X(B_1) \cong \pi^*(L)$ with $B_1 = \pi^{-1}(B)_{\mathrm{red}}$, $\pi^*(B) = nB_1$, $\omega_X \cong \pi^*(\omega_Y \otimes L^{n-1})$, and $\pi_*(\mathcal{O}_X) \cong \mathcal{O}_Y \oplus L^{-1} \oplus \cdots \oplus L^{-(n-1)}$, cf. [BPV*, I, (17.1)]. Alternatively, prove these properties as an exercise.]

Exercise 10.6. Let $f : X \to B$ be an elliptic fibration of the complex Enriques surface X over the curve B. Prove that $B \cong \mathbb{P}^1$, f has exactly two multiple fibers $2F_1$ and $2F_2$ (each with multiplicity 2), and $\omega_X \cong \mathcal{O}_X(F_1 - F_2) \cong \mathcal{O}_X(F_2 - F_1)$. Show that $\dim H^0(X, \mathcal{O}_X(F_i)) = 1$, $i = 1, 2$. If X is a $K3$ surface admitting an elliptic fibration $f : X \to B$, prove that $B \cong \mathbb{P}^1$ and f has no multiple fibers at all. [*Hint:* Use (8.3), (8.13), and (7.15).]

Exercise 10.7. Let C be an elliptic curve on a complex Enriques surface X. Prove that either $\dim |C| = 1$ and $|C|$ has no base points, or $\dim |C| = 0$, in which case $\dim |2C| = 1$ and $|2C|$ has no base points. [*Hint:* Observe that the normal bundle N of C in X is a 2-torsion element in $\mathrm{Pic}(C)$, and distinguish the cases $N \cong \mathcal{O}_C$ and $N \not\cong \mathcal{O}_C$, but $N^2 \cong \mathcal{O}_C$.]

Exercise 10.8. Let X be an Enriques surface, and let $C_1, \ldots, C_r$ be non-singular irreducible curves on X such that $C_1 \cup \cdots \cup C_r$ is connected and the intersection matrix $\|(C_i \cdot C_j)\|_{i,j=1,\ldots,r}$ is negative definite. Show that $C = C_1 \cup \cdots \cup C_r$ can be contracted to a rational double point z of a normal projective surface Z. Show also that the dualizing sheaf ω_Z is invertible, $\omega_Z \not\cong \mathcal{O}_Z$, $\omega_Z^2 \cong \mathcal{O}_Z$, and $H^1(Z, \mathcal{O}_Z) = 0$.

Exercise 10.9. Show that there are surfaces of degree 4 in $\mathbb{P}^3$ that do not have elliptic fibrations. [*Hint:* According to a theorem of M. Noether, for a general surface S of degree $d \geq 4$ in $\mathbb{P}^3$ (over $\mathbb{C}$) we have $\mathrm{Pic}(S) = \mathbb{Z}[\mathcal{O}_S(1)]$, cf. [DK].) Show that every surface of degree 4 in $\mathbb{P}^3$ that contains a line admits an elliptic fibration.]

Bibliographic References. Theorems 10.2, 10.3, 10.9, 10.19, 10.25, and the Bagnera–DeFranchis list are presented after [BM1]. The elementary properties of Enriques surfaces (10.11), (10.12), (10.13), (10.14), and (10.15) are from [BM2], and Example (10.16) is from [Bea]. The proof of Theorem 10.17 is adapted from Chapter X of [Sha3]. Theorem 10.21 and its consequence (10.22) can be found in various forms in [Sha2], [Sha3], and [Bea].

11
Ruled Surfaces. The Noether–Tsen Criterion

Definition 11.1. A surface X is a *ruled surface* if there exists a nonsingular projective curve B such that X is birationally isomorphic to $\mathbb{P}^1 \times B$.

Definition 11.2. Let X be a surface, and let B be a nonsingular projective curve. A *pencil of curves with base B on X* is a dominant rational map $\pi : X \dashrightarrow B$ such that $\Bbbk(B)$ is algebraically closed in $\Bbbk(X)$.

Then π is defined on the complement of a finite set of points $x_1, \ldots, x_n$ ($n \geq 0$). If these points are fundamental points for π (i.e., π is not regular at these points) then $x_1, \ldots, x_n$ are called the *base points* of the pencil π. For every point $y \in B$, $\pi^{-1}(y)$ (which by definition is the closure in X of the fiber of $\pi|_{X \setminus \{x_1, \ldots, x_n\}}$ over y) is a curve on X, and $\{\pi^{-1}(y) \mid y \in B\}$ is called the family of curves of the pencil π. If $y = \eta$ is the generic point of B, then $\pi^{-1}(\eta)$ is called the generic curve (or the generic fiber) of the pencil π. Let $\phi : X' \to X$ be a composite of quadratic transformations such that $\pi \circ \phi : X' \to B$ is a (surjective) morphism. Then $\pi^{-1}(\eta) \cong (\pi \circ \phi)^{-1}(\eta)$.

Theorem 11.3 (Noether–Tsen). *Let X be a surface that has a pencil of curves $\pi : X \dashrightarrow B$ with base B, such that the generic fiber has arithmetic genus zero. Then X is birationally isomorphic to $\mathbb{P}^1 \times B$. In particular, X is a ruled surface.*

The proof of this theorem uses several facts that are important in themselves. We begin with a definition.

Definition 11.4. Let K be a field. We say that K has *property C_i* ($i \geq 0$) if for every $n \geq 1$, every hypersurface $H \subset \mathbb{P}^n(K)$ of degree $d \geq 1$ with $d^i <$

$n + 1$ has a rational point over K. That is, every homogeneous polynomial of degree d in $n + 1$ variables with coefficients in K has a nonzero solution in K^{n+1} if $d^i < n + 1$ (with $d \geq 1$ and $n \geq 1$).

Remarks 11.5. (a) K has property C_0 if and only if K is algebraically closed.

(b) If K has property C_i for some $i \geq 0$, then K has property C_j for all $j \geq i$.

(c) All finite fields have property C_1 (this is known as the Chevalley–Warning theorem; see [Ser5, I, §2]).

(d) The real number field does not have property C_i for any $i \geq 0$.

Theorem 11.6. (a) *Let* $\Bbbk$ *be an algebraically closed field. Then the fraction field* $\Bbbk(T)$ *of the polynomial ring* $\Bbbk[T]$ *in one variable* T *has property* C_1.

(b) *If the field* K *has property* C_1, *then so does every algebraic extension* $E \supset K$ *of* K.

The proof of Theorem 11.6 is based on the following two lemmas.

(11.6.1) Lemma. *Let* K *be a field that is not algebraically closed. Then there exists an integer* $n > 1$ *and a homogeneous polynomial of degree* n *in* n *variables that has no nontrivial solution in* K^n.

Proof of Lemma 11.6.1. Since K is not algebraically closed, there exists a finite extension $L \supset K$ of degree $n > 1$. Let $N = N_{L/K} : L \to K$ be the norm map. Then $N(xy) = N(x)N(y)$, $\forall x, y \in L$, and $N(x) \neq 0$ for all $x \in L \setminus \{0\}$. Let $e_1, \ldots, e_n$ be a vector space basis of L over K. We will construct a homogeneous polynomial $P \in K[X_1, \ldots, X_n]$, of degree n, such that $N(x) = P(\xi_1, \ldots, \xi_n)$ for all $x = \sum \xi_i e_i \in L$, where $\xi_1, \ldots, \xi_n \in K$. To do that, consider the elements $\varepsilon_{ij}^r \in K$ such that

$$e_i e_j = \sum_{r=1}^{n} \varepsilon_{ij}^r e_r;$$

then put $P(X_1, \ldots, X_n) = \det |a_{ij}|$, where $a_{ij} = \sum_{r=1}^{n} X_r \varepsilon_{ri}^j$. To see that $N(x) = P(\xi_1, \ldots, \xi_n)$ for all $x = \sum \xi_i e_i \in L$, we note that

$$x \cdot e_i = \left(\sum_{r=1}^{n} \xi_r e_r \right) \cdot e_i = \sum_{r=1}^{n} \xi_r (e_r e_i)$$

$$= \sum_{r=1}^{n} \xi_r \sum_{s=1}^{n} \varepsilon_{ri}^s e_s = \sum_{s=1}^{n} \left(\sum_{r=1}^{n} \xi_r \varepsilon_{ri}^s \right) e_s,$$

and therefore

$$N(x) = \det \left| \sum_{r=1}^{n} \xi_r \varepsilon_{ri}^s \right| = \det |a_{is}(\xi)| = P(\xi_1, \ldots, \xi_n).$$

Since $P(\xi_1, \ldots, \xi_n) = N(x) \neq 0$ for all $(\xi_1, \ldots, \xi_n) \in K^n \setminus \{0\}$ (where $x = \sum \xi_i e_i$), the proof of the lemma is complete. $\square$

(11.6.2) Lemma. *Let K be a field that has property C_1, and let $f_1, \ldots, f_r$ $(r \geq 1)$ be r homogeneous forms of degree d in n variables over K. If $n > rd$, then these forms have a common nonzero solution in K^n.*

Proof of Lemma 11.6.2. If K is algebraically closed, then the result is well known. Therefore we'll assume that K is not algebraically closed.

By Lemma 11.6.1, if K is not algebraically closed, then there exists a homogeneous polynomial $P(X_1, \ldots, X_n)$ of degree $n \geq 2$ in n variables that has no nontrivial solutions in K^n. Put

$$Q(X_{11}, \ldots, X_{1n}, \ldots, , X_{n1}, \ldots, X_{nn})$$
$$= P(P(X_{11}, \ldots, X_{1n}), \ldots, , P(X_{n1}, \ldots, X_{nn})).$$

Then Q is a homogeneous polynomial of degree n^2 in n^2 variables, and it has no nontrivial solutions in K^{n^2}. Repeating the same argument, we see that we can find a homogeneous polynomial of degree e in e variables over K that has no nontrivial solutions in K^e, with e arbitrarily large.

Turning now to the proof of Lemma 11.6.2, let $\Phi(X_1, \ldots, X_e)$ be a homogeneous polynomial of degree $e \geq r$ that has no nontrivial solutions in K^e. Assume also that e is sufficiently large so that $(n - dr)[\frac{e}{r}] > dr$, where $[\frac{e}{r}]$ is the integral part (or round-down) of $\frac{e}{r}$. Consider the homogeneous polynomial

$$\Phi'(X_{11}, \ldots, X_{n1}, \ldots, X_{1[\frac{e}{r}]}, \ldots, X_{n[\frac{e}{r}]}),$$

obtained from Φ as follows:

- first replace the variables $X_1, \ldots, X_n$ with $f_1(X_{11}, \ldots, , X_{n1}), \ldots,$ $f_r(X_{11}, \ldots, X_{n1})$;

- then replace the variables $X_{n+1}, \ldots, X_{2n}$ with $f_1(X_{12}, \ldots, X_{n2}), \ldots,$ $f_r(X_{12}, \ldots, X_{n2})$, and so on;

- finally, if there are any variables left after performing these substitutions $[\frac{e}{r}]$ times, then set the remaining variables equal to 0.

Then Φ' is a homogeneous polynomial of degree de in $n[\frac{e}{r}]$ variables. We have $dr([\frac{e}{r}] + 1) > de$, and on the other hand $n[\frac{e}{r}] > dr([\frac{e}{r}] + 1)$ by the choice of e; therefore $n[\frac{e}{r}] > de$. Then Φ' has a nontrivial solution in K. But since Φ has no nontrivial solutions, we conclude that $f_1, \ldots, f_r$ have a common nonzero solution in K. $\square$

Proof of Theorem 11.6.

(a) Let $f(X_1, \ldots, X_n)$ be a homogeneous polynomial of degree $d < n$ in the variables $X_1, \ldots, X_n$, with coefficients in $\Bbbk(T)$. We may assume that the coefficients of f are, in fact, in $\Bbbk[T]$. In that case, we'll show that f has a nontrivial solution in $\Bbbk[T]$ (or, more exactly, in $\Bbbk[T]^n$).

Let

$$f(X_1, \ldots, X_n) = \sum_{i_1 + \cdots + i_n = d} c_{i_1 \ldots i_n} X_1^{i_1} \cdots X_n^{i_n}, \quad c_{i_1 \ldots i_n} \in \Bbbk[T].$$

Since the coefficients $c_{i_1 \ldots i_n}$ are polynomials in T, we can define

$$r = \max\{\deg c_{i_1 \ldots i_n} \mid (i_1, \ldots, i_n)\}.$$

Then we can write f in the form

$$f(X_1, \ldots, X_n) = f_0(X_1, \ldots, X_n) + T f_1(X_1, \ldots, X_n) + \cdots$$
$$+ T^r f_r(X_1, \ldots, X_n).$$

Consider new variables:

$$
\begin{array}{cccc}
Y_{10} & Y_{11} & \cdots & Y_{1s} \\
\multicolumn{4}{c}{\cdots\cdots\cdots\cdots\cdots} \\
Y_{n0} & Y_{n1} & \cdots & Y_{ns}
\end{array}
$$

where s is an integer such that

$$(n - d)s > (r + 1) - n. \tag{$*$}$$

Now let $\phi(Y_{10}, \ldots, Y_{ns})$ be the homogeneous polynomial of degree d in $n(s+1)$ variables Y_{ij} with coefficients in $\Bbbk[T]$, obtained from $f(X_1, \ldots, X_n)$ by the following substitutions:

$$X_1 \text{ replaced with } Y_{10} + Y_{11}T + \cdots + Y_{1s}T^s,$$
$$\cdots\cdots\cdots\cdots\cdots\cdots\cdots\cdots\cdots\cdots\cdots\cdots\cdots$$
$$X_n \text{ replaced with } Y_{n0} + Y_{n1}T + \cdots + Y_{ns}T^s;$$

thus

$$\phi(Y_{10}, \ldots, Y_{ns}) = f\left(\sum_{j=0}^{s} Y_{1j}T^j, \ldots, \sum_{j=0}^{s} Y_{nj}T^j\right).$$

The degree of ϕ in the variable T is clearly equal to $ds + r$. Therefore we can write

$$\phi(Y_{10}, \ldots, Y_{ns}) = \phi_0(Y_{10}, \ldots, Y_{ns})$$
$$+ T\phi_1(Y_{10}, \ldots, Y_{ns}) + T^{ds+r}\phi_{ds+r}(Y_{10}, \ldots, Y_{ns}).$$

We now have $ds + r + 1$ homogeneous polynomials $\phi_0, \ldots, \phi_{ds+r}$ in $n(s + 1)$ variables over the algebraically closed field $\Bbbk$. From $(*)$ we have that

$ds + r + 1 < n(s + 1)$, and therefore these forms have a common nonzero solution. This shows that f has a nonzero solution in $\Bbbk[T]$.

(b) It is enough to prove (b) for a finite extension E/K; indeed, if $f \in E[X_1, \ldots, X_n]$ is a homogeneous polynomial of degree $d < n$, then there exists a finite extension E_1/K ($E_1 \subset E$) such that $f \in E_1[X_1, \ldots, X_n]$, and then f has a nontrivial solution in E_1 (if (b) is proved for finite extensions).

Let $m = [E : K]$, and let $f \in E[X_1, \ldots, X_n]$ as before. Choose a vector space basis of E over K, say, $e_1, \ldots, e_m$. Introduce new variables

$$
\begin{matrix}
Y_{11} & \ldots & Y_{1m} \\
\cdots\cdots\cdots\cdots\cdots \\
Y_{n1} & \ldots & Y_{nm}
\end{matrix}
$$

and consider the homogeneous polynomial

$$\Phi(Y_{11}, \ldots, Y_{1m}, \ldots, Y_{n1}, \ldots, Y_{nm}) = f\left(\sum_{i=1}^{m} e_i Y_{1i}, \ldots, \sum_{i=1}^{m} e_i Y_{ni}\right).$$

Write Φ in the form $\Phi(Y) = \sum_{i=1}^{m} e_i \Phi_i(Y)$, with $\Phi_i(Y) \in K[Y_{11}, \ldots, Y_{nm}]$ homogeneous polynomials of degree d in nm variables over K. As $md < nm$ and K satisfies C_1, Lemma 11.6.2 shows that the forms $\Phi_1(Y), \ldots, \Phi_m(Y)$ have a nontrivial solution in K, say, $(\xi_{11}, \ldots, \xi_{nm}) \in K^{nm}$. Then

$$(\eta_1, \ldots, \eta_n) \in E^n, \quad \eta_j = \sum_{i=1}^{m} e_i \xi_{ji},$$

is a nontrivial solution of f in E. □

Lemma 11.7. *Let K be a field, and let $\bar{K}$ be the algebraic closure of K. Let X be an algebraic curve, proper over K, such that $\bar{X} = X \otimes_K \bar{K}$ is K-isomorphic to the projective line $\mathbb{P}^1(\bar{K})$. Then X is K-isomorphic to $\mathrm{Proj}(K[X_0, X_1, X_2]/(Q))$, where $Q \in K[X_0, X_1, X_2]$ is a nondegenerate quadratic form.*

Proof. From the hypotheses it follows that X is geometrically integral, and therefore the rational function field $K(X)$ of X is a regular extension of K, cf. [EGA IV, (4.5.9) and (4.6.1)]. Since X is also geometrically smooth, the sheaf $\Omega_{\bar{X}} = \Omega^1_{\bar{X}/\bar{K}}$ can be obtained by base change from Ω_X; that is, we have $\Omega_{\bar{X}} \cong \Omega_X \otimes_K \bar{K}$. Taking duals, we get a canonical isomorphism

$$(\Omega_{\bar{X}})\check{\ } \cong (\Omega_X)\check{\ } \otimes_K \bar{K}. \qquad (**)$$

But $(\Omega_{\bar{X}})\check{\ } = \mathcal{O}_{\mathbb{P}^1(\bar{K})}(2)$, and therefore $\dim H^0(\bar{X}, (\Omega_{\bar{X}})\check{\ }) = 3$. If $\sigma_0, \sigma_1, \sigma_2$ is a basis of $H^0(\bar{X}, (\Omega_{\bar{X}})\check{\ })$ over $\bar{K}$, then the rational map $\bar{X} \to \mathbb{P}^2(\bar{K})$ given by $x \mapsto (\sigma_0(x), \sigma_1(x), \sigma_2(x))$ is defined everywhere and is a closed embedding, because $\mathcal{O}(2)$ is a very ample invertible sheaf on $\bar{X}$. But the isomorphism $(**)$ shows that we can obtain a basis of $H^0(\bar{X}, (\Omega_{\bar{X}})^{\check{}})$ if we start with any basis $\sigma_0, \sigma_1, \sigma_2$ of $H^0(X, (\Omega_X)^{\check{}})$ over K. In this case the embedding is even a K-embedding (or more precisely is induced by a K-embedding). Therefore X is K-isomorphic to a closed subscheme of $\mathbb{P}^2(K)$ of the form $\mathrm{Proj}(K[X_0, X_1, X_2]/(Q))$, with Q a nondegenerate quadratic form in the variables X_0, X_1, X_2; Q is nondegenerate over K because it is so over $\bar{K}$. $\qquad\square$

Lemma 11.8. *In the hypotheses of Lemma 11.7, assume also that X has a rational point over K. Then X is K-isomorphic to $\mathbb{P}^1(K)$.*

Proof. Let $x \in X$ be a K-rational point. Let $I \subset \mathcal{O}_X$ be the coherent ideal sheaf that defines the closed subscheme $\{x\}$ of X. Then $\bar{I}^{-1} = I^{-1} \otimes_K \bar{K}$ is an invertible $\mathcal{O}_{\bar{X}}$-module of degree 1; since $\bar{X} \cong \mathbb{P}^1(\bar{K})$, we have $\dim_{\bar{K}} H^0(\bar{X}, \bar{I}^{-1}) = 2$. Since $H^0(\bar{X}, \bar{I}^{-1}) \cong H^0(X, I^{-1}) \otimes_K \bar{K}$, we see that there exists a nonconstant rational function $f \in K(X)$ that has a simple pole at x (and no other poles). Then f, regarded as a map $f : X \to \mathbb{P}^1(K)$, is a K-isomorphism. $\qquad\square$

Corollary 11.9. *Let K be a field that has property C_1, and let X be a geometrically integral, proper curve over K, of arithmetic genus 0. Then X is K-isomorphic to $\mathbb{P}^1(K)$.*

Proof. We have

$$p_a(X) = 1 - \chi(\mathcal{O}_X) = 1 - \dim_K H^0(\mathcal{O}_X) + \dim_K H^1(\mathcal{O}_X)$$
$$= 1 - \dim_{\bar{K}} H^0(\mathcal{O}_{\bar{X}}) + \dim_{\bar{K}} H^1(\mathcal{O}_{\bar{X}}).$$

Since X is geometrically integral, $\dim_{\bar{K}} H^0(\mathcal{O}_{\bar{X}}) = 1$, whence $H^1(\mathcal{O}_{\bar{X}}) = 0$, and therefore $\bar{X} \cong \mathbb{P}^1(\bar{K})$. By Lemma 11.7, X is K-isomorphic to a nondegenerate conic $Q(X_0, X_1, X_2) = 0$ in $\mathbb{P}^2(K)$. Since K has property C_1, this conic has at least one rational point over K, and therefore Lemma 11.8 gives $X \cong \mathbb{P}^1(K)$. $\qquad\square$

Proof of Theorem 11.3. Let η be the generic point of B. By Theorem 11.6, the field $\Bbbk(\eta) = \Bbbk(B)$ has property C_1. By hypothesis, the curve $X_\eta = \pi^{-1}(\eta)$ has arithmetic genus zero (over $\Bbbk(\eta)$). Let $\phi : X' \to X$ be a composite of quadratic transformations such that $\pi \circ \phi$ becomes a morphism. Since $\Bbbk(B)$ is algebraically closed in $\Bbbk(X') = \Bbbk(X)$, and since the generic fiber does not change with these quadratic transformations, we see that $X_\eta \cong (\pi \circ \phi)^{-1}(\eta)$ is geometrically integral (using (7.3)). By Corollary 11.9 X_η is therefore $\Bbbk(\eta)$-isomorphic to $\mathbb{P}^1(\Bbbk(\eta))$. So $\Bbbk(X_\eta) \cong \Bbbk(\eta)(t)$, with t an independent variable over $\Bbbk(\eta)$. Since $\Bbbk(X_\eta) = \Bbbk(X)$, we have proved that $\Bbbk(X) \cong \Bbbk(B)(t)$, with t an independent variable over $\Bbbk(B)$; in other words, X is birationally isomorphic to $\mathbb{P}^1 \times B$. $\qquad\square$

Theorem 11.10. *Let $\pi : X \to B$ be a surjective morphism from a surface X to a (nonsingular projective) curve B, such that there exists a closed point $b \in B$ with $\pi^{-1}(b) \cong \mathbb{P}^1$. Then there exists a section $\sigma : B \to X$ of π (that is, a morphism σ such that $\pi \circ \sigma = \mathrm{id}_B$), and an open subset $U \subseteq B$ that contains b and an isomorphism $f : \pi^{-1}(U) \overset{\cong}{\longrightarrow} \mathbb{P}^1 \times U$ such that the diagram*

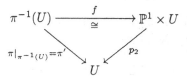

is commutative, where p_2 is the projection of $\mathbb{P}^1 \times U$ onto U.

Proof. Since B is a nonsingular curve and $\pi_*(\mathcal{O}_X)$ is a torsionfree coherent $\mathcal{O}_B$-module, $\pi_*(\mathcal{O}_X)$ is locally free of finite rank. Since π is flat and $H^1(\pi^{-1}(b), \mathcal{O}_{\pi^{-1}(b)}) = 0$, by the base change theorem [Har1, III.12] we have that the map $b' \mapsto \dim H^1(\pi^{-1}(b'), \mathcal{O}_{\pi^{-1}(b')})$ is identically zero in a neighborhood V of b in B, and therefore the natural homomorphism

$$\pi_*(\mathcal{O}_X) \otimes \Bbbk(b') \to H^0(\pi^{-1}(b'), \mathcal{O}_{\pi^{-1}(b')})$$

is an isomorphism for all $b' \in V$. We have $\dim H^0(\pi^{-1}(b), \mathcal{O}_{\pi^{-1}(b)}) = 1$ because $\pi^{-1}(b) \cong \mathbb{P}^1$, and therefore $\pi_*(\mathcal{O}_X)$ is locally free of rank 1. Then $\pi_*(\mathcal{O}_X) = \mathcal{O}_B$, and therefore $\Bbbk(B)$ is algebraically closed in $\Bbbk(X)$. From (7.3) we have that there exists an open neighborhood $U_1 \subseteq V$ of b in B such that the fiber $F_{b'} = \pi^{-1}(b')$ is geometrically integral for every $b' \in U_1$. As $F_b \cong \mathbb{P}^1$ and the arithmetic genus of $F_{b'}$ does not depend on b', we have that the generic fiber F_η has arithmetic genus zero, and $F_{b'} \cong \mathbb{P}^1$ for every closed point $b' \in U_1$. From Theorem 11.3 and its proof we see that $F_\eta \cong \mathbb{P}^1(\Bbbk(\eta)) = \mathbb{P}^1(\Bbbk(B))$, and therefore F_η has a rational point over $\Bbbk(B) = \mathcal{O}_{B,\eta}$. That is, there exists a $\Bbbk(B)$-morphism $\sigma' : \mathrm{Spec}(\Bbbk(B)) \to F_\eta$, and therefore a morphism $\sigma : \mathrm{Spec}(\mathcal{O}_{B,\eta}) \to X$ whose composite with π is precisely the canonical inclusion $\mathrm{Spec}(\mathcal{O}_{B,\eta}) \hookrightarrow B$. Using [EGA I, (6.5.1)] we conclude the existence of a rational section $\sigma : B \dashrightarrow X$ of π. Since B is a nonsingular curve and X is a projective variety, σ is in fact an everywhere regular morphism $\sigma : B \to X$.

Now put $D = \sigma(B)$. We have $(D \cdot F_{b'}) = 1$ for all $b' \in B$. Let $X' = \pi^{-1}(U_1)$ and $\pi' = \pi_{X'}$. Since all the fibers of π' are projective lines and $\mathcal{O}_{X'}(D) \otimes \mathcal{O}_{F_{b'}} \cong \mathcal{O}_{F_{b'}}(1)$ for all $b' \in U_1$, we have

$$\dim_{\Bbbk(b')} H^0(\mathcal{O}_X(D) \otimes \Bbbk(b')) = 2 \text{ for all } b' \in U_1.$$

From this and the base change theorems we get that $E = \pi'_*\mathcal{O}_{X'}(D)$ is a locally free $\mathcal{O}_{U_1}$-module of rank 2 and the canonical homomorphism

$$\pi'_*\mathcal{O}_{X'}(D) \otimes \Bbbk(b') \to H^0(\mathcal{O}_{X'}(D) \otimes \mathcal{O}_{F_{b'}})$$

is an isomorphism for every $b' \in U_1$. This immediately implies that the canonical homomorphism $\pi'^* \pi'_* \mathcal{O}_{X'}(D) = \pi'^*(E) \to \mathcal{O}_{X'}(D)$ is surjective. By the universal property of $\mathbb{P}(E)$, the projective bundle associated to E, we have that there exists a unique U_1-morphism $u : X' \to \mathbb{P}(E)$ such that $u^* \mathcal{O}_{\mathbb{P}(E)}(1) \cong \mathcal{O}_{X'}(D)$. From the definition of E we see that u is an isomorphism, because for every $b' \in U_1$ the morphism $u_{b'} : F_{b'} \to \mathbb{P}^1(\Bbbk(b'))$ induced by u is obviously an isomorphism. $\qquad\square$

Corollary 11.11. *In the hypotheses of Theorem 11.10, assume also that all the closed fibers of the morphism π are isomorphic to $\mathbb{P}^1$. Then there exists a locally free $\mathcal{O}_B$-module E of rank 2 and a B-isomorphism $u : X \to \mathbb{P}(E)$. Furthermore, if E and E' are two locally free $\mathcal{O}_B$-modules of rank 2, then there exists a B-isomorphism $v : \mathbb{P}(E) \to \mathbb{P}(E')$ if and only if there exists an invertible $\mathcal{O}_B$-module L such that $E' \cong E \otimes_{\mathcal{O}_B} L$.*

Proof. Everything except the last statement is clear from the proof of Theorem 11.10. If $E' \cong E \otimes L$, then $\mathbb{P}(E) \cong \mathbb{P}(E')$ by [EGA II, (4.1.4)]. Conversely, assume that there exists a B-isomorphism $v : \mathbb{P}(E) \xrightarrow{\cong} \mathbb{P}(E')$. Then, in view of Proposition 11.12 below, the invertible $\mathcal{O}_{\mathbb{P}(E)}$-module $v^* \mathcal{O}_{\mathbb{P}(E')}(1)$ can be written uniquely in the form

$$v^* \mathcal{O}_{\mathbb{P}(E')}(1) \cong p^*(L) \otimes \mathcal{O}_{\mathbb{P}(E)}(s),$$

with s a suitable integer, $p : \mathbb{P}(E) \to B$ (resp. $p' : \mathbb{P}(E') \to B$) the canonical projection of $\mathbb{P}(E)$ (resp. $\mathbb{P}(E')$) onto B, and L a suitable invertible $\mathcal{O}_B$-module. By the projection formula we get

$$p_* v^* \mathcal{O}_{\mathbb{P}(E')}(1) \cong L \otimes p_* \mathcal{O}_{\mathbb{P}(E)}(s). \qquad (*)$$

But the explicit computation of the cohomology of a projective bundle, cf. [EGA III, (2.1.15)], gives $p_* v^* \mathcal{O}_{\mathbb{P}(E')}(1) \cong p'_* \mathcal{O}_{\mathbb{P}(E')}(1) \cong E'$ and $p_* \mathcal{O}_{\mathbb{P}(E)}(s) = S^s(E)$, where $S^s(E) = 0$ if $s < 0$ and $S^s(E)$ is the sth symmetric power of E if $s \geq 0$. From $(*)$ we get $s = 1$, and therefore $E' \cong L \otimes E$. $\qquad\square$

Proposition 11.12. *Let $f : \mathbb{P}(E) \to Z$ be a projective fiber bundle, with Z an arbitrary algebraic variety and E a locally free $\mathcal{O}_Z$-module of finite rank. Then $\operatorname{Pic}(\mathbb{P}(E)) \cong \mathbb{Z}[\mathcal{O}_{\mathbb{P}(E)}(1)] \oplus f^*(\operatorname{Pic}(Z))$, where f^* is the (injective) group homomorphism induced by f on the level of Picard groups, and $[\mathcal{O}_{\mathbb{P}(E)}(1)]$ is the class of $\mathcal{O}_{\mathbb{P}(E)}(1)$ in $\operatorname{Pic}(\mathbb{P}(E))$.*

This proposition is well known; it is an easy consequence of the base change theorems.

Definition 11.13. A *geometrically ruled* surface is a surface X together with a morphism $f : X \to B$ onto a nonsingular curve B (called the *base curve*) such that the fiber $F_b = f^{-1}(B)$ is isomorphic to $\mathbb{P}^1$ for every closed point $b \in B$.

From Proposition 11.10 and Corollary 11.11 it follows that if $f : X \to B$ is a geometrically ruled surface, then there exists a locally free $\mathcal{O}_B$-module E of rank 2 and a B-isomorphism $u : X \xrightarrow{\cong} \mathbb{P}(E)$. Furthermore, there exists a section $\sigma : B \to X$ of f.

Corollary 11.14. *Let $f : X \to B$ be a geometrically ruled surface, and let $\sigma : B \to X$ be a section of f. Put $C = \sigma(B)$, and let F denote a closed fiber F_b of f. Then*

$$\operatorname{Pic}(X) \cong \mathbb{Z}[\mathcal{O}_X(C)] \oplus f^*(\operatorname{Pic}(B))$$

and

$$\operatorname{Num}(X) \cong \mathbb{Z}\{\mathcal{O}_X(C)\} \oplus \mathbb{Z}\{\mathcal{O}_X(F)\},$$

where $[L]$ (resp. $\{L\}$) is the class of the invertible $\mathcal{O}_X$-module L in $\operatorname{Pic}(X)$ (resp. in $\operatorname{Num}(X)$).

Proof. The first isomorphism follows immediately from Proposition 11.12, noting that if $X = \mathbb{P}(E)$, with E a locally free $\mathcal{O}_B$-module of rank 2, then

$$\mathcal{O}_X(C) \cong \mathcal{O}_{\mathbb{P}(E)}(1) \otimes f^*(L'),$$

with L' a suitable invertible $\mathcal{O}_B$-module. The second isomorphism follows from the first and from the observation that $\operatorname{Num}(B) = \mathbb{Z}\{b\}$, where $b \in B$ is an arbitrary closed point, and therefore $f^*(\operatorname{Num}(B)) \cong \mathbb{Z}\{\mathcal{O}_X(F)\}$. □

Proposition 11.15. *If X is a ruled surface, then $\kappa(X) = -\infty$; that is, $p_n = 0$ for all $n \geq 1$. In particular, $p_g = 0$.*

Proof. Since the p_n ($n \geq 1$) are birational invariants and X is birationally isomorphic to $\mathbb{P}^1 \times B$ (with B a nonsingular projective curve), the proposition follows from (5.8)(ii). □

Proposition 11.16. *Let B be a nonsingular projective curve, and let E be a locally free $\mathcal{O}_B$-module of rank $r \geq 1$. Then there exists a chain of locally free subsheaves of E,*

$$0 = E_0 \subset E_1 \subset \cdots \subset E_r = E,$$

with $\operatorname{rank}(E_i) = i$ and $E_i/E_{i-1} = L_i$ an invertible $\mathcal{O}_B$-module for every $i = 1, \ldots, r$.

Proof. Using induction, it is enough to show that E has an invertible subsheaf L such that E/L is locally free (necessarily of rank $r - 1$).

Let $P = \mathbb{P}(E^{\vee})$, and let $p : P \to B$ be the canonical projection. Then we have the universal embedding $\mathcal{O}_P(-1) \hookrightarrow p^*(E)$, where the quotient $E' = p^*(E)/\mathcal{O}_P(-1)$ is locally free.

Let $s \in \Gamma(U, E)$ be a nonzero section of E over a suitable nonempty open subset $U \subseteq B$. Since E is locally free, the section s induces a morphism $i_s : V \to P$, with $V \subseteq U$ a suitable nonempty open subset, such that $p \circ i_s$ coincides with the inclusion of V in B. Since i_s is a rational section of p, B is a nonsingular curve, and P is a projective variety, we see that in fact i_s is a regular section of p (that is, i_s is regular everywhere on B).

Finally, from the exact sequence of locally free $\mathcal{O}_P$-modules

$$0 \to \mathcal{O}_P(-1) \to p^*(E) \to E' \to 0$$

we get an exact sequence of locally free $\mathcal{O}_B$-modules

$$0 \to i_s^*(\mathcal{O}_P(-1)) = L_s \to i_s^* p^*(E) = E \to i_s^*(E') \to 0.$$

$\square$

Definition 11.17. With the hypotheses and notation as in Proposition 11.16, the chain

$$0 = E_0 \subset E_1 \subset \cdots \subset E_r = E,$$

with E_i locally free of rank i and $E_i/E_{i-1} = L_i$ an invertible $\mathcal{O}_B$-module for every $i = 1, \ldots, r$, is called a *splitting* of E. Sometimes we denote this splitting by $(L_1, \ldots, L_r)$, indicating only the sequence of quotients in the chain.

If E is a locally free $\mathcal{O}_B$-module of rank $r \geq 1$, we write $\det(E)$ for $\wedge^r(E)$, and $\deg(E)$ for $\deg(\det(E))$. It is easy to see that $\deg$ is an additive map on the category of locally free $\mathcal{O}_B$-modules of finite rank, that is, $\deg(E) = \deg(E_1) + \deg(E_2)$ for every exact sequence

$$0 \longrightarrow E_1 \longrightarrow E \longrightarrow E_2 \longrightarrow 0$$

of locally free $\mathcal{O}_B$-modules. Indeed, if $r_i = \operatorname{rank}(E_i), i = 1, 2$, and $r = \operatorname{rank}(E)$, then $r = r_1 + r_2$, and the exact sequence above induces an isomorphism $\det(E) \cong \det(E_1) \otimes_{\mathcal{O}_B} \det(E_2)$. In particular, if $(L_1, \ldots, L_r)$ is a splitting of E, then $\deg(E) = \sum_{i=1}^r \deg(L_i)$.

The following result is an immediate consequence of the existence of splittings.

Corollary 11.18 (Riemann–Roch). *If E is a locally free $\mathcal{O}_B$-module of rank r on a nonsingular projective curve B, then*

$$\chi(E) = \deg(E) + r(1 - g), \quad \text{where } g = p_a(B).$$

Proof. Induction on r. For $r = 1$, this is the usual Riemann–Roch theorem for invertible modules. Assume therefore that $r > 1$. By Definition 11.17, there exists an exact sequence of locally free $\mathcal{O}_B$-modules,

$$0 \to L \to E \to E' \to 0,$$

with $\text{rank}(L) = 1$ and $\text{rank}(E') = r - 1$. Using the inductive hypothesis, we have:

$$\begin{aligned}
\chi(E) &= \chi(L) + \chi(E') \\
&= \deg(L) + (1 - g) + \deg(E') + (r - 1)(1 - g) \\
&= \deg(E) + r(1 - g).
\end{aligned}$$

$\square$

Proposition 11.19. *Let $f : X \to B$ be a geometrically ruled surface, with B a nonsingular projective curve, $X = \mathbb{P}(E)$ for a locally free $\mathcal{O}_B$-module of rank 2, and f the canonical projection of $\mathbb{P}(E)$. Then:*

(a) $q = p_a(B)$;

(b) $(L \cdot L) = \deg(E)$, where $L = \mathcal{O}_{\mathbb{P}(E)}(1)$; and

(c) $(K^2) = 8(1 - q)$ and $b_2 = 2$.

Proof. Since $X = \mathbb{P}(E)$ and f is the canonical projection, [EGA III (2.1.15)] gives that the Leray spectral sequence

$$E_2^{pq} = H^p(B, R^q f_* \mathcal{O}_X(n)) \implies H^{p+q}(X, \mathcal{O}_X(n))$$

degenerates for every $n \geq 0$. Since $f_* \mathcal{O}_X(n) = S^n(E)$ (loc. cit.), we have that

$$H^p(B, S^n(E)) = H^p(X, \mathcal{O}_X(n)) \quad \text{for all } p \geq 0 \text{ and all } n \geq 0.$$

In particular, the case $p = 1, n = 0$ gives part (a) of the proposition.

For arbitrary $n \geq 0$, we have in particular that

$$\chi(X, \mathcal{O}_X(n)) = \chi(B, S^n(E)). \tag{$*$}$$

But Corollary 11.18 gives

$$\chi(B, S^n(E)) = \deg(S^n(E)) + \text{rank}(S^n(E))(1 - p_a(B)). \tag{$**$}$$

As $\text{rank}(E) = 2$, we have $\text{rank}(S^n(E)) = n + 1$. On the other hand, let (L_1, L_2) be a splitting of E, cf. Proposition 11.16. Since the map deg is additive, we have:

$$\begin{aligned}
\deg(S^n(E)) = \deg(S^n(L_1 \oplus L_2)) &= \deg\left(\bigwedge^{n+1} \left(\bigoplus_{i=0}^{n} (L_1^i \otimes L_2^{n-i}) \right) \right) \\
&= \deg\left(\bigotimes_{i=1}^{n} (L_1^i \otimes L_2^{n-i}) \right) = \frac{n(n+1)}{2} \cdot (\deg(L_1) + \deg(L_2)) \\
&= \frac{n(n+1)}{2} \cdot \deg(E).
\end{aligned}$$

This computation, together with Equations $(*)$ and $(**)$, gives

$$\chi(X, \mathcal{O}_X(n)) = \frac{n(n+1)}{2} \cdot \deg(E) + (n+1)(1 - p_a(B)),$$

and this proves part (b) of the proposition, cf. (1.21).

To prove part (c), we use the well known (and elementary) formula:

$$\omega_X \cong f^*(\omega_B \otimes \det(E)) \otimes \mathcal{O}_X(-2),$$

whence

$$\begin{aligned}
(K^2) &= (\mathcal{O}_X(-2) \cdot \mathcal{O}_X(-2)) + 2(\mathcal{O}_X(-2) \cdot f^*(\omega_B \otimes \det(E))) \\
&= 4(L \cdot L) - 4(L \cdot f^*(\omega_B \otimes \det(E))) \\
&= 4\deg(E) - 4\deg(\omega_B \otimes \det(E)) = -4\deg(\omega_B) \\
&= 8(1 - p_a(B)) = 8(1 - q).
\end{aligned}$$

Finally, since $p_g = 0$ by Proposition 11.15, Theorem 5.1 gives $\Delta = 0$, and therefore Noether's formula becomes

$$10 - 8q = (K^2) + b_2.$$

Since $(K^2) = 8 - 8q$, we get $b_2 = 2$. $\square$

11.20. Let Z be a complete algebraic variety, and let E_1 and E_2 be two locally free $\mathcal{O}_Z$-modules of finite rank. An *extension of E_2 by E_1* is an exact sequence of the form:

$$0 \longrightarrow E_1 \overset{\phi}{\longrightarrow} E \overset{\psi}{\longrightarrow} E_2 \longrightarrow 0, \qquad (\mathbf{E})$$

with E locally free of finite rank (E being locally free of finite rank follows automatically from the exact sequence, for E_1 and E_2 are locally free of finite rank).

If

$$0 \longrightarrow E_1 \overset{\phi'}{\longrightarrow} E' \overset{\psi'}{\longrightarrow} E_2 \longrightarrow 0$$

is another extension of E_2 by E_1, we say that the two extensions are *isomorphic* if there exists a homomorphism of $\mathcal{O}_Z$-modules, $u : E \to E'$, such that the following diagram is commutative:

$$\begin{array}{ccccccccc}
0 & \longrightarrow & E_1 & \overset{\phi}{\longrightarrow} & E & \overset{\psi}{\longrightarrow} & E_2 & \longrightarrow & 0 \\
& & \| & & \downarrow u & & \| & & \\
0 & \longrightarrow & E_1 & \underset{\phi'}{\longrightarrow} & E' & \underset{\psi'}{\longrightarrow} & E_2 & \longrightarrow & 0
\end{array}$$

The extension $(\mathbf{E})$ is *trivial* if $E \cong E_1 \oplus E_2$, ϕ is the canonical embedding $e_1 \mapsto (e_1, e_2)$, and ψ is the canonical projection $(e_1, e_2) \mapsto e_2$.

Proposition 11.21. *The isomorphism classes of extensions of E_2 by E_1 are in one-to-one correspondence with the elements of the group*

$$H^1(Z, \mathbf{Hom}(E_2, E_1)) = H^1(Z, \check{E}_2 \otimes E_1),$$

with the trivial extension of E_2 by E_1 corresponding to the zero element of $H^1(Z, \check{E}_2 \otimes E_1)$.

Proof. The proof is standard. We show only how to associate an element of $H^1(Z, \mathbf{Hom}(E_2, E_1))$ to an extension

$$0 \longrightarrow E_1 \longrightarrow E \longrightarrow E_2 \longrightarrow 0. \tag{E}$$

From this exact sequence we get another exact sequence

$$0 \to \mathbf{Hom}(E_2, E_1) \to \mathbf{Hom}(E_2, E) \to \mathbf{Hom}(E_2, E_2) \to 0,$$

and from this we get a canonical boundary map in cohomology:

$$\delta : \mathrm{Hom}(E_2, E_2) = H^0(Z, \mathbf{Hom}(E_2, E_2)) \to H^1(Z, \mathbf{Hom}(E_2, E_1)).$$

Then the element of $H^1(Z, \mathbf{Hom}(E_2, E_1))$ that corresponds to the extension (**E**) is the element $\delta(\mathrm{id}_{E_2})$. $\square$

Theorem 11.22. *Every locally free $\mathcal{O}_{\mathbb{P}^1}$-module of rank 2 is isomorphic to $\mathcal{O}_{\mathbb{P}^1}(m) \oplus \mathcal{O}_{\mathbb{P}^1}(n)$ for suitable integers m and n.*

Proof. Let E be a locally free sheaf of rank 2 on $\mathbb{P}^1$, and let M be any invertible $\mathcal{O}_{\mathbb{P}^1}$-module. Then we have

$$\deg(E \otimes M) = \deg(E) + 2 \deg(M).$$

Choose M as follows:

$$M = \begin{cases} \mathcal{O}_{\mathbb{P}^1}(-\frac{d}{2}), & \text{if } d = \deg(E) \text{ is even,} \\ \mathcal{O}_{\mathbb{P}^1}(-\frac{d+1}{2}), & \text{if } d \text{ is odd,} \end{cases}$$

and put $E' = E \otimes M$; then

$$\deg(E') = 0 \text{ or } -1.$$

It is clearly enough to prove the theorem for E', and therefore we may assume that $d = \deg(E) = 0$ or -1.

From Corollary 11.18 we get $\chi(E) = d + 2 \geq 1$, whence $\dim H^0(E) \geq 1$. Let $0 \neq s \in H^0(E)$. Using the notation introduced in the proof of Proposition 11.16, s is a global section of the invertible subsheaf L_s. Therefore $\deg(L_s) \geq 0$. Moreover, the quotient $L' = E/L_s$ is also invertible. Since L_s and L' are invertible on $\mathbb{P}^1$, there are two integers a and b such that

$L_s \cong \mathcal{O}_{\mathbb{P}^1}(a)$ (with $a \geq 0$ because $\deg(L_s) \geq 0$) and $L' \cong \mathcal{O}_{\mathbb{P}^1}(b)$. Then $a + b = \deg(E) = d$, and we have an exact sequence

$$0 \to \mathcal{O}_{\mathbb{P}^1}(a) \to E \to \mathcal{O}_{\mathbb{P}^1}(d - a) \to 0, \qquad (*)$$

with $d = \deg(E) = 0$ or -1. We have thus shown that E is an extension of $\mathcal{O}_{\mathbb{P}^1}(d - a)$ by $\mathcal{O}_{\mathbb{P}^1}(a)$. By Proposition 11.21, the group $H^1(\mathbb{P}^1, \mathcal{O}_{\mathbb{P}^1}(2a - d))$ classifies the isomorphism classes of extensions of $\mathcal{O}_{\mathbb{P}^1}(d-a)$ by $\mathcal{O}_{\mathbb{P}^1}(a)$. But $2a-d \geq -1$, for $a \geq 0$ and $d = 0$ or -1, and therefore $H^1(\mathbb{P}^1, \mathcal{O}_{\mathbb{P}^1}(2a-d)) = 0$. Therefore the extension $(*)$ is trivial, so that $E \cong \mathcal{O}_{\mathbb{P}^1}(a) \oplus \mathcal{O}_{\mathbb{P}^1}(d-a)$. $\square$

Remarks 11.23.

(a) Theorem 11.22 is a special case of a theorem of Grothendieck, cf. [Gro5] (see also [Ati]), which says that every locally free $\mathcal{O}_{\mathbb{P}^1}$-module E of finite rank has the form $\mathcal{O}_{\mathbb{P}^1}(m_1) \oplus \cdots \oplus \mathcal{O}_{\mathbb{P}^1}(m_r)$ for suitable integers $m_1, \ldots, m_r$, where r is the rank of E. However, we shall only need the special case $r = 2$, whose proof (given above) is easier.

(b) The construction of the invertible subsheaf L_s in the proof of Theorem 11.22 can also be obtained directly (i.e., without reference to the proof of Proposition 11.16), as follows. The section $s \in H^0(B, E) = \mathrm{Hom}(\mathcal{O}_B, E)$ yields a homomorphism $s^\vee : E^\vee \to \mathcal{O}_B$. Since B is a smooth curve, the image of $s^\vee$ is an ideal sheaf of the form $\mathcal{O}_B(-D)$, with D an effective divisor on B. In particular, $\mathcal{O}_B(-D)$ is invertible. It follows that the kernel $L'^\vee$ of the surjective map $E^\vee \to \mathcal{O}_B(-D)$ is also invertible. Dualizing, we get an exact sequence

$$0 \to L_s = \mathcal{O}_B(D) \to E \to L' \to 0,$$

with $\deg(L_s) \geq 0$ and $\deg(L') = d - \deg(L_s)$.

Note that, by construction, the effective divisor D is the zero locus of the section $s \in H^0(B, E)$.

EXERCISES

Exercise 11.1. Let $m, n \geq 1$ be two positive integers. Show that there exist exact sequences of sheaves on $\mathbb{P}^1$ of the form

$$0 \to \mathcal{O}_{\mathbb{P}^1} \to \mathcal{O}_{\mathbb{P}^1}(m) \oplus \mathcal{O}_{\mathbb{P}^1}(n) \to \mathcal{O}_{\mathbb{P}^1}(m + n) \to 0,$$

and none of these exact sequences splits.

Exercise 11.2. Let E be a locally free sheaf of rank 2 on a nonsingular projective curve B. Prove that there exists an invertible sheaf L on B such that the locally free sheaf $E' = E \otimes L$ has the following properties: $H^0(B, E') \neq 0$ and $H^0(B, E' \otimes M) = 0$ for all invertible sheaves M on

B with $\deg(M) < 0$. According to the terminology of [Har1], a locally free sheaf E' of rank 2 with these properties is called *normalized*. If E' is normalized, let $s \in H^0(B, E')$ be any nonzero section; then s corresponds to an injective map of sheaves $0 \to \mathcal{O}_B \to E'$. Prove that the quotient $E'/\mathcal{O}_B$ is invertible. [*Hint:* Observe that $H^0(B, E \otimes L) \neq 0$ for all invertible sheaves L with $\deg(L) \gg 0$, and $H^0(B, E \otimes L) = 0$ for all invertible sheaves L with $\deg(L) \ll 0$. For the last part use the fact that a rank 1 sheaf on a nonsingular curve is invertible if and only if it has no torsion.]

Exercise 11.3. A locally free sheaf E of finite rank on a nonsingular projective curve B is *stable* (resp. *semistable*) if for every quotient locally free sheaf $E \to F \to 0$ with $0 \neq F \neq E$, one has

$$\frac{\deg(F)}{\operatorname{rank}(F)} > \frac{\deg(E)}{\operatorname{rank}(E)}$$

(resp. if in this inequality $>$ is replaced by $\geq$). Prove that a locally free sheaf E of the form $E = E_1 \oplus E_2$, with E_i nonzero locally free sheaves, $i = 1, 2$, is never stable.

Exercise 11.4. Let E be a normalized locally free sheaf of rank 2 on a nonsingular projective curve B. Show that E is stable (resp. semistable) if and only if $\deg(E) > 0$ (resp. $\deg(E) \geq 0$).

Exercise 11.5. Let E be a normalized locally free sheaf of rank 2 on a nonsingular projective curve B of genus g such that $\deg(E) < 2 - 2g$. Show that $E \cong \mathcal{O}_B \oplus L$, with L an invertible sheaf on B with $\deg(L) = \deg(E)$. [*Hint:* Use Exercise 11.2 and Proposition 11.21.]

Exercise 11.6. Let E be a locally free sheaf of rank 2 on a nonsingular projective curve B and let $\pi : X \to B$ be the canonical projection of the geometrically ruled surface $X = \mathbb{P}(E)$. For every $n \geq 1$, show that there is a bijective correspondence between:

(a) the set of all effective divisors $D > 0$ on X with no fibers of π as components, of degree n over B and

(b) the set of all invertible subsheaves $L \subset S^n(E)$ such that $S^n(E)/L$ is locally free.

The correspondence is given by $D \to \pi_*(\mathcal{O}_X(n) \otimes \mathcal{O}_X(-D))$, and $L \to D_L$, where D_L is the subscheme of X defined by the homogeneous ideal $LS(E)$ of the symmetric $\mathcal{O}_B$-algebra $S(E)$. Show also that under this correspondence $(\mathcal{O}_X(1) \cdot D) = n \deg(E) - \deg(L)$. If $n = 1$, show that the normal bundle of D in X is $(E/L) \otimes L^{-1}$, and in particular, that $(D^2) = \deg(E) - 2 \deg(L)$.

Exercise 11.7. Let

$$0 \to \mathcal{O}_B \to E \to L \to 0$$

be a nonsplit exact sequence of locally free sheaves on a nonsingular projective curve B over $\Bbbk$, with char($\Bbbk$) $= 0$, L invertible, and $\deg(L) > 0$. Prove that $\mathcal{O}_{\mathbb{P}(E)}(1)$ is ample on the geometrically ruled surface $\mathbb{P}(E)$. [*Hint:* Use Proposition 11.19 to deduce that $(\mathcal{O}_{\mathbb{P}(E)}(1)\cdot\mathcal{O}_{\mathbb{P}(E)}(1)) > 0$. Let C be an integral curve on $\mathbb{P}(E)$. If C is a fiber of the canonical projection $\pi : \mathbb{P}(E) \to B$, then clearly $(\mathcal{O}_{\mathbb{P}(E)}(1)\cdot C) > 0$. If not, then the morphism $f = \pi|_C : C \to B$ is finite and surjective. Use char($\Bbbk$) $= 0$ to show that the exact sequence

$$0 \to \mathcal{O}_C \to f^*(E) \to f^*(L) \to 0$$

does not split. This will imply that $(\mathcal{O}_{\mathbb{P}(E)}(1) \cdot C) > 0$. Then conclude by the Nakai–Moishezon criterion.]

Exercise 11.8. Let C be an elliptic curve and let $X = C^{(2)}$ be the 2-fold symmetric product of C (char($\Bbbk$) $= 0$). Let $p : C \times C \to C$ be the morphism given by $p(x, y) = x + y$, $\forall x, y \in C$ (with the structure of Abelian variety on C by fixing a point c_0 as the origin). Prove that the morphism p yields a unique morphism $f : X \to C$ such that $p^{-1}(c) \cong \mathbb{P}^1$, $\forall c \in C$ (whence, by Corollary 11.11, $X \cong \mathbb{P}(E)$ for some rank two locally free $\mathcal{O}_C$-module E). [*Hint:* Show that $C \cong \{(x, c - x) \mid x \in C\} \to f^{-1}(c)$ is a finite morphism of degree 2 ramified at four points, and use Hurwitz's formula.] In this way, using Exercise 7.9, one gets an example of a geometrically ruled surface over an elliptic curve, which admits an elliptic fibration.

Bibliographic References. The proof of Theorem 11.6 (which is essential for the proof of the Noether–Tsen criterion) is taken from [Buc1]. Theorem 11.10 is a variant of the Noether–Tsen criterion; this variant, via its Corollary 11.11, makes the connection between geometrically ruled surfaces and rank 2 vector bundles on nonsingular projective curves. The facts regarding vector bundles on curves are presented after [Ati], and the simple proof of Grothendieck's theorem in the special case of vector bundles of rank 2, cf. Theorem 11.22, is from [Bea].

12
Minimal Models of Ruled Surfaces

12.1. Let B be a nonsingular projective curve, and let E be a locally free $\mathcal{O}_B$-module of rank 2. Let $X = \mathbb{P}(E)$ be the projective bundle associated with E, and let $f : X \to B$ be its canonical projection. Let $x \in X$ be a closed point on the fiber $F_b = f^{-1}(b)$, $b = f(x)$, and let $\pi : \tilde{X} \to X$ be the quadratic transformation of X with center x. Then the proper transform F' of F_b on $\tilde{X}$ has $p_a(F') = 0$ and $(F'^2) = -1$, because $p_a(F_b) = 0$ and $(F_b^2) = 0$. In other words, F' is an exceptional curve of the first kind. By (3.30), there exists a unique contraction $\pi' : \tilde{X} \to X'$ of F' to a nonsingular point; we shall denote this contraction by $\mathrm{cont}_{F'} : \tilde{X} \to X'$. As f is a morphism and F' is a component of a fiber of $f \circ \pi$, we get a commutative diagram:

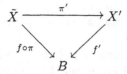

in which f' is a morphism. Moreover, all the fibers of f' are integral curves of arithmetic genus zero, and therefore $f' : X' \to B$ is a geometrically ruled surface. The birational map $\pi' \circ \pi^{-1} : X \dashrightarrow X'$ is called the *elementary transformation of the geometrically ruled surface* $f : X \to B$ *with center the point* x; we shall denote this elementary transformation by $\mathrm{elem}_x : X \dashrightarrow X'$.

Theorem 12.2. *Let B be a nonsingular projective curve with $p_a(B) > 0$. Then the following hold:*

(a) *If $f : X \to B$ is a geometrically ruled surface with base B, then X is a minimal model.*

(b) *For every minimal model X of the field $\Bbbk(B)(t)$, with t an independent variable over the rational function field $\Bbbk(B)$, there exists a morphism $f : X \to B$ such that X is a geometrically ruled surface with base B via the morphism f.*

(c) *Every birational map between two minimal models X, X' of $\Bbbk(B)(t)$ is a composite of an automorphism of X and a finite number of elementary transformations like the one described in (12.1).*

Proof. (a) By contradiction, assume that X contains an exceptional curve C of the first kind. Since C is a rational curve and $p_a(B) > 0$ (and B is nonsingular), C must be contained in a fiber of f. Since all the fibers of f are integral curves, C must in fact coincide with a fiber of f. But then $(C^2) = 0$, and therefore C cannot be an exceptional curve of the first kind.

Parts (b) and (c) of Theorem 12.2 follow from the next lemma.

Lemma 12.3. *Let E' be a locally free sheaf of rank 2 on the nonsingular projective curve B with $p_a(B) > 0$, X a surface, and $f' : X \dashrightarrow \mathbb{P}(E')$ a birational map. Then there exists a finite product $g : \mathbb{P}(E') \dashrightarrow \mathbb{P}(E)$ of elementary transformations such that $g \circ f'$ is a morphism.*

Proof. Let $g : \mathbb{P}(E') \dashrightarrow \mathbb{P}(E)$ be an arbitrary finite product of elementary transformations, $f = g \circ f'$, and $\lambda = \lambda(f)$ the minimal number of quadratic transformations of X needed to get a surface Y that dominates X (via the product π of those quadratic transformations) such that $f \circ \pi$ is a morphism.

Now choose g such that $\lambda = \lambda(f)$ is the smallest possible. We'll show that $\lambda = 0$, and therefore this f is a morphism.

Assume by contradiction that $\lambda > 0$. Then Y contains an exceptional curve C of the first kind, such that the surface $Y' = \mathrm{cont}_C(Y)$ dominates X. If the image of C in $\mathbb{P}(E)$ were a point, then the map from Y' to $\mathbb{P}(E)$ would be a morphism, and this would contradict the minimality of λ. Therefore the image D of C in $\mathbb{P}(E)$ is a curve. Since C is a rational curve, it follows that D is also a rational curve; and since $p_a(B) > 0$, D must be a fiber of $\mathbb{P}(E)$. Thus $(D^2) = 0$.

If the inverse of the birational morphism $Y \to \mathbb{P}(E)$ had no fundamental points on D, then this morphism would be a biregular isomorphism of an open neighborhood of C onto an open neighborhood of D; but this is impossible, because $(C^2) = -1$, while $(D^2) = 0$. Therefore there exists a point x on D that is a fundamental point of the inverse of the morphism $Y \to \mathbb{P}(E)$. Then Y dominates the surface $\mathrm{bl}_x(\mathbb{P}(E))$ obtained by blowing up the point x on $\mathbb{P}(E)$, and therefore Y dominates $\mathrm{elem}_x(\mathbb{P}(E))$. Since C is mapped to a point in $\mathrm{elem}_x(\mathbb{P}(E))$, we see that Y' maps regularly to $\mathrm{elem}_x(\mathbb{P}(E))$, and therefore $\lambda(\mathrm{elem}_x \circ g \circ f') < \lambda(g \circ f') = \lambda$, thus contradicting the choice of g. $\qquad\square$

Definition 12.4. A surface X is a *rational* surface if it is birationally isomorphic to the projective plane $\mathbb{P}^2$.

Since $\mathbb{P}^2$ is birationally isomorphic to $\mathbb{P}^1 \times \mathbb{P}^1$, every rational surface is a ruled surface with $q = 0$. Conversely, let X be a ruled surface with $q = 0$. Then X is birationally isomorphic to $\mathbb{P}^1 \times B$ with B a smooth projective curve. Moreover, $p_a(B) = q = 0$, by Proposition 11.19(a), so that $B \cong \mathbb{P}^1$, and therefore X is a rational surface.

12.5. Now we investigate the geometrically ruled surfaces with base $B = \mathbb{P}^1$.

By Theorem 11.22, every locally free $\mathcal{O}_{\mathbb{P}^1}$-module E of rank 2 is of the form $\mathcal{O}_{\mathbb{P}^1}(a) \oplus \mathcal{O}_{\mathbb{P}^1}(b)$ for suitable integers $a \le b$. Thus

$$
\begin{aligned}
E &= \mathcal{O}_{\mathbb{P}^1}(a) \oplus \mathcal{O}_{\mathbb{P}^1}(b) \\
&= \mathcal{O}_{\mathbb{P}^1}(a) \otimes \left(\mathcal{O}_{\mathbb{P}^1} \oplus \mathcal{O}_{\mathbb{P}^1}(b-a)\right) \\
&= L \otimes \left(\mathcal{O}_{\mathbb{P}^1} \oplus \mathcal{O}_{\mathbb{P}^1}(n)\right),
\end{aligned}
$$

with L an invertible $\mathcal{O}_{\mathbb{P}^1}$-module and $n \ge 0$ an integer. Then $\mathbb{P}(E) \cong \mathbb{P}(\mathcal{O}_{\mathbb{P}^1} \oplus \mathcal{O}_{\mathbb{P}^1}(n))$, by Corollary 11.11.

Let $\mathbb{F}_n$ denote the geometrically ruled surface $\mathbb{P}(\mathcal{O}_{\mathbb{P}^1} \oplus \mathcal{O}_{\mathbb{P}^1}(n))$, with base $\mathbb{P}^1$, where $n \ge 0$. Let $f_n : \mathbb{F}_n \to \mathbb{P}^1$ denote the canonical projection. Then $\mathbb{F}_n$ can be interpreted as the compactification of the vector bundle $\mathbb{V}(\mathcal{O}_{\mathbb{P}^1}(n)) \to \mathbb{P}^1$ associated to the invertible $\mathcal{O}_{\mathbb{P}^1}$-module $\mathcal{O}_{\mathbb{P}^1}(n)$ (compactifying each fiber of $\mathbb{V}(\mathcal{O}_{\mathbb{P}^1}(n))$ with the point at infinity). Then we have two sections $\sigma_0 : \mathbb{P}^1 \to \mathbb{F}_n$ and $\sigma_\infty : \mathbb{P}^1 \to \mathbb{F}_n$ of f_n, namely, the zero section and the section at infinity. The conormal bundle of $D_0 = \sigma_0(\mathbb{P}^1)$ in $\mathbb{F}_n$ is isomorphic to $\mathcal{O}_{\mathbb{P}^1}(n)$, and the conormal bundle of $D = \sigma_\infty(\mathbb{P}^1)$ in $\mathbb{F}_n$ is isomorphic to $\mathcal{O}_{\mathbb{P}^1}(-n)$. Thus

$$
(D_0^2) = -n \quad \text{and} \quad (D^2) = n.
$$

By (11.14) every divisor Δ on $\mathbb{F}_n$ is numerically equivalent to $aD + bF$, where (a, b) is a uniquely determined pair of integers and F is a closed fiber of f_n. Then

$$
(\Delta^2) = a^2(D^2) + 2ab(D \cdot F) = a^2 n + 2ab,
$$

for $(D \cdot F) = 1$ and $(F^2) = 0$. If Δ is an effective divisor, then $(\Delta \cdot F) \ge 0$. But $(\Delta \cdot F) = a(D \cdot F) + b(F^2) = a$. Hence if Δ is effective and $\Delta \equiv aD + bF$, then $a \ge 0$.

Let Δ be an effective divisor, numerically equivalent to $aD + bF$. From $p_a(D) = p_a(F) = 0$, $(D^2) = n$ and $(F^2) = 0$ we get $(K \cdot D) = -2 - (D^2) = -2 - n$ and $(K \cdot F) = -2$. Therefore $(K \cdot \Delta) = a(-2 - n) - 2b$, and

$$
\begin{aligned}
2p_a(\Delta) - 2 &= (\Delta^2) + (K \cdot \Delta) \\
&= a^2 n + 2ab - 2a - an - 2b.
\end{aligned}
$$

Since $\mathrm{Supp}(D_0) \cap \mathrm{Supp}(D) = \varnothing$, we have $(D_0 \cdot D) = 0$. Let $D_0 \equiv aD + bF$. Then $a = 1$ and $b = -n$. Indeed, $a = (D_0 \cdot F) = 1$, and on the other hand $0 = (D_0 \cdot D) = (D^2) + b(D \cdot F) = n + b$, whence $b = -n$. We have thus proved that

$$D_0 \equiv D - nF.$$

If Δ is an effective divisor that does not contain D_0 as a component, and $\Delta \equiv aD + bF$, then $b \geq 0$ as well. Indeed, in this case we have

$$0 \leq (\Delta \cdot D_0) = a(D \cdot D_0) + b(F \cdot D_0) = b.$$

Therefore if Δ is an integral curve on $\mathbb{F}_n$, we cannot have $(\Delta^2) < 0$ unless $n > 0$ and $\Delta = D_0$. Indeed, if $\Delta \neq D_0$, then $\Delta \equiv aD + bF$ with $a \geq 0$ and $b \geq 0$, and therefore $(\Delta^2) = a^2 n + 2ab \geq 0$; and for $\Delta = D_0$ we have $(D_0^2) = -n$. Since $p_a(D_0) = 0$, we have proved the following lemma:

Lemma 12.6. *The surfaces $\mathbb{F}_n$ are minimal models of $\mathbb{P}^2$ for all $n \geq 0, n \neq 1$, and if $n > 0$ then the only integral curve Δ on $\mathbb{F}_n$ such that $(\Delta^2) < 0$ is D_0. The surface $\mathbb{F}_1$ contains exactly one exceptional curve of the first kind, D_0, and $\mathrm{cont}_{D_0}(\mathbb{F}_1)$ is biregularly isomorphic to $\mathbb{P}^2$. Finally, the surface $\mathbb{F}_0$ is biregularly isomorphic to $\mathbb{P}^1 \times \mathbb{P}^1$.* $\quad\square$

So far we have used the following notation: if E is an exceptional curve of the first kind on a surface X, then $\mathrm{cont}_E(X)$ denotes the surface obtained from X by contracting E to a nonsingular point, cf. (3.30); and if $x \in X$ is a closed point on a surface X, then $\mathrm{bl}_x(X)$ denotes the surface obtained by blowing up the point x on X. If E is an exceptional curve of the first kind on X, and if E contracts to the point $y \in Y = \mathrm{cont}_E(X)$, then the birational maps $\mathrm{cont}_E : X \to Y$ and $\mathrm{bl}_y : Y \dashrightarrow X$ are inverse to each other. We have also introduced the elementary transformation elem_x, which transforms a geometrically ruled surface with base B into another geometrically ruled surface with base B. elem_x is called the elementary transformation with center x. Finally, we shall denote by $\mathrm{refl} : \mathbb{F}_0 = \mathbb{P}^1 \times \mathbb{P}^1 \to \mathbb{F}_0$ the reflection morphism $(x, y) \mapsto (y, x)$.

Now let $x \in \mathbb{F}_0$. It is very easy to see that $\mathrm{elem}_x(\mathbb{F}_0) \cong \mathbb{F}_1$. For $n \geq 1$, let $x \in \mathbb{F}_n$, and let F be the fiber of f through x. We distinguish two cases:

(a) $x \in F \cap D_0$. Then $\mathrm{elem}_x(\mathbb{F}_n) = \mathbb{F}_{n+1}$. Indeed, using the notation introduced in Figure 12.1, we have:
$(D_0'^2) = -n - 1, (D')^2 = n, (D_0''^2) = -n - 1, (D''^2) = n + 1$, and since $\mathrm{elem}_x(\mathbb{F}_n)$ is one of the surfaces $\mathbb{F}_m$, we must have $m = n + 1$.

(b) $x \notin F \cap D_0$. Without loss of generality, we may assume that $x \in D$ (because we can change the point at infinity on $\mathbb{P}^1$ while keeping the point $0 \in \mathbb{P}^1$ unchanged). Then $\mathrm{elem}_x(\mathbb{F}_n) = \mathbb{F}_{n-1}$. Indeed, using the notation from Figure 12.2, we have:
$(D'^2) = n - 1, (D_0'^2) = -n, (D''^2) = n - 1, (D_0''^2) = -n + 1$. Since $\mathrm{elem}_x(\mathbb{F}_n)$ is one of the surfaces $\mathbb{F}_m$, we must have $m = n - 1$.

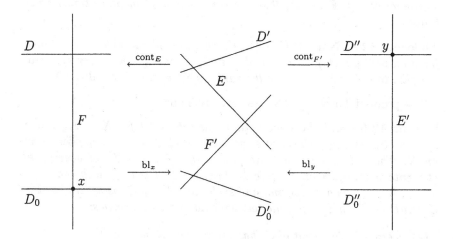

FIGURE 12.1.

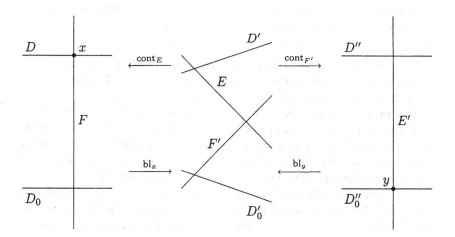

FIGURE 12.2.

Thus we have proved the following proposition:

Proposition 12.7. *If $m \neq n$ are nonnegative integers, then there exists a birational map $g : \mathbb{F}_m \to \mathbb{F}_n$ that is a composite of elementary transformations.*

Theorem 12.8 (Nagata [Nag]). *Every minimal model of the rational function field $\Bbbk(t_1, t_2)$, with t_1 and t_2 independent variables over $\Bbbk$, is isomorphic to either $\mathbb{P}^2$ or one of the surfaces $\mathbb{F}_n$ with $n \geq 0$ and $n \neq 1$.*

The proof of Theorem 12.8 uses several lemmas.

Lemma 12.9. *With the same notation as before, let $f_0 : X \to \mathbb{F}_{n_0}$ ($n_0 \geq 0$) be a birational morphism of surfaces, and let C be a nonsingular curve on X. Then there exists a finite product $g : \mathbb{F}_{n_0} \dashrightarrow \mathbb{F}_n$ of generalized elementary transformations (see Definition 12.10 below), such that $f = g \circ f_0 : X \to \mathbb{F}_n$ is a morphism, and such that $C' = f(C)$ is a curve on $\mathbb{F}_n$, $C' \equiv aD + bF$ on $\mathbb{F}_n$, with the following additional properties:*

(a) *every singular point of C' has multiplicity at most $a/2$;*

(b) *if $n = 0$, then $a \leq b$;*

(c) *if $n = 1$, every singular point of C' has multiplicity at most b;*

(d) *if $a = 1$, then $(C'^2) = (C^2)$; and*

(e) *if $n = 1$, $a = 2$ and $b = 0$, then $(C'^2) = (C^2)$.*

Definition 12.10. A *generalized elementary transformation* is one of the following birational transformations: elem_x, $\text{refl} : \mathbb{F}_0 \to \mathbb{F}_0$, and $\text{int}_x : \mathbb{F}_1 \dashrightarrow \mathbb{F}_1$, where $x \notin D_0$ and $\text{int}_x = \text{cont}_{D_0} \circ \text{bl}_x$.

Proof of Lemma 12.9. If $f_0(C)$ is a point $x \in \mathbb{F}_{n_0}$, then f_0 dominates $\text{bl}_x(\mathbb{F}_{n_0})$, and therefore it dominates $\text{elem}_x(\mathbb{F}_{n_0})$ as well. If $f_1(C)$ is still a point, where $f_1 = \text{elem}_x \circ f_0$, then repeating the same procedure we eventually find a finite product $g : \mathbb{F}_{n_0} \dashrightarrow \mathbb{F}_n$ of elementary transformations such that $g \circ f_0$ is a morphism and $(g \circ f_0)(C)$ is a curve on $\mathbb{F}_n$. (By contradiction, if $(g \circ f_0)(C)$ were a point for every finite product g of elementary transformations, then X would dominate an arbitrarily large number of consecutive blow-ups of points, contradicting Theorem 6.2.)

From among all the products g such that $g \circ f_0$ is a morphism and $(g \circ f_0)(C)$ is a curve, choose a g such that $a = a((g \circ f_0)(C))$ is minimal, where $(g \circ f_0)(C) \equiv aD + bF$. Such a g exists because the coefficient a is always a nonnegative integer. Furthermore, from among all the g that minimize $a((g \circ f_0)(C))$, choose one that minimizes the self-intersection of $(g \circ f_0)(C)$. This is possible because $((g \circ f_0)(C)^2) \geq (C^2)$ for every such g, for $g \circ f_0$ is a birational morphism. With g so chosen, we shall now prove that g has all the required properties. Put $f = g \circ f_0 : X \to \mathbb{F}_n$.

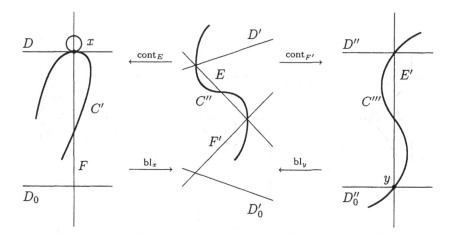

FIGURE 12.3.

(a) If C' has a singularity of multiplicity $r > a/2$ at a point x, then compose g to the left with elem_x. As C is nonsingular and C' is singular at x, f dominates $\text{bl}_x(\mathbb{F}_n)$, so that $\text{elem}_x \circ f$ is still a morphism. The situation is illustrated in Figure 12.3.

Computing the intersection numbers, cf. [Sha1, part I, ch. IV, §3], we have:

$$(C''^2) = (C'^2) - r^2 = (C'''^2) - r'^2,$$

where r' is the multiplicity of C''' at y (note that r' may be zero). Since $r = (E \cdot C'')$, $r' = (F' \cdot C'')$, and $(\text{cont}_E)^*(F) = E + F'$, we have by the projection formula:

$$r + r' = (E + F' \cdot C'') = (F \cdot C') = a,$$

and therefore $r' < r$, for $r > a/2$. Thus from the equalities above we have $(C''^2) < (C'^2)$, contradicting the minimality of (C'^2); note that $a(C''') = (E \cdot C''') = (E + F' \cdot C'') = a = a(C')$.

(b) If $n = 0$, note that the coefficients a and b can be switched by considering $\text{refl} \circ f$; therefore $a \leq b$ by the minimality of a.

(c) If $n = 1$, let x be a singular point of C', with multiplicity r. If $x \in D_0$, then $r \leq (C' \cdot D_0) = (aD + bF \cdot D_0) = b$, as required. If, however, $x \notin D_0$, consider $f_1 = \text{int}_x \circ f$. Let F be the fiber of $\mathbb{F}_1 \to \mathbb{P}^1$ through x, and let $y = D_0 \cap F$. The situation is illustrated in Figure 12.4.

We have: $(C' \cdot F) = a = a(C')$, $(C' \cdot D_0) = b$; $(C'' \cdot E) = r$, $(C'' \cdot E + F') = (C' \cdot F) = a$, and therefore $(C'' \cdot F') = a - r$; $(C'' \cdot D_0') = (C' \cdot D_0) = b$; and finally, $a' = a(C''') = (C''' \cdot F'') = (C'' \cdot F' + E) = a - r + b$. Since $a' \geq a$ by the minimality of a, we get that $r \leq b$, as required.

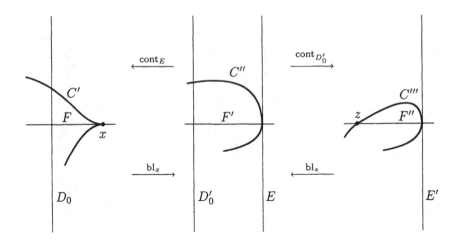

FIGURE 12.4.

(d) If $a = 1$, assume by contradiction that $(C^2) < (C'^2)$. Then there exists a point $x \in C'$ where f^{-1} is not defined. Applying the elementary transformation elem_x, we find that $f_1 = \mathrm{elem}_x \circ f$ is a morphism, $a_1 = a$, and $(C_1^2) = (C'^2) - 1$, contradicting the minimality of (C'^2).

(e) If $n = 1$, $a = 2$ and $b = 0$, assume by contradiction that $(C^2) < (C'^2)$; then there exists a point $x \in C'$ where f^{-1} is not defined. The point x cannot be on D_0, for $b = (C' \cdot D_0) = 0$ (and therefore C' does not intersect D_0). Applying int_x, we find that $a' = a + b - 1 = 1$ (because the curve C' is nonsingular by part (c) of the lemma), and this contradicts the minimality of a.   □

Lemma 12.11. *Let C be an irreducible curve on X, and let $f : X \to \mathbb{F}_n$ be a birational morphism that satisfies the properties* (a), (b), (c), (d), *and* (e) *in Lemma 12.9. Then:*

(i) *If $a = 2$, then C is nonsingular of genus g and $(C^2) \leq 4g + 4$.*

(ii) *If $a = 3$, then C is nonsingular of genus g and $(C^2) \leq 3g + 6$.*

(iii) *If $a \geq 4$ and $n \neq 1$, or if $a \geq 4, n = 1$ and $b \geq a/2$, then $(C^2) \leq \frac{2a}{a-2} \cdot (p_a(C) - 1)$.*

(iv) *If $a \geq 4, n = 1$ and $b < a/2$, then $(C^2) < \frac{2k}{k-3} \cdot (p_a(C) - 1)$, where $k = a + b$.*

Proof. First we prove the lemma for $X = \mathbb{F}_n$ and $f = \mathrm{id}_{\mathbb{F}_n}$. In view of (12.5), we have:

$$(C^2) = a^2 n + 2ab$$

and

$$2p_a(C) - 2 = a^2n + 2ab - 2a - an - 2b.$$

(i) If $a = 2$, then C is nonsingular by (a), and we have $(C^2) = 4n + 4b$ and $2p_a(C) - 2 = 2n + 2b - 4$, and therefore $(C^2) = 4p_a(C) + 4 = 4g + 4$.

(ii) If $a = 3$, then C is nonsingular by (a), and we have $(C^2) = 9n + 6b$ and $2p_a(C) - 2 = 6n + 4b - 6$, and therefore $(C^2) = 3p_a(C) + 6 = 3g + 6$.

(iii) In view of the explicit formulas for (C^2) and $p_a(C)$, the inequality $(C^2) \leq \frac{2a}{a-2} \cdot (p_a(C) - 1)$ for $a > 2$ is equivalent (by trivial algebraic manipulation) to $2a \leq an + 2b$. For $n = 0$ this inequality is true by part (b) of Lemma 12.9. For $n = 1$ the inequality is true exactly when $a \leq 2b$, that is, when $b \geq a/2$. Finally, for $n \geq 2$ the inequality is obviously true for all $a \geq 0$ and $b \geq 0$. But a is always nonnegative, and $b > 0$ except for the curve D_0; but our curve C is not equal to D_0, because $a(D_0) = 1$, while $a(C) = a \geq 4$.

(iv) Note that $k = a + b > 3$. Since $n = 1$, we have $(C^2) = a^2 + 2ab$ and $2p_a(C) - 2 = a^2 + 2ab - 3a - 2b$. The inequality $(C^2) < \frac{2k}{k-3} \cdot (p_a(C) - 1)$ is therefore equivalent (by trivial algebraic manipulation) to $2b^2 < ab$. This inequality, in turn, follows from $b > 0$ (which holds for the same reason as in (iii)) and $b < a/2$.

In the general case, since f is a birational morphism, it is a finite product of quadratic transformations. We shall prove the various statements in Lemma 12.11 by induction on the number of quadratic transformations in the decomposition of f. Thus the proof is reduced to showing that, if $X = \mathrm{bl}_x(X')$ with $x \in X'$, D is the image of C by $(\mathrm{bl}_x)^{-1}$, and the lemma holds for D, then it holds for C as well.

(i) and (ii) Since D is nonsingular, we have that C is also nonsingular, having the same genus as D, and $(C^2) \leq (D^2)$.

(iii) Let $r \geq 0$ be the multiplicity of D at x; $r = 0$ means that $x \notin D$. Then

$$(C^2) = (D^2) - r^2 \quad \text{and} \quad p_a(C) = p_a(D) - \tfrac{1}{2}r(r-1),$$

and therefore

$$(C^2) \leq \frac{2a}{a-2} \cdot (p_a(C) - 1) + \frac{a}{a-2}r(r-1) - r^2.$$

But $r \leq a/2$ by condition (a), and therefore

$$\frac{a}{a-2}r(r-1) - r^2 = \frac{r(2r-a)}{a-2} \leq 0,$$

and therefore the desired inequality holds for C.

(iv) We have

$$(C^2) = (D^2) - r^2 < \frac{2k}{k-3}(p_a(D) - 1) - r^2$$
$$= \frac{2k}{k-3}(p_a(C) - 1) + \frac{k}{k-3}r(r-1) - r^2.$$

But $r \le b$ by condition (c), and therefore $r \le k/3$ and

$$\frac{k}{k-3}r(r-1) - r^2 = \frac{r(3r-k)}{k-3} \le 0,$$

and this proves the desired inequality for C.   □

To complete the proof of Theorem 12.8, it suffices to prove the following lemma.

Lemma 12.12. *Let X be a surface, and let $f : X \dashrightarrow \mathbb{F}_{n_0}$ be a birational map. Then either there exists a finite product $g : \mathbb{F}_{n_0} \dashrightarrow \mathbb{F}_n$ of generalized elementary transformations such that $f = g \circ f_0$ is a morphism, or $X \cong \mathbb{P}^2$ and there exists a finite product $g : \mathbb{F}_{n_0} \dashrightarrow \mathbb{F}_1$ of generalized elementary transformations such that, if $f = g \circ f_0$, then $f = \mathrm{bl}_x \circ h$, with $x \in X$ and h a suitable biregular automorphism of X.*

Proof. Consider all the birational maps $f = g \circ f_0$, with $g : \mathbb{F}_{n_0} \to \mathbb{F}_n$ a finite product of generalized elementary transformations. For every such f, let $\lambda(f)$ be the smallest number of blow-ups of X needed to obtain a surface that dominates $\mathbb{F}_n$. Choose g such that $\lambda(f)$ is as small as possible. Then we'll show that either $\lambda(f) = 0$ (i.e., that f is a morphism), or $\lambda(f) = 1$, $X = \mathbb{P}^2$, and $n = 1$.

By way of contradiction, assume that $\lambda(f) > 0$. Let Y be a surface obtained by blowing up X $\lambda(f)$ times and such that Y dominates $\mathbb{F}_n$. Let C be an exceptional curve of the first kind on Y such that $\mathrm{cont}_C(Y)$ dominates X. By Lemma 12.9 we may assume that the morphism $\phi : Y \to \mathbb{F}_n$ obtained by composing f to the right with the $\lambda(f)$ blow-ups satisfies the properties (a), (b), (c), (d), and (e). Let $C' = \phi(C)$.

If $a = 0$, then $C' \equiv bF$ with $b > 0$, and C' is nonsingular and $(C'^2) = 0$. Since $(C^2) = -1$, there exists a point $x \in C'$ where ϕ^{-1} is not defined. Then $\mathrm{cont}_C(Y)$ dominates $\mathrm{elem}_x(\mathbb{F}_n)$, contradicting the minimality of $\lambda(f)$.

If $a = 1$, then $(C^2) = (C'^2) = -1$ by property (d), and therefore we must have $n = 1$ and $C' = D_0$, by Lemma 12.6. If ϕ is an isomorphism, then $X = \mathbb{P}^2$ and $\lambda(f) = 1$. If ϕ is not an isomorphism, then there exists a point $x \in \mathbb{F}_1 \setminus D_0$ such that $\mathrm{cont}_C(Y)$ dominates $\mathrm{int}_x(\mathbb{F}_1)$, which again contradicts the minimality of $\lambda(f)$.

If $a = 2$, then C' is nonsingular by condition (a); and since C is a rational curve, we have $p_a(C') = 0$. From the formula expressing $p_a(C)$ in terms of a, b and n, we have that $b + n = 1$; then either $n = 0$ and $b = 1$, which is impossible by condition (b), or $n = 1$ and $b = 0$. But in the latter case $(C'^2) = 4 > 0$, and we reach a contradiction with condition (e).

If $a = 3$, we have $p_a(C') = 0$, just as in the previous case, and then $3n + 2b = 2$; thus $n = 0$ and $b = 1$, but this contradicts (b).

Finally, if $a \geq 4$, then we are either in case (iii) or (iv) of Lemma 12.11. Furthermore, here we have $p_a(C) = 0$. Therefore in case (iii) of Lemma 12.11 we get $(C^2) \leq -\frac{2a}{a-2}$, and in case (iv) we get $(C^2) \leq -\frac{2k}{k-3}$. In particular, in both cases we get that $(C^2) < -1$, which is a contradiction.

$\square$

This concludes the proof of Theorem 12.8.

EXERCISES

Exercise 12.1. On the surface $\mathbb{F}_n$ $(n \geq 0)$ consider a divisor $\Delta \in \mathrm{Div}(\mathbb{F}_n)$ that is linearly equivalent to $aD_0 + bF$, with $a, b \geq 0$. Show that Δ is ample if and only if $a > 0$ and $b > an$. (Use the Nakai–Moishezon criterion.) If $\Delta \sim D_0 + bF$, show that $|\Delta|$ has no base points if and only if $b \geq n$.

Exercise 12.2. Let Δ be a divisor on $\mathbb{F}_n$ such that $\Delta \sim D_0 + bF$ with $b > n$. Prove that Δ is very ample. Then, more generally, prove that Δ is very ample if $\Delta \sim aD_0 + bF$ with $a > 0$ and $b > an$. In particular, every ample divisor on $\mathbb{F}_n$ is very ample.

Exercise 12.3. Prove that the linear system $|D_0 + nF|$ on $\mathbb{F}_n$ has no base points and defines a morphism $\phi : \mathbb{F}_n \to \mathbb{P}^{n+1}$ that is an embedding off D_0 and blows D_0 down to a point y. Show that $\phi(\mathbb{F}_n)$ is the cone in $\mathbb{P}^{n+1}$ with vertex y over the normal rational curve of degree n in $\mathbb{P}^n$.

Exercise 12.4. Let X be a nondegenerate closed irreducible subvariety of positive dimension of $\mathbb{P}^N$. Prove the following inequality:

$$\deg(X) \geq \mathrm{codim}_{\mathbb{P}^N}(X) + 1.$$

If we have equality in this inequality, we say that X is of *minimal degree* in $\mathbb{P}^N$.

Exercise 12.5. Let X be a (nonsingular projective) surface and let Y be a nonsingular rational curve on X that is an ample divisor on X. Prove that X is one of the following:

(i) $X \cong \mathbb{P}^2$ and Y is a line or a conic in $\mathbb{P}^2$, or

(ii) $X \cong \mathbb{F}_n$ for some $n \geq 0$ and Y is embedded in X as a section of the canonical projection $\mathbb{F}_n \to \mathbb{P}^1$.

If X is assumed to be only normal and projective and Y is an ample Cartier divisor on X, show that besides these two possibilities, there is also the following:

(iii) X is isomorphic to the cone in $\mathbb{P}^{n+1}$ over the normal rational curve of degree n in $\mathbb{P}^n$ $(n \geq 2)$ and Y is the intersection of X with the hyperplane at infinity, $H = \mathbb{P}^n$.

Exercise 12.6. Let $Y' \hookrightarrow X$ and $Y' \hookrightarrow X'$ be inclusions of curves Y and Y' into surfaces X and X', respectively. We say that they are *Zariski equivalent* if there exist two Zariski open subsets $U \subset X$ and $U' \subset X'$ containing Y and Y', respectively, and an isomorphism $\phi : U \cong U'$ such that $\phi(Y) = Y'$.

Let Y be a nonsingular rational (irreducible) curve on a surface X, with $(Y^2) > 0$. Prove that $Y \hookrightarrow X$ is Zariski equivalent to one of the embeddings of case (i), (ii), or (iii) of the preceding exercise. [*Hint:* Use Exercises 1.8 and 12.5.]

Exercise 12.7. Let Y be an elliptic curve that is an ample divisor on a (nonsingular projective) surface X over $\Bbbk$, with char($\Bbbk$) = 0. Prove that either $-Y$ is a canonical divisor on X (whence ω_X^{-1} is ample; the surfaces X with ω_X^{-1} ample are called *Del Pezzo surfaces*), or $X \cong \mathbb{P}(E)$ for some rank 2 vector bundle E on Y, and Y is embedded in X as a section of the canonical projection $\mathbb{P}(E) \to Y$.

Exercise 12.8. Consider the very ample linear system $|D_0 + bF|$ on the surface $\mathbb{F}_n$, with $b \geq n+1$ (cf. Exercise 12.2), and let $\phi = \phi_{|D_0+bF|} : X \hookrightarrow \mathbb{P}(H^0(\mathbb{F}_n, \mathcal{O}_{\mathbb{F}_n}(D_0+bF))^\vee) = \mathbb{P}^N$, with $N = 2b-n+1$, be the corresponding embedding. Show that for every fiber F of the ruled fibration $\mathbb{F}_n \to \mathbb{P}^1$, $\phi(F)$ is a straight line in $\mathbb{P}^N$ (i.e., $X := \phi(\mathbb{F}_n)$ is a scroll in $\mathbb{P}^N$ in the classical terminology), and the degree of X in $\mathbb{P}^N$ is $N - 1 = 2b - n$. (In other words, X is a surface of minimal degree in $\mathbb{P}^N$ in the sense of Exercise 12.4.)

Exercise 12.9. Conversely, prove the following theorem of Del Pezzo: let X be a nonsingular nondegenerate surface in $\mathbb{P}^N$ $(X \neq \mathbb{P}^N)$ of minimal degree $N - 1$. Then, up to projective equivalence, X is either the scroll described in Exercise 12.8 (for some $n \geq 0$ and $b \geq n+1$), or the Veronese surface in $\mathbb{P}^5$ (i.e., the image of the 2-fold Veronese embedding $\mathbb{P}^2 \hookrightarrow \mathbb{P}^5$). [*Hint:* Use the fact that a general hyperplane section of X is a rational normal curve of degree $N - 1$ in $\mathbb{P}^{N-1}$, and then, Exercise 12.5.]

Exercise 12.10. Let C be a curve on a nonsingular projective surface X.

(i) If $X \setminus C$ is isomorphic to the affine plane $\mathbb{A}^2$ and C is irreducible, prove that $X \cong \mathbb{P}^2$ and C is a line.

(ii) Prove that there are no pairs (X, C) with C irreducible and $X \setminus C$ isomorphic to the affine plane $\mathbb{A}^2$ minus the two coordinate axes.

(iii) Classify all nonsingular projective surfaces X containing a curve C with two irreducible components, such that $X \setminus C \cong \mathbb{A}^2$.

[*Hint:* Use Theorem 12.8.]

We emphasize that by the term *isomorphism* we mean isomorphism of algebraic varieties. For example, the problem over $\mathbb{C}$ becomes a lot more difficult if $X \setminus C$ is only assumed to be analytically isomorphic to $\mathbb{A}^2$, or to $\mathbb{A}^2$ minus the two coordinate axes; see, for example, [Har1, p.231, example 3.2].

Bibliographic References. This chapter is based on [Har4]. Although Theorems 12.2 and 12.8 will be proved again in the next chapter by a different argument (via the classification of surfaces), we wanted to give also the direct and elementary proofs included in this chapter.

13

Characterization of Ruled and Rational Surfaces

In view of (9.3) and the proof of (8.3)(a), we have the following.

Proposition 13.1. *If X is a minimal surface, then the following conditions are equivalent:*

(a) *There exists an integral curve C on X such that $(K \cdot C) < 0$; in other words, X belongs to class (a), cf. (8.2).*

(b) *Adjunction terminates on X; that is, for every divisor $D \in \mathrm{Div}(X)$ there exists an integer n_D such that $|D + nK| = \varnothing$ for all $n \geq n_D$.*

(c) *$\kappa(X) = -\infty$, that is, $p_n = 0$ for all $n \geq 1$.*

(d) *$p_{12} = 0$.*

This chapter is devoted to the study of minimal surfaces X satisfying the equivalent conditions (a), (b), (c), and (d) in Proposition 13.1. We have already seen examples of such surfaces, namely, the minimal ruled surfaces, cf. Proposition 11.15.

The main results proved in this chapter are the following two theorems:

Theorem 13.2 (Enriques). *Let X be a minimal surface. Then X is a ruled surface if and only if it satisfies the equivalent conditions (a), (b), (c), and (d) in Proposition 13.1.*

Theorem 13.3 (Castelnuovo). *Let X be a minimal surface. Then X is a rational surface if and only if $q = p_2 = 0$.*

The study of ruled surfaces is closely related to the study of families of rational curves. If D is an effective divisor on X, then the isomorphism $\mathcal{O}_X(K+D) \cong \omega_D$ suggests that the effective divisors D with the property that $|K+D| = \varnothing$ must be related to families of rational curves on X, because an integral curve D on X is rational if and only if $|\omega_D| = \varnothing$. On a minimal surface X that satisfies the equivalent conditions (a), (b), (c), and (d) in Proposition 13.1, the existence of effective divisors D such that $|K+D| = \varnothing$ is guaranteed by condition (b). We begin by studying these divisors.

Proposition 13.4. *Let X be a minimal surface, and let D be an effective divisor on X such that $|K+D| = \varnothing$. Then $p_g = 0$ and the natural morphism $\underline{\mathrm{Pic}}^0(X) \to \underline{\mathrm{Pic}}^0(D)$ is surjective, and $\underline{\mathrm{Pic}}^0(D)$ is an Abelian variety.*

Proof. Since $|K| \subseteq |K+D| = \varnothing$ (D is an effective divisor), we have $p_g = 0$. On the other hand, the condition $|K+D| = \varnothing$ is equivalent, by Serre duality, to the vanishing $H^2(\mathcal{O}_X(-D)) = 0$; from the exact sequence

$$0 \to \mathcal{O}_X(-D) \to \mathcal{O}_X \to \mathcal{O}_D \to 0$$

we get that the natural homomorphism $H^1(\mathcal{O}_X) \to H^1(\mathcal{O}_D)$ is surjective. Since $p_g = 0$, the Picard scheme $\underline{\mathrm{Pic}}^0(X)$ is reduced, cf. [Mum1, Lecture 27], so that $\underline{\mathrm{Pic}}^0(X)$ is an Abelian variety. Since $H^1(\mathcal{O}_X)$ (resp. $H^1(\mathcal{O}_D)$) is canonically identified with the tangent space to $\underline{\mathrm{Pic}}^0(X)$ (resp. to $\underline{\mathrm{Pic}}^0(D)$) at the origin, we see that $\underline{\mathrm{Pic}}^0(X) \to \underline{\mathrm{Pic}}^0(D)$ is surjective, and because $\underline{\mathrm{Pic}}^0(X)$ is an Abelian variety, we conclude that $\underline{\mathrm{Pic}}^0(D)$ is also an Abelian variety. $\qquad\square$

Proposition 13.5. *Let X be a minimal surface, and let $D = \sum_{i=1}^r n_i E_i$ be an effective divisor such that $|K+D| = \varnothing$, where $n_i \geq 1$ for all i and the E_i are mutually distinct integral curves on X. Then:*

(i) *Every E_i is a nonsingular curve, and $\sum_{i=1}^r p_a(E_i) \leq q$.*

(ii) *$\{E_i\}$ is a configuration of curves with no loops.*

(iii) *If $n_i \geq 2$, then E_i has at least one of the following properties:*

(a) *E_i is a rational curve; or*

(b) *$(E_i^2) < 0$; or*

(c) *E_i is an elliptic curve with $(E_i^2) = 0$ and the normal bundle of E_i in X is nontrivial.*

Proof. (i) Since $|K+E_i| \subseteq |K+D| = \varnothing$, by Proposition 13.4 we have that $\underline{\mathrm{Pic}}^0(X) \to \underline{\mathrm{Pic}}^0(E_i)$ is a surjective morphism and $\underline{\mathrm{Pic}}^0(E_i)$ is an Abelian variety. Therefore E_i is nonsingular. Indeed, if E_i has singularities, let $\tilde{E}_i$ be the normalization of E_i. By the structure theorem of the Picard scheme of a singular curve, we have that $\underline{\mathrm{Pic}}^0(E_i)$ is an extension of $\underline{\mathrm{Pic}}^0(\tilde{E}_i)$ by a nontrivial affine subgroup, combination of additive groups G_a and/or multiplicative groups G_m; see [Ser3, Chapter V] or [Oor]. But $\underline{\mathrm{Pic}}^0(E_i)$ is an Abelian variety, and therefore it cannot have any nontrivial affine subgroups. Therefore $\tilde{E}_i = E_i$.

Similarly, $|K + D'| \subseteq |K + D| = \varnothing$, where $D' = \sum_{i=1}^r E_i$, and by Proposition (13.4 we have that $\underline{\mathrm{Pic}}^0(D')$ is an Abelian variety. Therefore any two components E_i and E_j of D' with $i \neq j$ and $E_i \cap E_j \neq \varnothing$ intersect transversely, for otherwise $\underline{\mathrm{Pic}}^0(D')$ would contain affine subgroups isomorphic to G_m (see, for example, the discussion in (3.4.4)). The exact sequence (3.4.2)

$$0 \to \mathcal{O}_{D'} \to \prod_{i=1}^r \mathcal{O}_{E_i} \to F \to 0$$

gives $\dim H^1(\mathcal{O}_{D'}) \geq \sum_{i=1}^r \dim H^1(\mathcal{O}_{E_i}) = \sum_{i=1}^r p_a(E_i)$. But since $|K + D'| = \varnothing$, we have $H^2(\mathcal{O}_X(-D')) = 0$, and therefore the restriction homomorphism $H^1(\mathcal{O}_X) \to H^1(\mathcal{O}_{D'})$ is surjective, whence

$$\sum_{i=1}^r p_a(E_i) \leq \dim H^1(\mathcal{O}_X) = q.$$

(ii) If D contains a loop, then $\underline{\mathrm{Pic}}^0(D')$ contains a subgroup isomorphic to G_m; but this is not possible, for $\underline{\mathrm{Pic}}^0(D')$ is an Abelian variety.

(iii) If $n_i \geq 2$, then $|K + 2E_i| \subseteq |K + D| = \varnothing$, and therefore $\underline{\mathrm{Pic}}^0(2E_i)$ is an Abelian variety, by Proposition 13.4. Hence the natural morphism $\underline{\mathrm{Pic}}^0(2E_i) \to \underline{\mathrm{Pic}}^0(E_i)$ is an isomorphism.

Indeed, consider the exact sequences

$$0 \to I = \mathcal{O}_{E_i}(-E_i|_{E_i}) \to \mathcal{O}_{2E_i} \to \mathcal{O}_{E_i} \to 0$$

and

$$0 \to I \xrightarrow{\alpha} \mathcal{O}^*_{2E_i} \to \mathcal{O}^*_{E_i} \to 1$$

($I^2 = 0$ and $\alpha(x) = 1 + x$). The second exact sequence induces the cohomology exact sequence

$$0 \to H^1(I) \to \mathrm{Pic}(2E_i) \to \mathrm{Pic}(E_i),$$

and therefore the exact sequence

$$0 \to H^1(I) \to \underline{\mathrm{Pic}}^0(2E_i) \to \underline{\mathrm{Pic}}^0(E_i).$$

If $H^1(I)$ were nonzero, then the Abelian variety $\underline{\mathrm{Pic}}^0(2E_i)$ would contain an affine subgroup isomorphic to G_a, which is impossible. Therefore

$H^1(I) = 0$. Then the cohomology exact sequence associated to the first exact sequence gives

$$0 = H^1(I) \to H^1(\mathcal{O}_{2E_i}) \to H^1(\mathcal{O}_{E_i}) \to 0,$$

so that the tangent map to $\underline{\operatorname{Pic}}^0(2E_i) \to \underline{\operatorname{Pic}}^0(E_i)$ at the origin is an isomorphism, and then the map $\underline{\operatorname{Pic}}^0(2E_i) \to \underline{\operatorname{Pic}}^0(E_i)$ is an isomorphism, as claimed.

In particular, we have also shown that $H^1(\mathcal{O}_{E_i}(-E_i|_{E_i})) = 0$. If $(E_i^2) \geq 0$, then since $\deg_{E_i} \mathcal{O}_{E_i}(-E_i|_{E_i}) = -(E_i^2) \leq 0$, we have that E_i is either a rational curve (in which case $(E_i^2) = 0$ or 1) or an elliptic curve (in which case $(E_i^2) = 0$). If E_i is elliptic then $(E_i^2) = 0$ and $\mathcal{O}_{E_i}(-E_i|_{E_i}) \not\cong \mathcal{O}_{E_i}$, for $H^1(\mathcal{O}_{E_i}(-E_i|_{E_i})) = 0$.   □

Proposition 13.6. *Let X be a surface, $x_1, \ldots, x_n$ n distinct closed points on X, and $D \in \operatorname{Div}(X)$ a divisor such that $\dim |D| \geq 3n$. Then there exists an effective divisor $D' \in |D|$ such that every x_i is on D' and is a multiple point of D'.*

Proof. Consider the exact sequence

$$0 \to \mathcal{O}_X(D) \otimes I_1 \otimes \cdots \otimes I_n \to \mathcal{O}_X(D) \to \bigoplus_{i=1}^n \Bbbk(x_i)^3 \to 0,$$

where $(I_i)_z = \mathcal{O}_{X,z}$ if $z \neq x_i$ and $(I_i)_{x_i} = \mathfrak{m}_{X,x_i}^2$ ($\dim \mathcal{O}_{X,x_i}/\mathfrak{m}_{X,x_i}^2 = 3$). From the associated cohomology exact sequence,

$$0 \to H^0(\mathcal{O}_X(D) \otimes I_1 \otimes \cdots \otimes I_n) \to H^0(\mathcal{O}_X(D)) \to \bigoplus_{i=1}^n \Bbbk(x_i)^3,$$

we get $H^0(\mathcal{O}_X(D) \otimes I_1 \otimes \cdots \otimes I_n) \neq 0$, because $\dim H^0(\mathcal{O}_X(D)) > 3n$ by hypothesis, and $\dim \oplus_{i=1}^n \Bbbk(x_i)^3 = 3n$. Let s be a nonzero section in $H^0(\mathcal{O}_X(D) \otimes I_1 \otimes \cdots \otimes I_n)$. Then $D' = \operatorname{div}_X(s)$ satisfies all the requirements.   □

Proposition 13.7. *Let X be a minimal surface with $q = 0$, and let E be a rational curve on X such that $(K \cdot E) < 0$ and $|K + E| = \varnothing$ (then E is necessarily nonsingular, by (13.5)(i)). Then X is either a geometrically ruled surface over $\mathbb{P}^1$ or isomorphic to $\mathbb{P}^2$. Furthermore, in the geometrically ruled case the ruling is given by a morphism $\phi_{|C|}$ for a (nonsingular) rational curve C on X such that $(K \cdot C) < 0$ and $|K + C| = \varnothing$.*

Proof. Since X is minimal and $(K \cdot E) < 0$, we have $(E^2) \geq 0$. Now let

$$a = \min\{(C^2) \mid C \text{ a rational curve with } (C^2) \geq 0 \text{ and } |K + C| = \varnothing\}.$$

Let $\mathcal{C}$ denote the set of all (integral) rational curves C with $(C^2) = a$ and $|K + C| = \emptyset$. Fix a hyperplane section H on X, and put

$$b = \min\{(C \cdot H) \mid C \in \mathcal{C}\}.$$

Fix a curve $C \in \mathcal{C}$ with $(C \cdot H) = b$. First we prove that every element of $|C|$ is a nonsingular rational curve.

Let $D = \sum_{i=1}^{r} n_i E_i \in |C|$, with $n_i \geq 1$ and E_i mutually distinct integral curves on X. We have

$$\sum_{i=1}^{r} n_i (K \cdot E_i) = (K \cdot D) = (K \cdot C) < 0,$$

and therefore $(K \cdot E_i) < 0$ for at least one i. Fix such an index i. Since X is minimal, we must have $(E_i^2) \geq 0$. Since $|K + D| = |K + C| = \emptyset$, E_i is a nonsingular rational curve (by Proposition 13.5(i)) and $|K + E_i| = \emptyset$. Therefore $(E_i^2) \geq a$ by the choice of a, and we have

$$a = (D^2) = \sum_{j=1}^{r} n_j (C \cdot E_j) \geq n_i (C \cdot E_i)$$

$$= n_i \sum_{h=1}^{r} n_h (E_h \cdot E_i) \geq n_i^2 (E_i^2).$$

Since $n_i \geq 1$ we get $(E_i^2) \leq a$, and therefore $(E_i^2) = a$, by the minimality of a. Therefore $E_i \in \mathcal{C}$. Furthermore,

$$(H \cdot E_i) \geq b = (H \cdot C) = \sum_{h=1}^{r} n_h (H \cdot E_h).$$

As $(H \cdot E_h) > 0$ for all h, we get $r = 1$ and $n_i = 1$; in other words, $D = E_i$, and we have proved that every $D \in |C|$ is a nonsingular rational curve. In particular, from Proposition 13.6 we see that $\dim |C| \leq 2$.

From the exact sequence

$$0 \to \mathcal{O}_X \to \mathcal{O}_X(C) \to \mathcal{O}_X(C) \otimes \mathcal{O}_C \to 0$$

we get the cohomology exact sequence

$$0 \to H^0(\mathcal{O}_X) \to H^0(\mathcal{O}_X(C)) \to H^0(\mathcal{O}_X(C) \otimes \mathcal{O}_C) \to H^1(\mathcal{O}_X) = 0.$$

Since $\deg_C(\mathcal{O}_X(C) \otimes \mathcal{O}_C) = (C^2)$ and $C \cong \mathbb{P}^1$, we have $\dim H^0(\mathcal{O}_X(C) \otimes \mathcal{O}_C) = 1 + (C^2)$, and therefore $\dim H^0(\mathcal{O}_X(C)) = 2 + (C^2)$, or $\dim |C| = 1 + (C^2)$. We have seen earlier that $\dim |C| \leq 2$, so that (C^2) must be either 0 or 1 (recall also that $(C^2) \geq 0$). Also, the cohomology exact sequence shows that $|C|$ is base point free.

If $(C^2) = 0$, then dim $|C| = 1$. Hence the morphism $\phi_{|C|}$ maps X to $\mathbb{P}^1$, $\phi_{|C|} : X \to \mathbb{P}(H^0(\mathcal{O}_X(C))^{\check{}}) \cong \mathbb{P}^1$. The fibers of this morphism are the elements of $|C|$, which are all nonsingular rational curves, so that X is a geometrically ruled surface with base $\mathbb{P}^1$.

If $(C^2) = 1$, let D be an arbitrary integral curve on X, and let $x \in D$ be a closed point. Since dim $|C| = 2$ in this case, there exists a curve $C' \in |C|$ such that $x \in C'$. If $C' = D$, then $(C \cdot D) = (C \cdot C') = (C^2) = 1 > 0$, and if $C' \neq D$, then also $(C \cdot D) = (C' \cdot D) > 0$. Therefore the morphism $\phi_{|C|} : X \to \mathbb{P}(H^0(\mathcal{O}_X(C))^{\check{}}) \cong \mathbb{P}^2$ is a finite (surjective) morphism. Since C is the inverse image by $\phi_{|C|}$ of a straight line in $\mathbb{P}^2$ and $(C^2) = 1$, $\phi_{|C|}$ must be an isomorphism. □

Proposition 13.8. *Let X be a minimal surface with $\kappa(X) = -\infty$, $q > 0$, and such that through every point $x \in X$ there is a nonsingular rational curve $C \subset X$. Then X is a geometrically ruled (and therefore ruled) surface, via the morphism $f : X \to B$ obtained from the Stein factorization of the Albanese morphism $\alpha : X \to \mathrm{Alb}(X)$.*

Proof. From $\kappa(X) = -\infty$ we have, in particular, that $p_g = 0$; therefore, by Theorem (5.3) we have dim $\mathrm{Alb}(X) = q > 0$. Let $\alpha : X \to \mathrm{Alb}(X)$ be the canonical morphism. Since $\alpha(X)$ generates $\mathrm{Alb}(X)$, it cannot be a point. That is, dim $\mathrm{Alb}(X) \geq 1$. If $x \in X$ is a closed point and $C \subset X$ is a nonsingular rational curve passing through x, then $\alpha(C)$ is a point on $\mathrm{Alb}(X)$, and therefore dim $\alpha^{-1}\alpha(x) = 1$. Hence dim $\alpha(X) = 1$, so that the normalization B of $\alpha(X)$ in the field $\Bbbk(X)$ of rational functions on X is a nonsingular projective curve. If $f : X \to B$ is the morphism obtained by Stein factorization from α, then $f(C)$ is a point for every nonsingular rational curve C on X. Since the general fiber of f is an integral curve, by (7.3), it follows that in our case the general fiber of f is a nonsingular rational curve. From (12.2) we get that f is a geometric ruling of X, for X is a minimal surface. (The same conclusion, that f is a geometric ruling, can also be proved directly, without reference to Chapter 12, as follows: If $D = \sum_{i=1}^r n_i E_i$ is a fiber of f, with $n_i \geq 1$ and the E_i distinct integral curves, then $(D^2) = 0$ and $(K \cdot D) = -2$, because $p_a(D) = 0$. If $r > 1$, then $(E_i^2) < 0$ for all i, cf. (2.6). On the other hand, since $(K \cdot D) < 0$, we have $(K \cdot E_i) < 0$ for at least one i. But then E_i is an exceptional curve of the first kind, contradicting the minimality of X. Thus $r = 1$, that is, $D = nE$ with $n \geq 1$ and E an integral curve. But $-2 = (K \cdot D) = n(K \cdot E)$; if $n \geq 2$ then $(K \cdot E) = -1$, and since $(E^2) = 0$ this is not possible, due to the genus formula. Therefore $n = 1$, and $D = E$ is a nonsingular rational curve). □

Lemma 13.9. *If X is a minimal surface with $\kappa(X) = -\infty$, then there exists a hyperplane section H on X such that $(K \cdot H) < 0$.*

Proof. Let C be an integral curve on X such that $(K \cdot C) < 0$. As X is minimal, we have $(C^2) \geq 0$. Let H_1 be a hyperplane section on X. For every $n \geq 0$ the divisor $nC + H_1$ is ample, because

$$((nC + H_1)^2) = n^2(C^2) + 2n(C \cdot H_1) + (H_1^2) \geq (H_1^2) > 0$$

and

$$(nC + H_1 \cdot D) = n(C \cdot D) + (H_1 \cdot D) \geq (H_1 \cdot D) > 0$$

for every integral curve D on X, so that $nC + H_1$ is ample by the Nakai–Moishezon criterion (1.25). On the other hand, $(K \cdot nC + H_1) < 0$ for all $n \gg 0$. □

Next we analyze the minimal surfaces X with $\kappa(X) = -\infty$ and $(K^2) \leq 0$.

Proposition 13.10. *Let X be a minimal surface with $\kappa(X) = -\infty$ and $(K^2) \leq 0$. Then for every $n \geq 0$ there exists a divisor $D \in \mathrm{Div}(X)$ such that $|K + D| = \varnothing$, $(D \cdot K) < 0$ and $\dim |D| \geq n$.*

Proof. By Lemma 13.9 there is a hyperplane section H on X such that $(K \cdot H) \leq -2n - 2q$. From $\kappa(X) = -\infty$ and (13.1) we have that adjunction on X terminates. Thus there exists $m \geq 0$ such that $|H + mK| \neq \varnothing$ and $|H + (m + 1)K| = \varnothing$. Let $D' \in |H + mK|$. We cannot have $D' = 0$; if we did, then we'd have $H \sim -mK$, and therefore $(H^2) = m^2(K^2) \leq 0$, which is a contradiction. Therefore $D' > 0$. Write $D' = D + D''$, where D consists of the components E of D' such that $(K \cdot E) < 0$, and D'' consists of the components E of D' with $(K \cdot E) \geq 0$. By construction,

$$|K + D| \subseteq |K + D'| = |H + (m + 1)K| = \varnothing.$$

Since $|K - D| \subseteq |K| = \varnothing$, we have by the Riemann–Roch theorem:

$$\begin{aligned}
\dim |D| &\geq \tfrac{1}{2}(D^2) - \tfrac{1}{2}(D \cdot K) - q \\
&\geq \tfrac{1}{2}(D^2) - \tfrac{1}{2}(D + D'' \cdot K) - q \\
&= \tfrac{1}{2}(D^2) - \tfrac{1}{2}(D' \cdot K) - q \\
&= \tfrac{1}{2}(D^2) - \tfrac{1}{2}(H \cdot K) - \tfrac{1}{2}m(K^2) - q \\
&\geq -\tfrac{1}{2}(H \cdot K) - q \\
&\geq (n + q) - q \\
&= n.
\end{aligned}$$

We used the following inequalities: $(D'' \cdot K) \geq 0$, $(K^2) \leq 0$, and $(D^2) \geq 0$; the last inequality is proved as follows: if E is a component of D, then since $(K \cdot E) < 0$ and X is minimal we have $(E^2) \geq 0$; then $(D^2) \geq 0$ follows.

Since $D \neq 0$, we have $(K \cdot D) < 0$. Also, $|K + D| \subseteq |K + D'| = |H + (m + 1)K| = \varnothing$. The proof of Proposition 13.10 is complete. □

Proposition 13.11. *Let X be a minimal surface with $p_2 = q = 0$. Then $\kappa(X) = -\infty$. If in addition $(K^2) \leq 0$, then there exists a nonsingular rational curve C on X with $(K \cdot C) < 0$ and $|K + C| = \varnothing$; in particular, in this case the surface X is rational.*

Proof. If $(K^2) < 0$, then $\kappa(X) = -\infty$. Indeed, assume by contradiction that there exists $D \in |nK|$ with $n \geq 1$. Then $(K \cdot D) = n(K^2) < 0$, so that $\kappa(X) = -\infty$ by (13.1), contradiction.

Now assume that $p_2 = q = 0$ and $(K^2) \geq 0$. We have

$$\dim |-K| + \dim |2K| \geq (K^2) - 1$$

by the Riemann–Roch theorem, and since $\dim |2K| = -1$ (for $p_2 = 0$) we get that $\dim |-K| \geq (K^2) \geq 0$. Thus there exists an effective divisor $D \in |-K|$. If $D = 0$ then $K \sim 0$, so that $p_g = 1$, contradiction; therefore $D > 0$. Then for every divisor $D' \in \mathrm{Div}(X)$ there exists an integer $n_{D'} \geq 1$ such that $|D' + nK| = |D' - nD| = \varnothing$ for all $n \geq n_{D'}$ (see (7.11.3)). Therefore adjunction on X terminates, so that $\kappa(X) = -\infty$ by (13.1).

Finally, assume that $(K^2) \leq 0$ (with $p_2 = q = 0$). By Proposition 13.10 there exists an effective divisor E on X such that $(K \cdot E) < 0$ and $|K + E| = \varnothing$. Then some irreducible component C of E still has the properties $(K \cdot C) < 0$ and $|K+C| = \varnothing$. By Proposition 13.5(i) we have $p_a(C) \leq q = 0$, so that C is a nonsingular rational curve. Now the last statement in the proposition follows from Proposition 13.7. $\qquad\square$

Proposition 13.12. *Let X be a minimal surface with $p_g = 0$ and $q > 0$. Then $(K^2) \leq 0$, and if $q > 1$ then $(K^2) < 0$.*

Proof. Since $p_g = 0$, Noether's formula becomes

$$10 - 8q = (K^2) + b_2$$

(see (5.1)).

If $q > 1$, then we see at once that $(K^2) < 0$. If $q = 1$, we must show that $b_2 \geq 2$. But $\dim \mathrm{Alb}(X) = 1$ in this case, by Theorem 5.3, and therefore the Albanese morphism cannot be constant. Hence X is a fibration over $\mathrm{Alb}(X)$, and by Lemma 8.7 and the Igusa–Severi inequality we get $b_2 \geq 2$. $\qquad\square$

Proposition 13.13. *Let X be a minimal surface with $q > 0$. Then $\kappa(X) = -\infty$ if and only if X is a geometrically ruled surface.*

Proof. By (11.15), if X is a ruled surface, then $\kappa(X) = -\infty$.

Now assume that X is a minimal surface with $q > 0$ and $\kappa(X) = -\infty$. We must prove that X is geometrically ruled. By Proposition 13.8 it is enough to show that for every point $x \in X$ there is a nonsingular rational curve that passes through x.

By way of contradiction, assume that this is not the case. Then there are at most finitely many nonsingular rational curves on X. Indeed, let $f : X \to B$ be the morphism obtained from the Albanese morphism $\alpha :$ $X \to \mathrm{Alb}(X)$ by Stein factorization, as in the proof of Proposition 13.8. If there were infinitely many nonsingular rational curves on X, then each of them would be contained in a fiber of f. Therefore there are infinitely many points $b \in B$ with $\dim f^{-1}(b) = 1$. Hence B cannot be a surface. So B is a nonsingular curve. Since the general fiber of f is integral, and since infinitely many fibers of f contain at least one nonsingular rational curve, at least one of the fibers of f is a nonsingular rational curve. But then the proof of Proposition 13.8 shows that all the fibers of f are smooth rational curves, which is a contradiction.

Now fix a projective embedding $X \hookrightarrow \mathbb{P}^n$ of X. For a fixed integer $d \geq 1$, there are only finitely many integral curves E on X with $\deg(E) = d$ and $H^0(N_E) = 0$, where N_E is the normal bundle of E in X. Indeed, if $C(d)$ is the Hilbert scheme of all curves of degree d on X, and if $e \in C(d)$ is the (closed) point corresponding to a curve E, then $H^0(N_E)$ is canonically isomorphic to the tangent space to $C(d)$ at e; see [Mum1]. If $H^0(N_E)=0$ then e is an isolated point of $C(d)$. But since $C(d)$ is a projective scheme (loc. cit.), it has only finitely many isolated points, and therefore X contains at most finitely many curves E of degree d with $H^0(N_E) = 0$. In particular, X contains at most finitely many nonsingular curves E of degree d having one of the following two additional properties:

(a) $(E^2) < 0$, or

(b) E is elliptic, $(E^2) = 0$, and $N_E \not\cong \mathcal{O}_E$.

Let $\mathcal{F}$ be the family of all nonsingular curves E on X that either are rational or satisfy condition (a) or (b).

First we prove Proposition 13.13 under the additional hypothesis that the field $\Bbbk$ is uncountable (and algebraically closed). Let $d = \deg(X)$ with respect to the embedding $X \hookrightarrow \mathbb{P}^n$ considered earlier. Since there are at most finitely many curves of degree d in $\mathcal{F}$, a general hyperplane H in $\mathbb{P}^n$ intersects X in a nonsingular, connected curve C (by Bertini's theorem), with $C \notin \mathcal{F}$. Since $\mathcal{F}$ is at most countable and C is uncountable (because $\Bbbk$ is uncountable), we have in particular that $C \setminus \bigcup_{E \in \mathcal{F}}(C \cap E)$ is infinite, and therefore $X \setminus \bigcup_{E \in \mathcal{F}} E$ is infinite.

Let $x_1, \dots, x_q$ be q distinct (closed) points on $X \setminus \bigcup_{E \in \mathcal{F}} E$, where q is the irregularity of X. By Proposition 13.12 we have $(K^2) \leq 0$, and therefore by Proposition 13.10 there exists an effective divisor $D' \in \mathrm{Div}(X)$ such that $(K \cdot D') < 0$, $|K + D'| = \varnothing$, and $\dim |D'| \geq 3q$. Then by Proposition 13.6 there exists a divisor $D = \sum_{i=1}^{r} n_i E_i \in |D'|$ such that $x_1, \dots, x_q$ are multiple points on D. If E_i is a component of D that passes through an x_j, then $E_i \notin \mathcal{F}$, and therefore $p_a(E_i) \geq 1$ and the corresponding coefficient n_i is 1, by Proposition 13.5(iii). Also, E_i is nonsingular, by Proposition 13.5(i).

Since x_j is a multiple point of D, this means that at least two components of D must pass through x_j. As $x_1, \ldots, x_q$ are all multiple points on D and $\{E_i\}$ is a configuration of curves without loops, there must be at least $q+1$ distinct components E_i of D, each passing through at least one x_j. Then all these components E_i are smooth curves with $p_a(E_i) \geq 1$, contradicting the inequality in Proposition 13.5(i).

Proposition 13.13 is therefore proved when $\Bbbk$ is uncountable. Specifically, in this case we have shown that if X is a minimal surface with $q > 0$ and $\kappa(X) = -\infty$, then the morphism $f : X \to B$ obtained from the Albanese morphism $\alpha : X \to \mathrm{Alb}(X)$ by Stein factorization is a geometric ruling of X.

If $\Bbbk$ is countable, let $\Bbbk \hookrightarrow \Bbbk'$ be an extension of $\Bbbk$ with $\Bbbk'$ uncountable and algebraically closed. Let $X' = X \otimes_{\Bbbk} \Bbbk'$. Then $\kappa(X') = \kappa(X) = -\infty$ and $q(X') = q(X) > 0$. Therefore the morphism $f' : X' \to B'$, obtained from the Albanese morphism $\alpha' : X' \to \mathrm{Alb}(X')$ by Stein factorization, has the general fiber isomorphic to $\mathbb{P}^1_{\Bbbk'}$. One can check easily that $B' = B \otimes_{\Bbbk} \Bbbk'$ and $f' = f \otimes_{\Bbbk} \Bbbk'$. Let F_b be an integral fiber of f over a closed point $b \in B$, and let $b' \in B'$ be a closed point such that $v(b') = b$, where $v : B' \to B$ is the morphism induced by $\mathrm{Spec}(\Bbbk') \to \mathrm{Spec}(\Bbbk)$. Then the fiber $(f')^{-1}(b')$ is isomorphic (over $\Bbbk'$) to $F_b \otimes_{\Bbbk} \Bbbk'$, and $f'^{-1}(b')$ is an integral curve of arithmetic genus zero. Hence F_b is itself an integral curve of arithmetic genus zero, that is, $F_b \cong \mathbb{P}^1_{\Bbbk}$. Since X is a minimal surface, we have in fact proved that f is a geometric ruling of X (as in the proof of Proposition 13.8, one shows that all the fibers of f are integral curves, thus isomorphic to $\mathbb{P}^1_{\Bbbk}$). $\qquad\square$

Remarks 13.14. (a) The proof of Proposition 13.13 gives a different, more complicated way of proving part (b) of Theorem 12.2.

(b) If X is a minimal surface with $p_g = 0$ and $q \geq 2$, then X is a ruled surface. Indeed, $(K^2) < 0$ by Proposition 13.12, and therefore $\kappa(X) = -\infty$ by Theorem 1.25. Now the fact that X is ruled follows from Proposition 13.13.

(c) Propositions 13.11 and 13.13 reduce the proof of Theorem 13.2 to Theorem 13.3 with the additional hypothesis that $(K^2) > 0$.

13.15. Next we prove Theorem 13.3 in the case $(K^2) > 0$. The proof in this case is much more delicate than in the case $(K^2) \leq 0$.

First we observe that if there exists an effective divisor D on X such that $(K \cdot D) < 0$ and $|K + D| = \varnothing$, then at least one component C of D has the same properties, that is, $(K \cdot C) < 0$ and $|K + C| = \varnothing$. Then $p_a(C) \leq q = 0$ by Proposition 13.5(i), and therefore C is a nonsingular rational curve. Then the rationality of X follows from Proposition 13.7.

Therefore there only remains to analyze the case when $(K \cdot D) \geq 0$ for every effective divisor D such that $|K + D| = \varnothing$.

Lemma 13.16. *If X is a minimal surface with $p_2 = q = 0$, $(K^2) > 0$, and $(K \cdot D) \geq 0$ for every effective divisor D on X such that $|K + D| = \varnothing$, then:*

(a) $\text{Pic}(X)$ *is generated by* $\omega_X = \mathcal{O}_X(K)$, *and* $\mathcal{O}_X(-K)$ *is ample. In particular, X does not contain any nonsingular rational curves.*

(b) *Every* $D \in |-K|$ *is an integral curve with* $p_a(D) = 1$.

(c) $(K^2) \leq 5$ *and* $b_2 \geq 5$.

Proof. First we show that every $D \in |-K|$ is an irreducible curve. By way of contradiction, assume that there exists a reducible $D \in |-K|$. We have $D > 0$ and $(K \cdot D) = -(K^2) < 0$. Therefore at least one component C of D has $(K \cdot C) < 0$. But $|K + C| = \varnothing$, for otherwise there exists $\Delta \in |K + C|$, and then $0 < (D - C) + \Delta \sim -K - C + K + C = 0$, which is a contradiction. So $|K + C| = \varnothing$. But this contradicts the hypothesis of the lemma, because $(K \cdot C) < 0$.

Therefore we have shown that every $D \in |-K|$ is irreducible. In fact, the same argument shows that every such D is also reduced, that is, an integral curve. Note also that

$$p_a(D) = \tfrac{1}{2}(D \cdot D + K) + 1 = 1,$$

since $D + K \sim 0$. Therefore (b) is proved.

Now we show that *the only effective divisor D such that $|K + D| = \varnothing$ is the zero divisor.* Indeed, assume by contradiction that $|K + D| = \varnothing$ for some $D > 0$. Fix a point $x \in D$. Since $\dim |-K| \geq (K^2) \geq 1$, there exists $C \in |-K|$ passing through x. By what we have shown earlier, C is an integral curve. C cannot be one of the components of D; if it were, then $|K + D| \supseteq |K + C| = |0| \neq \varnothing$, which is a contradiction. Since $D \cap C \neq \varnothing$ (they meet at least at x) and C is an integral curve that is not a component of D, we have $(C \cdot D) > 0$. But then $(K \cdot D) = (-C \cdot D) < 0$, which contradicts the hypothesis of the lemma.

By Proposition 13.11 we have $\kappa(X) = -\infty$; therefore adjunction terminates. Let $\Delta \in \text{Div}(X)$ be an arbitrary effective divisor. Then there exists an integer $n \geq 0$ such that $|\Delta + nK| \neq \varnothing$ and $|\Delta + (n+1)K| = \varnothing$. If we put $D = \Delta + nK$, then D is an effective divisor such that $|K + D| = \varnothing$. By what we have shown, we must have $D = 0$, and therefore $\Delta \sim -nK$. Since every divisor on X is the difference of two effective divisors, we have shown that $\text{Pic}(X)$ is cyclic, generated by $\omega_X = \mathcal{O}_X(K)$. Moreover, if H is a hyperplane section on X, then $H \sim -nK$ with $n \geq 0$, and then in fact $n > 0$, for $H \not\sim 0$; therefore $\mathcal{O}_X(-K)$ is ample.

If C is any integral curve on X, then we must have $C \sim -mK$ for some integer $m \geq 1$. Computing the arithmetic genus of C by the genus formula, we have

$$p_a(C) = p_a(-mK) = \tfrac{1}{2}m(m-1)(K^2) + 1 \geq 1,$$

and therefore C cannot be a nonsingular rational curve. Therefore (a) is proved.

Finally, assume by contradiction that $(K^2) \geq 6$. Then $\dim |-K| \geq (K^2) \geq 6$. Fix two points x and y on X. By Proposition 13.6 there exists $C \in |-K|$ so that x and y are multiple points of C. But we have already shown that C is an integral curve with $p_a(C) = 1$, so that C can have at most one singular point. Therefore we must have $(K^2) \leq 5$. Now Noether's formula, which in this case is $10 = (K^2) + b_2$, gives $b_2 \geq 5$.   □

To complete the proof of Theorem 13.3, there only remains to prove the following fundamental lemma.

Lemma 13.17. *There are no surfaces X with $q = 0$ satisfying conditions (a), (b), and (c) of Lemma 13.16.*

Note that condition (a) in Lemma 13.16 implies that the surface X is minimal, $(K^2) > 0$, and $p_n = 0$ for all $n \geq 1$.

First we prove Lemma 13.17 in the case when $\Bbbk$ is a field of characteristic zero. In this case we don't even need condition (b).

Lemma 13.18. *If $\Bbbk$ is an algebraically closed field of characteristic zero, then there are no surfaces X with $q = 0$ satisfying conditions (a) and (c) of Lemma 13.16.*

Proof. Assume, by contradiction, that such a surface X does exist.

By Lefschetz's principle, cf. [Buc2, p. 124], we may assume that $\Bbbk = \mathbb{C}$, the field of complex numbers. Then, using the notation from [Ser2], for every coherent $\mathcal{O}_X$-module F let F^h denote the analytic coherent sheaf associated to F. Then the exponential exact sequence

$$0 \to \mathbb{Z} \to \mathcal{O}_X^h \to (\mathcal{O}_X^h)^* \to 1$$

induces the cohomology exact sequence

$$H^1(\mathcal{O}_X^h) \to H^1((\mathcal{O}_X^h)^*) \to H^2(X, \mathbb{Z}) \to H^2(\mathcal{O}_X^h).$$

By [Ser2] we have $H^i(\mathcal{O}_X^h) \cong H^i(\mathcal{O}_X)$, $\forall i$, and $H^1((\mathcal{O}_X^h)^*) \cong H^1(\mathcal{O}_X^*) = \operatorname{Pic}(X)$. Since $q = p_g = 0$, we get $\operatorname{Pic}(X) \cong H^2(X, \mathbb{Z})$, so that $b_2 = \operatorname{rank} H^2(X, \mathbb{Z}) = \operatorname{rank} \operatorname{Pic}(X) = 1$, the last equality being given by condition (a) of Lemma 13.16, and this contradicts the inequality $b_2 \geq 5$ from condition (c) in the same lemma.   □

Remark 13.19. The proof of Castelnuovo's criterion in characteristic zero also gives a new proof of Theorem 12.8 (for this special case, when the characteristic of $\Bbbk$ is zero). Indeed, let X be a minimal rational surface. Then $p_2 = q = 0$. By Proposition 13.11 and the proof of Castelnuovo's criterion when $(K^2) > 0$ and $\operatorname{char}(\Bbbk) = 0$, there exists a nonsingular rational curve C on X such that $(C \cdot K) < 0$ and $|K + C| = \emptyset$. Then, by Proposition 13.7,

X is isomorphic to either $\mathbb{P}^2$ or a geometrically ruled surface over $\mathbb{P}^1$ (i.e., to one of the surfaces $\mathbb{F}_n$ with $n \geq 0$ and $n \neq 1$, cf. (12.5)).

We included the other, longer proof of (12.8) in Chapter 12 because it is elementary, does not depend on the classification of surfaces, and works in positive characteristic as well as in characteristic zero.

Now we proceed to prove the fundamental Lemma 13.17 (and therefore Castelnuovo's criterion of rationality of surfaces) in characteristic $p > 0$. The idea is to reduce the proof to characteristic zero.

Lemma 13.20. *Let X be a surface with $q = 0$ and with properties (a), (b), and (c) in Lemma 13.16. Then there exists a nonsingular curve $D \in |-K|$.*

Proof. By way of contradiction, assume that all the curves in the complete linear system $|-K|$ are singular. Since every $D \in |-K|$ is an integral curve with $p_a(D) = 1$, D must have exactly one singular point, which can be either an ordinary cuspidal point or a double point with distinct tangents, cf. the proof of Theorem 7.16 or [Ser3, Chapter IV]. We have $\dim |-K| \geq (K^2) \geq 1$. Let $L \subseteq |-K|$ be a one-dimensional linear subsystem. Then the fibers of the rational map $\phi = \phi_L : X \dashrightarrow \mathbb{P}^1$ are exactly the curves in L, because L has no fixed components; indeed, all the elements of L are integral curves.

Let Y be the set of all singular points on the curves in L, and let x be a base point of L (if any). Then $x \notin Y$.

Indeed, assume by contradiction that $x \in Y$. Let then $D \in L$ be a curve that has a singularity at x, and let $u_1 : X_1 \to X$ be the quadratic transformation of X with center x. Then the proper transform D' of D on X_1 is a nonsingular rational curve and a fiber of the rational map $\phi_1 = \phi \circ u_1 : X_1 \to \mathbb{P}^1$. After at most (K^2) quadratic transformations we get a surface $\tilde{X}$ such that $\tilde{\phi} = \phi \circ u$ is a morphism (where u is the composite of those quadratic transformations), and we have that a fiber of $\tilde{\phi}$ is a nonsingular rational curve. By (11.10) we get that $\tilde{X}$ (and therefore X) is a rational surface. But Theorem 12.8 shows, in particular, that there are no minimal models X of rational surfaces with the property that $\mathrm{Pic}(X) = \mathbb{Z}\omega_X$. Therefore we reach a contradiction.

We have thus shown that Y does not contain any base point of L. Let $u : \tilde{X} \to X$ be a composite of a minimal number of quadratic transformations such that $\tilde{\phi} = \phi \circ u$ becomes a morphism. Then $\tilde{\phi} : \tilde{X} \to \mathbb{P}^1$ is a quasielliptic fibration, whose fibers are all integral rational curves, each having exactly one singular point. By (7.18), each such singular point is an ordinary cuspidal point, and this situation is possible only if the characteristic of $\Bbbk$ is 2 or 3.

Let $\tilde{Y}$ be the set of singularities of the fibers of $\tilde{\phi}$. If $\mathrm{char}(\Bbbk) = 3$, from the proof of (7.18) we know that $\tilde{Y}$ is a nonsingular irreducible curve and the restriction $\tilde{\phi}|_{\tilde{Y}} : \tilde{Y} \to \mathbb{P}^1$ is a bijective, purely inseparable morphism of degree $p = 3$.

A similar conclusion holds when char($\Bbbk$) $= 2$. Indeed, using the notation in the proof of Theorem 7.18 (with $f = \tilde{\phi}$ and $B = \mathbb{P}^1$), we have

$$f(u, v) = U(u, v) \cdot (u^2 + v^3), \quad U(u, v) \in \Bbbk[[u, v]], \; U(0, 0) \neq 0.$$

We observe that $\frac{\partial f}{\partial v}(u, v) = U(0, 0)v^2 +$ terms of higher degree in v, and therefore by Weierstrass' preparation theorem we have:

$$\frac{\partial f}{\partial v}(u, v) = V(u, v) \cdot (v^2 + R_1(u)v + R_2(u)),$$

with $V(0, 0) \neq 0$ and $R_1(u)$ and $R_2(u)$ two noninvertible series in $\Bbbk[[u]]$. One can see easily that $R_1(u)$ is, in fact, zero. Moreover, we claim that the coefficients of the odd powers of u in $R_2(u)$ are all zero. If this weren't so, then since $\tilde{Y}$ is a curve and the power series $v^2 + R_2(u)$ is irreducible, we have that $v^2 + R_2(u)$ divides the power series $\frac{\partial f}{\partial u}$ (since $\tilde{Y}$ is exactly the set of points of X where $f = \tilde{\phi}$ is not a smooth morphism). But this is easily seen to be impossible.

We have shown that, if char($\Bbbk$) $= 2$, then $\frac{\partial f}{\partial v} = V(u, v) \cdot g^2$, with $V(0, 0) \neq 0$ and $g \in \Bbbk[[u, v]]$ a series of order 1. Since $g = 0$ is a local equation for $\tilde{Y}$, $\tilde{Y}$ is a nonsingular irreducible curve. Since $\tilde{Y}$ intersects each fiber F of $\tilde{\phi}$ in exactly one point and $(\tilde{Y} \cdot F) = p = 2$, we have that $\tilde{\phi}|_{\tilde{Y}} : \tilde{Y} \to \mathbb{P}^1$ is a bijective, purely inseparable morphism, with $\tilde{Y}$ a nonsingular (irreducible) curve, just like in the case char($\Bbbk$) $= 3$.

Therefore we see that, if char($\Bbbk$) $= 2$ or 3, then $\tilde{Y}$ is a nonsingular rational curve. Since no base point of L is on Y, we have that $\tilde{Y}$ is isomorphic to Y via $u|_{\tilde{Y}}$, and therefore Y is also a nonsingular rational (irreducible) curve. But this contradicts the fact that X does not contain any nonsingular rational curves, cf. Lemma 13.16(a). $\qquad\square$

Lemma 13.21. *Let X be a surface with $q = 0$ and with properties (a), (b), and (c) in Lemma 13.16. Then $H^2(X, T_X) = 0$, where $T_X = (\Omega^1_{X/\Bbbk})^{\check{}}$ is the sheaf associated to the tangent bundle of X.*

Proof. By Lemma 13.20 there exists a nonsingular elliptic curve $D \in |-K|$. Consider the exact sequence

$$0 \to \mathcal{O}_X((n-1)D) \otimes T_X \to \mathcal{O}_X(nD) \otimes T_X \to \mathcal{O}_X(nD) \otimes T_X \otimes \mathcal{O}_D \to 0$$

$(n \geq 0)$, which induces the cohomology exact sequence

$$H^1(\mathcal{O}_X(nD) \otimes T_X \otimes \mathcal{O}_D) \to H^2(\mathcal{O}_X((n-1)D) \otimes T_X) \to H^2(\mathcal{O}_X(nD) \otimes T_X).$$

By Lemma 13.16(a) D is an ample divisor on X, and therefore by Serre's theorem, cf. [Ser1] or [EGA III, (2.2.2)], we have $H^2(\mathcal{O}_X(nD) \otimes T_X) = 0$ for all $n \gg 0$. To prove that $H^2(X, T_X) = 0$ it will suffice therefore to show that $H^1(\mathcal{O}_X(nD) \otimes T_X \otimes \mathcal{O}_D) = 0$ for all $n \geq 1$.

Consider the canonical exact sequence

$$0 \to T_D \to T_X \otimes \mathcal{O}_D \to N = \mathcal{O}_X(D) \otimes \mathcal{O}_D \to 0,$$

which induces the cohomology exact sequence

$$H^1(\mathcal{O}_X(nD) \otimes \mathcal{O}_D) \to H^1(\mathcal{O}_X(nD) \otimes T_X \otimes \mathcal{O}_D) \to H^1(\mathcal{O}_X(nD) \otimes N)$$

(by tensoring with $\mathcal{O}_X(nD)$ and observing that $T_D \cong \mathcal{O}_D$, because D is a nonsingular elliptic curve).

Since D is an elliptic curve and $(D^2) > 0$, we get $H^1(\mathcal{O}_X(nD) \otimes \mathcal{O}_D) = H^1(\mathcal{O}_X(nD) \otimes N) = 0$, and consequently $H^1(\mathcal{O}_X(nD) \otimes T_X \otimes \mathcal{O}_D) = 0$, for all $n \geq 1$. □

13.22. We return to the proof of the fundamental Lemma 13.17 in the case $p = \operatorname{char}(\Bbbk) > 0$. Let $A = W(\Bbbk)$ be the ring of Witt vectors of $\Bbbk$; A is a complete discrete valuation ring of characteristic zero, with maximal ideal $\mathfrak{m}$ generated by p and residual field $A/\mathfrak{m}$ isomorphic to $\Bbbk$.

The method of proof is to lift X to characteristic zero and to use the result already proved in characteristic zero. By Lemma 13.21 we have that $H^2(X, T_X) = 0$. Since in addition X is projective and $H^2(\mathcal{O}_X) = 0$, we can use Theorem (7.3), exposé III in [Gro7], which gives the existence of a smooth projective morphism $f : U \to V = \operatorname{Spec}(A)$, whose closed fiber is isomorphic to X. Let X' denote the generic fiber of f. Then X' is a nonsingular projective surface over the quotient field $\Bbbk'$ of A.

Since the fibers of f are two-dimensional, we have $R^i f_*(\mathcal{O}_U) = 0$ for all $i \geq 3$, by [EGA III, (4.2.2)]. Using the base change theorems, cf. [Mum2, II, §5], it follows that the canonical homomorphism

$$(R^2 f_*(\mathcal{O}_U)) \otimes_A (A/\mathfrak{m}) \to H^2(f^{-1}(\mathfrak{m}), \mathcal{O}_U/\mathfrak{m}\mathcal{O}_U) = H^2(X, \mathcal{O}_X) = 0$$

is an isomorphism. By Nakayama's lemma we get that $R^2 f_*(\mathcal{O}_U) = 0$. Repeating the argument we see that $R^1 f_*(\mathcal{O}_U) = 0$, and therefore

$$H^1(X', \mathcal{O}_{X'}) = 0. \tag{13.22.1}$$

Let $\Bbbk''$ be an algebraic closure of $\Bbbk'$, and let $\{\Bbbk'_i\}$ be the family of all finite subextensions $\Bbbk'_i \supseteq \Bbbk'$ of $\Bbbk''$. Put $X'' = X' \otimes_{\Bbbk'} \Bbbk''$ and $T_i = X' \otimes_{\Bbbk'} \Bbbk'_i$. Let A'' be the integral closure of A in $\Bbbk''$, and A_i the integral closure of A in $\Bbbk'_i$. Let $\mathfrak{m}''$ be a maximal ideal of A'' lying over $\mathfrak{m}$. Let B'' be the localization $A''_{\mathfrak{m}''}$, $\mathfrak{m}_i = \mathfrak{m}'' \cap A_i$, $B_i = (A_i)_{\mathfrak{m}_i}$, $\mathfrak{n}'' = \mathfrak{m}''B''$, and $\mathfrak{n}_i = \mathfrak{m}_i B_i$. Since $\Bbbk = A/\mathfrak{m}$ is an algebraically closed field, we have $B''/\mathfrak{n}'' = B_i/\mathfrak{n}_i = \Bbbk$.

Let $V_i = \operatorname{Spec}(B_i)$, $U_i = U \otimes_A B_i$, and $f_i = f \otimes_A B_i : U_i \to V_i$. Since $B_i/\mathfrak{n}_i = \Bbbk$, the closed fiber of f_i is canonically isomorphic to X. Also, the generic fiber of f_i is isomorphic to T_i. Moreover, since the extension $\Bbbk'_i \supseteq \Bbbk'$ is finite, B_i is a discrete valuation ring.

Since the finite subextensions $\Bbbk_i$ form an inductive system with inductive limit equal to $\Bbbk''$, the family of $\mathcal{O}_X$-algebras $\mathcal{O}_{X'} \otimes_{\Bbbk'} \Bbbk_i'$ (with the homomorphisms induced by the natural inclusions $\Bbbk_i' \subseteq \Bbbk_j'$) is an inductive system with inductive limit equal to the $\mathcal{O}_X$-algebra $\mathcal{O}_{X'} \otimes_{\Bbbk'} \Bbbk''$. As $T_i = \mathrm{Spec}(\mathcal{O}_{X'} \otimes_{\Bbbk'} \Bbbk_i')$ and $X'' = \mathrm{Spec}(\mathcal{O}_{X'} \otimes_{\Bbbk'} \Bbbk'')$, we have therefore that X'' is the projective limit of the schemes T_i, by [EGA IV, (8.2.3)]. In particular, the natural morphisms $X'' \to T_i$ induce group homomorphisms $\mathrm{Pic}(T_i) \to \mathrm{Pic}(X'')$, and therefore a canonical group homomorphism

$$a : \mathrm{inj}\lim_i \mathrm{Pic}(T_i) \to \mathrm{Pic}(X'').$$

Theorem (8.5.2) and Proposition (8.5.5) in [EGA IV] show that the homomorphism a is actually an isomorphism.

For the sake of simplicity, for an arbitrary index i, write $V' = V_i$, $U' = U_i$, $f' = f_i$, and $T = T_i$. With this notation, we have the following.

Lemma 13.23. *There exists a group isomorphism*

$$b : \mathrm{Pic}(T) \xrightarrow{\;\cong\;} \mathrm{Pic}(X),$$

defined as follows: Let L be an invertible $\mathcal{O}_T$-module, and let L_1 be an invertible $\mathcal{O}_{U'}$-module such that $L_1|_T \cong L$. Then $b([L]) = [L_1|X]$, where by $[\;]$ we denote the class of an invertible module in its corresponding Picard group.

Proof. First we show that the definition of $b([L])$ does not depend on the choice of the extension L_1 of L to U' (if such an extension exists at all). Indeed, if L_2 is another invertible $\mathcal{O}_{U'}$-module such that $L_2|_T \cong L$, then $(L_1 \otimes L_2^{-1})|_T \cong \mathcal{O}_T$. Since V' is the spectrum of a discrete valuation ring, T is the complement of the effective divisor X, whose normal bundle in U' is trivial. Indeed, if t is a local parameter for the ring B_i, then X has the global equation $t = 0$ on U'.

Since $L_1 \otimes L_2^{-1}$ is trivial in the complement of X and since X is irreducible, we have $L_1 \otimes L_2^{-1} \cong \mathcal{O}_{U'}(mX)$ for some integer m, and therefore $L_1 \otimes L_2^{-1}|_X \cong \mathcal{O}_{U'}(mX)|_X \cong \mathcal{O}_X$. Therefore if the definition of b is at all possible, it does not depend on the choice of an extension of L.

On the other hand, we can always find an extension L_1 of L to U', since U' is a nonsingular scheme (because the morphism f' is smooth and V' is nonsingular). Thus b is correctly defined, and it is clear that b is a group homomorphism.

Finally, we show that b is an isomorphism. Let $\omega_{U'/V'} = \bigwedge^2(\Omega^1_{U'/V'})$. Since $\Omega^1_{U'/V'}$ is locally free of rank two, $\omega_{U'/V'}$ is an invertible $\mathcal{O}_{U'}$-module. Also, $(\omega_{U'/V'})|_X \cong \omega_X$ and $(\omega_{U'/V'})|_T \cong \omega_T$. Thus $b([\omega_T]) = [\omega_X]$. By condition (a) of Lemma 13.16, $[\omega_X]$ generates $\mathrm{Pic}(X)$, and therefore the homomorphism b is surjective.

Now assume that $b([L]) = 0$. If L_1 is an extension of L to U', then we have an exact sequence

$$0 \longrightarrow L_1 \stackrel{p}{\longrightarrow} L_1 \longrightarrow \mathcal{O}_X \longrightarrow 0$$

(for $L_1|_X \cong \mathcal{O}_X$ by $b([L]) = 0$), and therefore a cohomology exact sequence

$$H^0(L_1) \to H^0(\mathcal{O}_X) \to H^1(L_1) \stackrel{p}{\to} H^1(L_1) \to H^1(\mathcal{O}_X) = 0.$$

Since $H^1(L_1)$ is a finitely generated B_i-module (for f' is a proper morphism) and the exact sequence shows, in particular, that $H^1(L_1) = p \cdot H^1(L_1)$, we have $H^1(L_1) = 0$ by Nakayama's lemma. Therefore from the same exact sequence we see that there exists a section $s \in H^0(L_1)$ whose restriction to X coincides with the section $1 \in H^0(\mathcal{O}_X)$. Since f' is a proper morphism and the set $\{z \in U' \mid s(z) \neq 0\}$ is open in U', we have $s(z) \neq 0$ for all $z \in U'$; that is, $L_1 \cong \mathcal{O}_{U'}$, and therefore $L \cong \mathcal{O}_T$.   □

Finally, we complete the proof of the fundamental Lemma 13.17. Noting that the isomorphisms $b_i : \mathrm{Pic}(T_i) \to \mathrm{Pic}(X)$ are compatible with the homomorphisms of the inductive system of groups $\mathrm{Pic}(T_i)$, we get the existence of a canonical isomorphism between $\mathrm{Pic}(X)$ and $\mathrm{inj\,lim}_i \mathrm{Pic}(T_i)$. Since the latter group is canonically identified with $\mathrm{Pic}(X'')$ by the isomorphism a, we have established the existence of a canonical isomorphism between $\mathrm{Pic}(X)$ and $\mathrm{Pic}(X'')$. From the construction of this isomorphism we have that ω_X is mapped to $\omega_{X''}$, and since $\mathrm{Pic}(X) = \mathbb{Z}\omega_X$, we have $\mathrm{Pic}(X'') = \mathbb{Z}\omega_{X''}$. Using [EGA III, (4.7.1)] and the fact that $\omega_{U/V}^{-1}|_X \cong \omega_X^{-1}$ is ample on X, we see that $\omega_{U/V}^{-1}$ is ample on U, and therefore $\omega_{X'}^{-1} = \omega_{U/V}^{-1}|_{X'}$ is ample on X'. Since ampleness is preserved under base change, we have proved that $\omega_{X''}^{-1}$ is ample on X''. Hence X'' satisfies condition (a) of Lemma 13.16.

By Equation (13.22.1) we have $H^1(X'', \mathcal{O}_{X''}) = H^1(X', \mathcal{O}_{X'}) \otimes_{\Bbbk'} \Bbbk'' = 0$.

Now we show that X'' satisfies condition (c) of Lemma 13.16 as well. Note that $(\omega_{X''} \cdot \omega_{X''}) = (\omega_{X'} \cdot \omega_{X'})$, because $\omega_{X''}$ is the inverse image of $\omega_{X'}$ by the canonical morphism $X'' \to X'$.

On the other hand, for every integer m the $\mathcal{O}_U$-module $\omega_{U/V}^m$ is f-flat, $\omega_{U/V}^m|_{X'} \cong \omega_{X'}^m$, and $\omega_{U/V}^m|_X \cong \omega_X^m$. Using the base change theorems, [Mum2, II, §5], we have

$$\chi(X', \omega_{X'}^m) = \chi(X, \omega_X^m), \quad \forall m \in \mathbb{Z}.$$

Referring to the definition of the intersection number, we see that $(\omega_{X'} \cdot \omega_{X'}) = (\omega_X \cdot \omega_X)$. Using also the equality observed earlier, we have $(\omega_{X''} \cdot \omega_{X''}) = (\omega_X \cdot \omega_X) \leq 5$ (because X satisfies condition (c) in Lemma 13.16). Using Noether's formula, we see also that $b_2(X'') \geq 5$.

Summing up, we have constructed a surface X'' over the field $\Bbbk''$ of characteristic zero, with $q(X'') = 0$ and satisfying conditions (a) and (c) in Lemma 13.16. But this contradicts Lemma 13.18. Thus the fundamental

Lemma 13.17, and with it Theorems 13.2 and 13.3, are completely proved in any characteristic.   □

Remark 13.24. The condition $p_2 = q = 0$ in Castelnuovo's criterion of rationality implies immediately the condition $p_g = q = 0$. One may naturally ask whether this weaker condition is already sufficient for rationality. The answer is negative, since the Enriques surfaces, cf. (10.10), and the Godeaux surfaces, cf. (9.6.2), satisfy $p_g = q = 0$ but are not rational.

Example 13.25. Using Castelnuovo's criterion, we can show very easily that every nonsingular surface of degree three in $\mathbb{P}^3$ is rational. Indeed, if X is such a surface, then $H^1(\mathcal{O}_X) = 0$ and $\omega_X = \mathcal{O}_X(-1)$, because X is a complete intersection. Therefore $p_2 = q = 0$, and X is rational by Theorem 13.3. The rationality of X can also be proved directly (by a much more elementary argument) using the fact that X contains a straight line in $\mathbb{P}^3$. (This line has self-intersection -1 on X, and therefore X is not a minimal surface.)

Definition 13.26. Let X be a nonsingular projective variety of dimension $n \geq 1$. X is a *unirational* variety if there exists a separable and dominant rational map $\phi : \mathbb{P}^n \to X$.

Clearly, every rational variety (i.e., every variety that is birationally isomorphic to $\mathbb{P}^n$) is unirational. On the other hand, if X is a unirational curve, then Hurwitz's formula [Har1, p. 301] shows immediately that $X \cong \mathbb{P}^1$.

For surfaces, Castelnuovo's criterion implies the following theorem.

Theorem 13.27. *Every unirational surface is rational.*

Proof. Let $\phi : \mathbb{P}^2 \to X$ be a separable and dominant rational map. After a suitable number of quadratic transformations of $\mathbb{P}^2$, we arrive at a separable, surjective (regular) morphism $\psi : Y \to X$, with Y a rational surface. For each $n \geq 1$, the separable morphism ψ induces an injective map $\psi^* : \Gamma(X, \omega_X^n) \to \Gamma(Y, \omega_Y^n)$ (the separability of ψ is absolutely essential here), and since $p_n(Y) = 0$ for all $n \geq 1$, we have $p_n(X) = 0$ for all $n \geq 1$. On the other hand, if $\alpha_X : X \to \text{Alb}(X)$ is the Albanese morphism of X, we have $q(X) = \dim \text{Alb}(X)$ by (5.3). But $\text{Alb}(X)$ is an Abelian variety generated by $\alpha_X(X) = (\alpha_X \circ \psi)(Y) = (\text{Alb}(\psi) \circ \alpha_Y)(Y) = 0$. Since $p_g(X) = 0$, we get $q(X) = 0$, using (5.3) again. In particular, we have shown that $p_2(X) = q(X) = 0$, and therefore X is rational, by Theorem 13.3.   □

Remarks 13.28.   (i) If the condition of unirationality is weakened by removing the separability requirement for the rational map ϕ, then Theorem 13.27 becomes false in general. Indeed, Zariski [Zar3] has constructed examples of unirational surfaces in this weaker sense, which are not rational.

(ii) The first proof of Castelnuovo's criterion in arbitrary characteristic was given by Zariski in [Zar2, Zar3]. Later M. Artin (cf. [BH]) and H. Kurke [Kur1] gave shorter proofs, entirely different from Zariski's, using étale cohomology instead of the classical cohomology with coefficients in $\mathbb{Z}$, and the Kummer exact sequence instead of the exponential exact sequence used in the proof of Lemma 13.18.

(iii) It was recently proved that there exist unirational varieties of dimension three, namely, certain hypersurfaces of degree 3 and 4 in $\mathbb{P}^4$, that are not rational; see [CG] and [IM].

EXERCISES

Exercise 13.1. Using Castelnuovo's criterion (Theorem 13.3), show that every Del Pezzo surface is rational (cf. Exercise 12.6 for the definition of Del Pezzo surfaces).

Exercise 13.2. Prove that every Del Pezzo surface X is isomorphic to either $\mathbb{F}_0 = \mathbb{P}^1 \times \mathbb{P}^1$ or $\mathbb{P}^2$ blown up at m distinct points in general position, with $0 \le m \le 8$. Recall that the points $P_1, \dots, P_m \in \mathbb{P}^2$ are said to be in general position if no three of them are on a line, no six of them are on a conic, and if $m = 8$, there is no cubic passing through seven of them and passing doubly through the eighth. [*Hint:* By Exercise 13.1 X is rational, and by Theorem 12.8 either $X \cong \mathbb{P}^2$ or X dominates $\mathbb{F}_n$ for some $n \ge 0$. The case $n \ge 2$ is ruled out by Exercise 5.2. Finally, if the points are not in general position, show that ω_X^{-1} cannot be ample.]

Exercise 13.3. Let X be the Fermat surface $x_0^3 + x_1^3 + x_2^3 + x_3^3 = 0$ in $\mathbb{P}^3$ (char($\mathbb{k}$) $\neq 3$). By Example 13.25 X is a rational surface, and by a general result [GH*, Har1, Sha1] every nonsingular cubic surface X in $\mathbb{P}^3$ contains exactly 27 lines. Find the 27 lines explicitly in this case.

Exercise 13.4. Let X be a surface containing infinitely many exceptional curves of the first kind (see Exercise 7.2 for the existence of such surfaces). Prove that X is rational. [*Hint:* Use Castelnuovo's criterion 13.3.]

Exercise 13.5. Let X be a surface with $\kappa(X) = -\infty$ and such that for every irreducible nonsingular curve C on X one has $(C^2) > 0$. Show that $X \cong \mathbb{P}^2$. Show by example that the hypothesis $\kappa(X) = -\infty$ is necessary.

Exercise 13.6. Let X be a surface with ample tangent sheaf T_X (see [Har3] for the definition and elementary properties of ample vector bundles). Prove that $\kappa(X) = -\infty$ and $(C^2) > 0$ for every irreducible nonsingular curve. In particular, by Exercise 13.5, T_X ample implies $X \cong \mathbb{P}^2$ (this last statement belongs to Hartshorne, cf. [Har3]).

Exercise 13.7. Let C be a line in $\mathbb{P}^2$. Denote by D the divisor $2C$ (the first infinitesimal neighborhood of C in $\mathbb{P}^2$). Prove that the restriction map $\text{Pic}(\mathbb{P}^2) \to \text{Pic}(D)$ is surjective. [*Hint:* Consider the truncated exponential sequence $0 \to \mathcal{O}_C(-1) \to \mathcal{O}_D^* \to \mathcal{O}_C^* \to 0$.]

In particular, every surface X that birationally dominates $\mathbb{P}^2$ contains nonsingular curves Y with $(Y^2) > 0$ such that the restriction map $\operatorname{Pic}(X) \to \operatorname{Pic}(2Y)$ is surjective.

Exercise 13.8. Conversely, let Y be a nonsingular curve on a surface X such that $(Y^2) > 0$ and the restriction map $\operatorname{Pic}(X) \to \operatorname{Pic}(2Y)$ is surjective. Show that there exists a birational morphism $f : X \to \mathbb{P}^2$ and an open neighborhood U of Y in X such that $f|_U$ defines an isomorphism $U \cong f(U)$ onto an open subset $f(U)$ of $\mathbb{P}^2$, and $f(Y)$ is a line in $\mathbb{P}^2$. [*Hint:* From the surjectivity of $\operatorname{Pic}(X) \to \operatorname{Pic}(2Y)$ one deduces that the Picard scheme $\underline{\operatorname{Pic}}^0(2Y)$ is an Abelian variety. Then the truncated exponential sequence immediately yields the exact sequence of algebraic groups

$$0 \to H^1(Y, N^{-1}) \to \underline{\operatorname{Pic}}^0(2Y) \to \underline{\operatorname{Pic}}^0(Y) \to 0,$$

from which it follows that $H^0(Y, N^{-1}) = 0$, where N is the normal bundle of Y in X. Since $(Y^2) > 0$, $\deg(N) > 0$, whence this vanishing implies $Y \cong \mathbb{P}^1$ and $(Y^2) = 1$. Then the morphism f is the morphism associated to the linear system $|Y|$, cf. the last part of the proof of Proposition 13.7.]

The statement of Exercise 13.8 was first found by J. D'Almeida [Alm*] (with a more complicated argument). For a higher dimensional analogue, when the situation is a lot more complicated, see [BBI*].

Exercise 13.9. Find examples of nonsingular curves Y on nonrational surfaces X with $(Y^2) \le 0$, such that the map $\operatorname{Pic}(X) \to \operatorname{Pic}(2Y)$ is surjective. [*Hint:* Let C be a nonsingular curve of genus $g > 0$. Take $X = \mathbb{P}(E)$, with $E = \mathcal{O}_C \oplus L$ and L an invertible sheaf of degree d on C such that $d \le 2 - 2g$ and $L \not\cong \omega_C^{-1}$. Take for Y the section of the canonical projection $\mathbb{P}(E) \to C$ corresponding to the projection $E = \mathcal{O}_C \oplus L \to L$.]

Exercise 13.10. Let C be a line in $\mathbb{P}^2$. Prove that $H^0(nC, \mathcal{O}_{nC}) \cong \Bbbk$ for every $n \ge 1$, where $\Bbbk$ is the ground field.

Exercise 13.11. Let C be a line in $\mathbb{P}^2$. Prove that the cokernel of the restriction map $\operatorname{Pic}(\mathbb{P}^2) \to \operatorname{Pic}(3C)$ is isomorphic to the underlying additive group of the ground field $\Bbbk$ (this is the algebraic formulation of the so-called *Reiss condition*, as explained in [GH*]).

Bibliographic References. The proof of Theorem 13.2 is presented after [Mum4], and the proof of Castelnuovo's criterion in characteristic zero included here is due to Kodaira; see [Sha1], or [BH], or [Bea]. The proof of Castelnuovo's criterion in positive characteristic is from [Isk].

14

Zariski Decomposition and Applications

In this chapter we present Zariski's theory of finite generation of the graded algebra $R(X, D)$ associated to a divisor D on a surface X, cf. [Zar1] and some more recent developments related to this theory.

14.1. First we introduce some more terminology. We shall continue to use the term *surface* for a nonsingular projective surface (over the fixed algebraically closed field $\Bbbk$). However, in this chapter we shall also work with singular surfaces, in which case we shall be explicit, depending on the context.

Let X be a normal, complete algebraic surface over $\Bbbk$. Denote by $\mathrm{Div}(X)$ the free Abelian group generated by all integral curves contained in X. An element of $\mathrm{Div}(X)$ is called a *Weil divisor* on X. A Weil divisor $D \in \mathrm{Div}(X)$ is said to be a *Cartier* divisor if for every point $x \in X$ there exists a Zariski open subset U_x of X containing x, and a nonzero rational function $f_x \in \Bbbk(X)$ such that the restriction $D|_{U_x}$ coincides with $(f_x)|_{U_x}$, where (f_x) is the Weil divisor on X associated to f_x (this makes perfect sense because X is normal). Of course, when X is smooth, there is no difference between Weil and Cartier divisors on X. A Weil divisor $D \in \mathrm{Div}(X)$ is said to be $\mathbb{Q}$-*Cartier* if there is a positive integer $n > 0$ such that nD is a Cartier divisor.

Set $\mathrm{Div}_{\mathbb{Q}}(X) = \mathrm{Div}(X) \otimes_{\mathbb{Z}} \mathbb{Q}$. The elements of $\mathrm{Div}_{\mathbb{Q}}(X)$ are called *Weil $\mathbb{Q}$-divisors* on X. Thus, a Weil $\mathbb{Q}$-divisor $D \in \mathrm{Div}_{\mathbb{Q}}(X)$ can be expressed in a unique way as a formal combination $D = \sum_{i=1}^{n} a_i C_i$, with $a_i \in \mathbb{Q}$ and $C_1, \ldots, C_n$ pairwise distinct integral curves on X. D is said to be *positive* (notation: $D \geq 0$) if $a_i \geq 0$, $\forall i = 1, \ldots, n$.

Let D be a Weil $\mathbb{Q}$-divisor on the normal complete surface X. Then one associates to D the sheaf $\mathcal{O}_X(D)$, defined as follows. For every open subset $U \subseteq X$ let

$$\Gamma(U, \mathcal{O}_X(D)) = \{f \in \mathbb{k}(X) \setminus \{0\} \mid [(f) + D]|_U \geq 0\} \cup \{0\},$$

where $\mathbb{k}(X)$ is the field of rational functions on X, (f) is the Weil divisor associated to the rational function f, and $\Delta|_U$ denotes the restriction to U of a Weil divisor $\Delta \in \mathrm{Div}_{\mathbb{Q}}(X)$. From the definition of $\mathcal{O}_X(D)$ it follows that $\mathcal{O}_X(D) = \mathcal{O}_X([D])$, where if $D = \sum_i a_i C_i$ as earlier, we put $[D] = \sum_i [a_i] C_i \in \mathrm{Div}(X)$; for every real number x, $[x]$ denotes the largest integer $\leq x$. In particular, since X is normal, $\mathcal{O}_X(D)$ is a reflexive sheaf of rank 1 on X.

14.2. D-dimension. Let $D \in \mathrm{Div}_{\mathbb{Q}}(X)$ be a $\mathbb{Q}$-divisor on the surface X. In the sequel we shall need Iitaka's definition of the D-dimension of X, denoted by $\kappa(X, D)$, which actually can be defined in a much more general situation (see [Iit1*]). Let $N(X, D) = \{m > 0 \mid H^0(X, \mathcal{O}_X(mD)) \neq 0\}$. If $N(X, D) \neq \varnothing$, note that for every $m, n \in N(X, D)$ we have $mn \in N(X, D)$. In other words, $N(X, D)$ is a subsemigroup of $\mathbb{N}$. By definition, if $N(X, D) = \varnothing$, we say that the D-dimension of X is $-\infty$, and we write $\kappa(X, D) = -\infty$. If $N(X, D) \neq \varnothing$, we can define the graded $\mathbb{k}$-algebra

$$R(X, D) = \bigoplus_{i=0}^{\infty} H^0(X, \mathcal{O}_X(iD)),$$

with multiplication defined as follows. Let $f \in H^0(X, \mathcal{O}_X(iD))$ and $g \in H^0(X, \mathcal{O}_X(jD))$ $(f, g \neq 0)$. Then $(f) + iD \geq 0$ and $(g) + jD \geq 0$, whence $(fg) + (i+j)D = ((f) + iD) + ((g) + jD) \geq 0$. This shows that the product fg of rational functions $f, g \in k(X)$ defines an element of degree $i + j$ of $R(X, D)$. As a ring $R(X, D)$ is a domain. In particular, it makes sense to consider the transcendence degree of $R(X, D)$ over $\mathbb{k}$, denoted by $\mathrm{tr.deg}_{\mathbb{k}}(R(X, D))$. Then by definition

$$\kappa(X, D) = \mathrm{tr.deg}_{\mathbb{k}}(R(X, D)) - 1 \text{ if } N(X, D) \neq \varnothing.$$

We have the following result (see [Iit1*, Iit2*]).

Theorem 14.3 (Iitaka). *With the preceding notation, let $D \in \mathrm{Div}(X)$ be a divisor on the surface X such that $\kappa(X, D) \geq 0$. Then $\kappa(X, D)$ coincides with the number κ of either of the following conditions:*

(i) $\kappa = \max_{n \in N(X,D)} \dim(\varphi_{|nD|}(X))$, *where $\varphi_{|nD|}(X)$ denotes the closure in $\mathbb{P}^{N(n)}$ of the image of the open subset of X on which the rational map $\varphi_{|nD|} : X \dashrightarrow \mathbb{P}^{N(n)} = \mathbb{P}(H^0(X, \mathcal{O}_X(nD))^{\check{}})$ is defined;*

(ii) *There is a unique nonnegative integer κ and integers $m_0, a, b > 0$ such that for all $m \gg 0$ one has the inequalities*

$$am^{\kappa} \leq \dim_{\mathbb{k}} H^0(X, \mathcal{O}_X(mm_0 D)) \leq bm^{\kappa}.$$

In particular, $\kappa(X, D) \leq \dim(X) = 2$.

If K is a canonical divisor on X, then $\kappa(X, K)$ is just the canonical dimension of the surface X studied in the previous chapters.

Definition 14.4. Let D be a divisor on a surface X. Then D is *pseudo-effective* if $(D \cdot H) \geq 0$ for every ample divisor H on X. D is *nef* if $(D \cdot C) \geq 0$ for every integral curve C on X.

Clearly, every effective divisor is pseudo-effective. More generally, every divisor $D \in \mathrm{Div}(X)$ such that $\kappa(X, D) \geq 0$ is pseudo-effective. The converse is not true in general, as the following example shows. Let C be an elliptic curve over $\mathbb{C}$, and set $X = C \times \mathbb{P}^1$. Let δ be a divisor of degree zero on C that is not a torsion element in $\mathrm{Pic}^0(C)$, and set $D = \delta \times \mathbb{P}^1$. Clearly, D is pseudo-effective and $\kappa(X, D) = -\infty$.

On the other hand, every divisor $D \geq 0$ on a surface X such that its linear system has no fixed components is nef. By Theorem 1.25, every nef divisor D satisfies $(D^2) \geq 0$. Every nef divisor is obviously pseudo-effective. Note that the notions of pseudo-effective and nef divisors extend easily to $\mathbb{Q}$-divisors. As an example, let X be a surface of canonical dimension ≥ 0. Then its canonical class K is nef if and only if X is minimal. Indeed, if X is not minimal, let C be an exceptional curve of the first kind on X. Then $(K \cdot C) = (C^2) = -1$, and in particular K cannot be nef. Assume therefore that X is minimal. By Theorems 9.4 and 13.2, K is nef if and only if the canonical dimension of X is ≥ 0, that is, if X is not ruled.

Lemma 14.5. *Let D be a $\mathbb{Q}$-divisor on the surface X. Then D is pseudo-effective if and only if $(D \cdot P) \geq 0$ for every nef divisor P on X.*

Proof. Multiplying D by a suitable positive integer, we may assume that D is an integral divisor. One implication is obvious. For the other, assume that D is pseudo-effective. Let P be a nef divisor and H an ample divisor on X. For every $n \geq 0$, the Nakai-Moishezon criterion (Corollary 1.24) shows that the divisor $H + nP$ is ample. Since D is pseudo-effective, it follows that $(D \cdot (H + nP)) \geq 0, \forall n \geq 0$, which forces $(D \cdot P) \geq 0$. $\square$

The following lemma generalizes Remark 7.11.3.

Lemma 14.6. *Let D be a divisor on the surface X. Assume that D is not pseudo-effective. Then for every $Z \in \mathrm{Div}(X)$ we have $H^0(X, \mathcal{O}_X(nD + Z)) = 0$ for every $n \gg 0$.*

Proof. Since D is not pseudo-effective, there exists an ample divisor H such that $(D \cdot H) < 0$. Then for every Z we have $(nD + Z \cdot H) < 0$ for $n \gg 0$, and therefore $|nD + Z| = \varnothing$. $\square$

Lemma 14.7. *Let D be a nef $\mathbb{Q}$-divisor on the surface X. Then $\kappa(X, D) = 2$ if and only if $(D^2) > 0$.*

Proof. Without loss of generality we may assume that D is an integral divisor (just multiply D by an appropriate positive integer). Assume first that $(D^2) > 0$. For every integer $m > 0$ we have

$$h^0(mD) + h^2(mD) \geq \frac{1}{2}m^2(D^2) - \frac{1}{2}m(D \cdot K) + \chi(\mathcal{O}_X)$$

by Riemann–Roch, where $h^i(D) = \dim_k H^i(X, \mathcal{O}_X(D))$. By Serre duality, $h^2(mD = h^0(K - mD)$. If the divisor $-D$ would be pseudo-effective, by Lemma 14.5 we would get $((-D) \cdot D) \geq 0$ (because D is nef), which would contradict the hypothesis $(D^2) > 0$. Thus $-D$ is not pseudo-effective, and by Lemma 14.6, $h^0(K - mD) = 0$ for all $m \gg 0$. Then the preceding inequality becomes

$$h^0(mD) \geq \frac{1}{2}m^2(D^2) - \frac{1}{2}m(D \cdot K) + \chi(\mathcal{O}_X), \ \forall m \gg 0,$$

which together with $(D^2) > 0$ implies that $\kappa(X, D) = 2$.

Conversely, assume that $\kappa(X, D) = 2$. Since D is nef, $(D^2) \geq 0$ (Theorem 1.25). Assume that $(D^2) = 0$. Since $\kappa(X, D) = 2$, $h^0(mD) \geq 2$ for $m \gg 0$. Set $|mD| = |M| + Z$, where Z is the fixed part of the linear system $|mD|$. Then we have $0 = ((mD)^2) = (M^2) + (M \cdot Z) + m(D \cdot Z)$. Since $|M|$ has no fixed components, M is nef, and in particular $(M^2) \geq 0$ (Theorem 1.25) and $(M \cdot Z) \geq 0$. Since D is nef by hypothesis, $(D \cdot Z) \geq 0$. These inequalities, with the above equality, imply $(M^2) = 0$, which in turn implies that $\varphi_{|mD|}(X)$ is a curve, contradicting the hypothesis that $\kappa(X, D) = 2$ (via Theorem 14.3(i)). □

Lemma 14.8. *Let H be a nef $\mathbb{Q}$-divisor and E an effective $\mathbb{Q}$-divisor on the surface X. If $((H + E) \cdot C) \geq 0$ for every component C of E then $H + E$ is nef.*

The proof is obvious.

Lemma 14.9. *Let $C_1, \ldots, C_q$ be integral curves on the surface X such that the intersection matrix $\|(C_i \cdot C_j)\|_{i,j}$ is negative definite. Let $D = \sum_{i=1}^q a_i C_i$ be a divisor such that $(D \cdot C_i) \leq 0$, $\forall i = 1, \ldots, q$. Then $D \geq 0$.*

Proof. Let $D = A - B$ be the decomposition of D with $A \geq 0$, $B \geq 0$, and A and B having no common components. Since $B \geq 0$ we have $0 \geq (D \cdot B) = (A \cdot B) - (B^2)$. Since A and B have no common components, $(A \cdot B) \geq 0$, whence $(B^2) \geq 0$. On the other hand, because the matrix $\|(C_i \cdot C_j)\|_{i,j}$ is negative definite, $(B^2) \leq 0$. It follows that $(B^2) = 0$, whence $B = 0$ (again because $\|(C_i \cdot C_j)\|_{i,j}$ is negative definite). □

Lemma 14.10. *Let $C_1, \ldots, C_q$ be integral curves on the surface X such that the intersection matrix $\|(C_i \cdot C_j)\|_{i,j}$ is negative definite. Let $D = \sum_{i=1}^q a_i C_i$ be a $\mathbb{Q}$-divisor and let D' be a pseudo-effective $\mathbb{Q}$-divisor such* that $((D' - D) \cdot C_i) \leq 0$, $\forall i = 1, \ldots, q$. Then the divisor $D' - D$ is pseudo-effective.

Proof. Let H be an arbitrary nef $\mathbb{Q}$-divisor on X. Since $\|(C_i \cdot C_j)\|_{i,j}$ is negative definite, there exists a $\mathbb{Q}$-divisor of the form $Y = \sum_{i=1}^{q} \lambda_i C_i$ such that $(Y \cdot C_i) = -(H \cdot C_i)$, $\forall i = 1, \ldots, q$. Since $(Y \cdot C_i) = -(H \cdot C_i) \leq 0$, $\forall i = 1, \ldots, q$, we have $Y \geq 0$ by Lemma 14.9. Then from the hypotheses we get $((D' - D) \cdot Y) \leq 0$. On the other hand, $((H + Y) \cdot C_i) = 0$, $\forall i = 1, \ldots, q$ implies that $H + Y$ is nef (Lemma 14.8), and since D' is pseudo-effective, $((H + Y) \cdot D') \geq 0$ (Lemma 14.5). Combining these inequalities we get $((D' - D) \cdot H) = (D' \cdot H) - (D \cdot H) = (D' \cdot H) + (D \cdot Y) \geq -(D' \cdot Y) + (D \cdot Y) = -((D' - D) \cdot Y) \geq 0$. $\square$

Lemma 14.11. *Let $C_1, \ldots, C_q$ be integral curves on the surface X such that the intersection matrix $I(C_1, \ldots, C_q) = \|(C_i \cdot C_j)\|_{i,j=1,\ldots,q}$ is negative semidefinite of signature $(0, r)$, with $r < q$. Assume also that the matrix $I(C_1, \ldots, C_r) = \|(C_i \cdot C_j)\|_{i,j=1,\ldots,r}$ is negative definite. Then for every $j > r$ there exists an effective $\mathbb{Q}$-divisor $D_j = \sum_{i=1}^{r} a_{ji} C_i$ such that $((D_j + C_j) \cdot C_i) = 0$, $\forall i = 1, \ldots, q$.*

Proof. For each $1 \leq p \leq q$ denote by $V(C_1, \ldots, C_p)$ the $\mathbb{Q}$-vector space spanned by $C_1, \ldots, C_p$ (i.e., the set of $\mathbb{Q}$-divisors of the form $\sum_{i=1}^{p} a_i C_i$, with $a_i \in \mathbb{Q}$). Let $V_1 = V(C_1, \ldots, C_r)$, and let V_2 be the singular $\mathbb{Q}$-subspace of the matrix $I(C_1, \ldots, C_q)$ in $V = V(C_1, \ldots, C_q)$, that is, $V_2 = \{D \in V \mid (D \cdot D') = 0 \ \forall D' \in V\}$. From our hypotheses it follows that $V = V_1 \oplus V_2$. In particular, for every $j > r$ there exist $D_j \in V_1$ and $E_j \in V_2$ such that $C_j = -D_j + E_j$. Since $E_j = C_j + D_j \in V_2$ we get $((D_j + C_j) \cdot C_i) = 0$, $\forall i = 1, \ldots, q$. Moreover, $(D_j \cdot C_i) = -(C_j \cdot C_i) \leq 0$, $\forall i = 1, \ldots, r$, so by Lemma 14.9, $D_j \geq 0$, $\forall j > r$. $\square$

Lemma 14.12. *Let $C_1, \ldots, C_q$ be integral curves on the surface X, and let D be a pseudo-effective $\mathbb{Q}$-divisor on X such that $(D \cdot C_i) \leq 0$, $\forall i = 1, \ldots, q$. Assume that there is an r such that $1 \leq r < q$ and $(D \cdot C_j) < 0$, $\forall j = r + 1, \ldots, q$. If the intersection matrix $I(C_1, \ldots, C_r) = \|(C_i \cdot C_j)\|_{i,j=1,\ldots,r}$ is negative definite, then the intersection matrix $I(C_1, \ldots, C_q) = \|(C_i \cdot C_j)\|_{i,j=1,\ldots,q}$ is also negative definite.*

Proof. Assume first that $I(C_1, \ldots, C_q)$ is not negative semidefinite. Then there exists a $\mathbb{Q}$-divisor $Z \in V(C_1, \ldots, C_q)$ (see the notation of the proof of the previous lemma) such that $(Z^2) > 0$. Multiplying Z by a suitable positive integer, we may assume that $Z \in \mathrm{Div}(X)$. We claim that we may also assume that Z is effective. Indeed, if $Z = A - B$ with $A \geq 0$, $B \geq 0$, and A and B with no common components, we have $0 < (Z^2) = (A^2) + (B^2) - 2(A \cdot B)$. Since $(A \cdot B) \geq 0$, it follows that $(A^2) > 0$ or $(B^2) > 0$. Since we can use A or B instead of Z, the claim is proved. So we may assume $Z > 0$. Then by Riemann-Roch we have

$$h^0(nZ) + h^0(K - nZ) \geq \frac{n^2}{2}(Z^2) - \frac{n}{2}(Z \cdot K) + \chi(\mathcal{O}_X).$$

Since Z is effective, $(H \cdot Z) > 0$ for every ample divisor H on X, whence $(H \cdot (-Z)) < 0$. In particular, $-Z$ cannot be pseudo-effective. Then by Lemma 14.6 (or by (7.11.3)) we deduce that $h^0(K - nZ) = 0$, $\forall n \gg 0$. From the previous inequality it follows that $\kappa(X, Z) = 2$. By Theorem 14.3 the image of X under the rational map $\varphi_{|nZ|}$ is a surface if $n \gg 0$. Replacing Z by a suitable multiple of it, we may therefore assume that $\varphi_{|Z|}(X)$ is a surface. Denote the fixed part of $|Z|$ by F, that is, the linear system $|Z'| \overset{\text{def}}{=} |Z| - F$ has no fixed components, $Z' \geq 0$ and $Z' \in V(C_1, \ldots, C_q)$. Clearly $\kappa(X, Z') = \kappa(X, Z) = 2$. Moreover, since Z' is effective and $|Z'|$ has no fixed components, Z' is nef. Therefore by Lemma 14.7 we get $(Z'^2) > 0$.

Since $Z' \in V(C_1, \ldots, C_q)$ and $Z' \geq 0$, from the hypotheses on D it follows that $(D \cdot Z') \leq 0$. Recalling that Z' is nef and D is pseudo-effective, Lemma 14.5 implies $(D \cdot Z') \geq 0$. Thus $(D \cdot Z') = 0$. From the hypotheses on D it again follows that $Z' \in V(C_1, \ldots, C_r)$. Since $(Z'^2) > 0$, this last fact contradicts the hypothesis that the matrix $I(C_1, \ldots, C_r)$ is negative definite.

Therefore we have proved that the matrix $I(C_1, \ldots, C_q)$ is negative semi-definite. Assume that $I(C_1, \ldots, C_q)$ is not negative definite. By Lemma 14.11, for every $j = r + 1, \ldots, q$, there is an effective $\mathbb{Q}$-divisor $D_j \in V(C_1, \ldots, C_r)$ such that $((D_j + C_j) \cdot C_i) = 0$, $\forall i = 1, \ldots, q$. By Lemma 14.8, $D_j + C_j$ is nef (take $H = 0$ and $E = D_j + C_j$). But from the hypotheses on D it follows that $(D \cdot (D_j + C_j)) \leq (D \cdot C_j) < 0$, which contradicts the fact that D is pseudo-effective. The proof of Lemma 14.12 is complete. $\square$

Corollary 14.13. *Let D be a pseudo-effective divisor on the surface X. Then there are only finitely many integral curves C on X such that $(D \cdot C) < 0$.*

Proof. Let $C_1, \ldots, C_q$ be integral curves on X such that $(D \cdot C_i) < 0$, $\forall i = 1, \ldots, q$. By Lemma 14.12, the intersection matrix $I(C_1, \ldots, C_q)$ is negative definite. This implies that the classes of the curves $C_1, \ldots, C_q$ define linearly independent elements of the Néron–Severi group $\mathrm{NS}(X) = \mathrm{Pic}(X)/\mathrm{Pic}^0(X)$. Then the corollary is a consequence of the fact that the rank of $\mathrm{NS}(X)$ is finite. $\square$

Theorem 14.14 (Zariski–Fujita). *Let D be a pseudo-effective $\mathbb{Q}$-divisor on the surface X. Then D can be uniquely written in the form $D = P + N$, where P is a nef $\mathbb{Q}$-divisor, $N \geq 0$, and if $N > 0$, then $(P \cdot C_i) = 0$, $\forall i = 1, \ldots, q$ and the intersection matrix $I(C_1, \ldots, C_q) = \|(C_i \cdot C_j)\|_{i,j=1,\ldots,q}$ is negative definite, where $C_1, \ldots, C_q$ are the (reduced) irreducible components of $\mathrm{Supp}(N)$.*

Proof. We shall first prove the existence of the decomposition.

If D is nef, put $P = D$ and $N = 0$. Otherwise, let $C_1, \ldots, C_{q_1}$ be all the integral curves C_i on X such that $(D \cdot C_i) < 0$ (by Corollary 14.13, there are only finitely many such curves). By the proof of Corollary 14.13, the intersection matrix $I(C_1, \ldots, C_{q_1})$ is negative definite. Therefore there exists a unique $\mathbb{Q}$-divisor $N_1 \in V(C_1, \ldots, C_{q_1})$ such that $(N_1 \cdot C_i) = (D \cdot C_i)$, $\forall i = 1, \ldots, q_1$. Since $(D \cdot C_i) < 0$, $\forall i = 1, \ldots, q_1$, we infer that $N_1 \geq 0$ (Lemma 14.9). If $D_1 \overset{\text{def}}{=} D - N_1$ is nef, we take $P = D_1$ and $N = N_1$. If not, from the definition of N_1 and Lemma 14.10 it follows that D_1 is pseudo-effective. Let then $C_{q_1+1}, \ldots, C_{q_2}$ be all the integral curves C_j on X such that $(D_1 \cdot C_j) < 0$ (Corollary 14.13). From Lemma 14.12 we deduce that the intersection matrix $I(C_1, \ldots, C_{q_2})$ is negative definite, and consequently, there exists a unique $\mathbb{Q}$-divisor $N_2 \in V(C_1, \ldots, C_{q_2})$ such that $(N_2 \cdot C_j) = (D_1 \cdot C_j)$, $\forall j = 1, \ldots, q_2$. As earlier, $N_2 \geq 0$, and $D_2 \overset{\text{def}}{=} D_1 - N_2$ is pseudo-effective. If D_2 is nef we conclude by taking $P = D_2$ and $N = N_1 + N_2$. If not, we continue the above procedure, which has to stop after finitely many steps because the rank of $\text{NS}(X)$ is finite. This proves the existence part of our theorem.

To prove the uniqueness part we need the following lemma.

Lemma 14.15. *Let $C_1, \ldots, C_q$ be integral curves on the surface X such that the intersection matrix $I(C_1, \ldots, C_q)$ is negative definite. Let D' be an effective $\mathbb{Q}$-divisor on X, and let $D \in V(C_1, \ldots, C_q)$ be a $\mathbb{Q}$-divisor such that $((D' - D) \cdot C_i) \leq 0$, $\forall i = 1, \ldots, q$. Then $D' - D \geq 0$.*

Proof. Write $D' = Z + D_1'$, with $Z \in V(C_1, \ldots, C_q)$ and $C_i \not\subseteq \text{Supp}(D_1')$, $\forall i = 1, \ldots, q$. Then $(D_1' \cdot C_i) \geq 0$, and hence $(Z \cdot C_i) = (D' \cdot C_i) - (D_1' \cdot C_i) \leq (D' \cdot C_i) \leq (D \cdot C_i)$, $\forall i = 1, \ldots, q$. Then by Lemma 14.9, $Z - D \geq 0$, and therefore $D' - D \geq 0$. $\qquad\square$

We can now prove the uniqueness part of Theorem 14.14. Let $D = P + N = P' + N'$ be two decompositions as in the theorem. If $\text{Supp}(N) = C_1 \cup \cdots \cup C_q$, we have $(P \cdot C_i) = 0$ and $(P' \cdot C_i) \geq 0$, $\forall i = 1, \ldots, q$ (because P' is nef). It follows that $(N \cdot C_i) = (D \cdot C_i) = (P' \cdot C_i) + (N' \cdot C_i) \geq (N' \cdot C_i)$, $\forall i = 1, \ldots, q$. Then by Lemma 14.9 we get $N' \geq N$. Similarly $N \geq N'$, whence $N = N'$. $\qquad\square$

Definition 14.16. The decomposition $D = P + N$ of a $\mathbb{Q}$-divisor D on the surface X as in Theorem 14.14 is called the *Zariski decomposition* of D. P is called the *positive part* and N the *negative part* of D.

Theorem 14.14 asserts that every pseudo-effective divisor has a unique Zariski decomposition. Zariski proved that every effective divisor D admits such a decomposition and used it as an essential tool in his investigation of the finite generation of the $\Bbbk$-algebras $R(X, D)$; cf. [Zar1]. Subsequently,

Fujita [Fuj*] showed that, more generally, every pseudo-effective divisor admits a Zariski decomposition.

Conversely, every divisor D admitting a Zariski decomposition $D = P + N$ is necessarily pseudo-effective. Indeed, if H is an arbitrary ample divisor on X then $(D \cdot H) = (P \cdot H) + (N \cdot H)$, and $(P \cdot H) \geq 0$ (because P is nef) and $(N \cdot H) \geq 0$ (because H is ample and N is effective). Note that even if D is a pseudo-effective $\mathbb{Z}$-divisor on X, the positive (resp. negative) part of D might have rational (nonintegral) coefficients. This is one of the important reasons it is natural to consider divisors with rational coefficients on a surface.

An important example of Zariski decomposition is the following. Let X' be a (not necessarily minimal) surface of nonnegative canonical dimension. Then we know that X' dominates a unique minimal model X. Let $f : X' \to X$ be the corresponding birational morphism. Let K be a canonical divisor of X. Since X is minimal of nonnegative canonical dimension, $(K \cdot C) \geq 0$ for every irreducible C on X, that is, K is nef. Then it is easy to see that $f^*(K)$ is also nef. On the other hand, since f is a composition of quadratic transformations and since we know the behavior of the canonical class via a quadratic transformation (see the proof of Proposition 5.7), we get that there is an effective divisor N on X', supported on the exceptional locus of f, such that $K' = f^*(K) + N$ is a canonical divisor of X'. This formula represents the Zariski decomposition of the canonical divisor K'. In fact, the whole idea of the Zariski decomposition of a divisor comes from this formula in connection with the theory of minimal models of surfaces.

Lemma 14.17. *Let $D = P + N$ be the Zariski decomposition of a pseudo-effective divisor D on the surface X (with P the positive part). Then for every $n \geq 0$ one has $H^0(X, \mathcal{O}_X(nD)) = H^0(X, \mathcal{O}_X(nP))$. In particular, $R(X, D) = R(X, P)$ and $\kappa(X, D) = \kappa(X, P)$.*

Proof. Since $nD = nP + nN$ is a Zariski decomposition of nD (if $n > 0$), it is sufficient to prove that $H^0(X, \mathcal{O}_X(D)) = H^0(X, \mathcal{O}_X(P))$. Since $N \geq 0$, the inclusion $H^0(X, \mathcal{O}_X(D)) \subseteq H^0(X, \mathcal{O}_X(P))$ is obvious. Conversely, let $f \in H^0(X, \mathcal{O}_X(D))$ ($f \neq 0$), whence $(f) + D \geq 0$. We must prove that $(f) + P \geq 0$. Set $D' = (f) + D$; since for every irreducible component C of $\text{Supp}(N)$, $(P \cdot C) = 0$, we get $((D' - N) \cdot C) = 0$. Therefore Lemma 14.15 implies $D' - N \geq 0$, that is, $(f) + P \geq 0$. $\square$

Corollary 14.18. *Let D be a pseudo-effective divisor on the surface X, and let $D = P + N$ be the Zariski decomposition of D (with P the positive part). Then $\kappa(X, D) = 2$ if and only if $(P^2) > 0$.*

The corollary is a consequence of Lemmas 14.17 and 14.7.

Zariski considered the following algebro-geometric analogue of Hilbert's fourteenth problem: if X is a nonsingular projective variety and if $D \in \text{Div}(X)$, find conditions under which the $\Bbbk$-algebra $R(X, D)$ is finitely generated. When X is a surface, Zariski solved this problem completely by proving the following fundamental result (see [Zar1]).

Theorem 14.19 (Zariski). *Let* $D \in \mathrm{Div}(X)$ *be a divisor on the surface* X.

(i) *If* $\kappa(X, D) \le 1$ *then the* $\Bbbk$*-algebra* $R(X, D)$ *is finitely generated.*

(ii) *Assume that* $\kappa(X, D) = 2$, *and let* $D = P + N$ *be the Zariski decomposition of* D *(with* P *the positive part of* D*). Then* $R(X, D)$ *is a finitely generated* $\Bbbk$*-algebra if and only if there exists a positive integer* $n > 0$ *such that* $nP \in \mathrm{Div}(X)$ *and the linear system* $|nP|$ *has no fixed components (by Theorem 9.17, the latter condition is also equivalent to the following: there is a positive integer* $n > 0$ *such that* $nP \in \mathrm{Div}(X)$ *and the linear system* $|nP|$ *has no base points).*

One of the main technical tools Zariski used in proving this theorem was the Zariski decomposition of the divisor D. By Lemma 14.17, one may replace D by P and therefore assume that D is nef. The proof of part (i) of Theorem 14.19 is much simpler than the proof of part (ii). Proposition 9.2 and Theorem 9.17 prove (by cohomological arguments) the harder implication of (ii). Instead of proving the converse implication of (ii) (for which we refer the reader to [Zar1] because that implication is quite readable there), in the rest of this chapter we present some applications of this fundamental result.

We shall need the following general contractability criterion due to Grauert–Artin (see [Gra*] in the complex-analytic case, and [Art3] for the generalization of Grauert's result to algebraic spaces).

Theorem 14.20 (Grauert–Artin). *Let* $C_1, \ldots, C_n$ *be integral curves on a surface* X, *such that the intersection matrix* $I(C_1, \ldots, C_n)$ *is negative definite. Then there exists a morphism* $f : X \to Y$, *with* Y *a normal complete two-dimensional algebraic space, such that* $f(C)$ *is a finite set of points of* Y, *and* $f|_{X \setminus C} : X \setminus C \to Y \setminus f(C)$ *is a biregular isomorphism, where* $C \overset{\mathrm{def}}{=} C_1 \cup \cdots \cup C_n$. *Moreover, if* $f' : X \to Y'$ *is another morphism with the same properties as* f, *and* Y' *a normal complete two-dimensional algebraic space, then there exists a unique biregular isomorphism* $u : Y \to Y'$ *such that* $u \circ f = f'$. □

The category of algebraic spaces of dimension 2 over $\Bbbk$ is a slightly larger category than the category of (possibly singular) surfaces over $\Bbbk$. The algebraic spaces over $\Bbbk$ form a category that contains as a full subcategory the category of algebraic varieties over $\Bbbk$. These objects were introduced independently by M. Artin [Art3] and B. Moishezon [Moi*], the main motivation being that certain natural functors in algebraic geometry are represented in the category of algebraic spaces but not necessarily in the category of algebraic varieties. For a systematic account on algebraic spaces see [Knu*].

If $k = \mathbb{C}$ is the field of complex numbers, by a result of Moishezon (loc. cit.), the category of complete irreducible algebraic spaces is equivalent to the category of compact irreducible complex analytic spaces X such that tr. $\deg_{\mathbb{C}} \mathcal{M}(X) = \dim(X)$, where $\mathcal{M}(X)$ is the field of all meromorphic functions on X (in general, if X is an irreducible complex analytic space, a result of Siegel shows that one has the inequality tr. $\deg_{\mathbb{C}} \mathcal{M}(X) \leq \dim(X)$; see, for example, [Sha1, Part III, Chapter 8]. By Theorem 14.20, Nagata's example from (3.1) yields in particular an example of a normal complete two-dimensional algebraic space over $\mathbb{C}$ that is not an algebraic variety. On the other hand, one can prove that every complete nonsingular two-dimensional algebraic space is an algebraic surface (this result extends Theorem 1.25).

The reader who is not familiar with algebraic spaces may assume that Y is a normal complete surface because the applications we have in mind will make use of the following two contractability criteria in the category of projective surfaces (see Theorems 14.21 and 14.23 below).

Theorem 14.21 (M. Artin). *In the hypotheses of Theorem 14.20, assume also that the ground field k is the algebraic closure of a finite field. Then the curve C can be contracted projectively, that is, there exists a morphism $f : X \to Y$, with Y a normal projective surface, such that $f(C)$ is a finite set of points and $f|_{X \setminus C} : X \setminus C \to Y \setminus f(C)$ is a biregular isomorphism. In other words, the algebraic space Y given by Theorem 14.20 is a projective surface.*

Proof. We may clearly assume that C is connected. There are two main steps in the proof. The first step is valid for any algebraically closed ground field k.

Step 1. There exists a positive divisor $E \in \mathrm{Div}(X)$ with $\mathrm{Supp}(E) = C$ such that for every positive divisor $Z > E$ with $\mathrm{Supp}(Z) = C$, the restriction map $\mathrm{Pic}(Z) \to \mathrm{Pic}(E)$ is an isomorphism.

Proof of Step 1. Choose a sufficiently large and divisible integer m with the property that for every $i = 1, \ldots, n$ and for every $L \in \mathrm{Pic}(C_i)$ such that $\deg_{C_i}(L) > m$ we have $H^1(L) = 0$. Since the matrix $I(C_1, \ldots, C_n)$ is negative definite, there is an integral divisor E with $\mathrm{Supp}(E) = C$ such that $(E \cdot C_i) < -m + (C_i^2)$. E is positive by Lemma 14.9. We claim that E satisfies the requirements of step 1.

Indeed, for every integral divisor $Z > E$ with $\mathrm{Supp}(Z) = C$, set $Z - E = \sum_{i=1}^n a_i C_i$, with $a_i \geq 0$. We shall proceed by induction on $\sum_{i=1}^n a_i$. First observe that $((Z - E)^2) = ((Z - E) \cdot \sum_{i=1}^n a_i C_i) < 0$, so that $((Z - E) \cdot C_i) < 0$ for some i, say, for $i = 1$. Then $(Z \cdot C_1) < -m + (C_1^2)$. Set $Z_1 = Z - C_1$. Clearly $Z_1 \geq E$. By induction we may assume that the restriction map $\mathrm{Pic}(Z_1) \to \mathrm{Pic}(E)$ is an isomorphism. Therefore it is sufficient to prove that the restriction map $\mathrm{Pic}(Z) = H^1(Z, \mathcal{O}_Z^*) \to \mathrm{Pic}(Z_1) = H^1(Z_1, \mathcal{O}_{Z_1}^*)$ is an isomorphism. The ideal sheaf of Z_1 in Z is isomorphic to $\mathcal{O}_X(-Z_1)/\mathcal{O}_X(-Z) \cong \mathcal{O}_{C_1}(-Z_1)$. Since this ideal is a square zero ideal on the scheme Z, it is also isomorphic to the kernel of the restriction map $\mathcal{O}_Z^* \to \mathcal{O}_{Z_1}^*$ (via the map $u \mapsto 1 + u$), whence we get the cohomology exact sequence

$$H^1(C_1, \mathcal{O}_{C_1}(-Z_1)) \to \operatorname{Pic}(Z) \to \operatorname{Pic}(Z_1) \to H^2(C_1, \mathcal{O}_{C_1}(-Z_1)) = 0.$$

To conclude the proof of Step 1 it suffices to show that $H^1(C_1, \mathcal{O}_{C_1}(-Z_1)) = 0$. But this is clear by the choice of m, because $\deg_{C_1}(\mathcal{O}_{C_1}(-Z_1)) = (-Z_1 \cdot C_1) = -((Z - C_1) \cdot C_1) > m$.

Step 2. We shall try to adapt the proof of Theorem 3.9. Let H be a very ample divisor on X such that $H^q(X, \mathcal{O}_X(H)) = 0$, $\forall q \geq 1$. Replacing H by a suitable multiple, we may assume that $(H \cdot C_i) = -(Z \cdot C_i)$ for a certain integral divisor $Z > 0$, with $\operatorname{Supp}(Z) = C$ (see the proof of Theorem 3.9). Replacing again Z by a suitable multiple, we may also assume that $Z > E$, with E the divisor from Step 1. The proof of Theorem 3.9 will go through if we show that we can choose H and Z such that $\mathcal{O}_X(H + Z) \otimes \mathcal{O}_Z \cong \mathcal{O}_Z$ and $H^1(\mathcal{O}_X(H)) = H^2(\mathcal{O}_X(H)) = 0$. By Step 1, this is equivalent to the fact that $\mathcal{O}_X(H + Z) \otimes \mathcal{O}_E \cong \mathcal{O}_E$.

Set $L \overset{\text{def}}{=} \mathcal{O}_X(H + Z) \otimes \mathcal{O}_E$. Considering the map $d : \operatorname{Pic}(E) \to \mathbb{Z}^n$ of Lemma 3.4, defined by $d(L) = (\deg_{C_1}(L|_{C_1}), \ldots, \deg_{C_n}(L|_{C_n}))$, we have $L \in \operatorname{Ker}(d)$. But $\operatorname{Ker}(d)$ is the underlying group of the Picard scheme $\underline{\operatorname{Pic}}^0(E)$ of the scheme E, which is a commutative algebraic group over $\Bbbk$.

Now, $\Bbbk$ being the algebraic closure of a finite field, $\Bbbk$ is the union $\cup \Bbbk_i$ of all finite subfields $\Bbbk_i$ of $\Bbbk$. Consequently, $\underline{\operatorname{Pic}}^0(E)(\Bbbk) = \cup_i \underline{\operatorname{Pic}}^0(E)(\Bbbk_i)$, where by $\underline{\operatorname{Pic}}^0(E)(\Bbbk')$ we denoted the group of all $\Bbbk'$-rational points of $\underline{\operatorname{Pic}}^0(E)$ for every subfield $\Bbbk'$ of $\Bbbk$. Since $\Bbbk_i$ is a finite field, the group $\underline{\operatorname{Pic}}^0(E)(\Bbbk_i)$ is also finite. This implies that every element of the group $\underline{\operatorname{Pic}}^0(Z)(\Bbbk)$ has finite order. In particular, $L \in \underline{\operatorname{Pic}}^0(E)(\Bbbk)$ has a finite order s. Replacing H by sH and Z by sZ we get the desired condition.

Therefore we may assume that $L \cong \mathcal{O}_E$, and then the last part of the proof of Theorem 3.9 applies; we get a morphism $f' : X \to Y'$, with Y' a normal projective surface, such that $f(C)$ is a point (because C is connected) and the restriction $f|_{X \setminus C} : X \setminus C \to Y' \setminus f(C)$ a biregular isomorphism. On the other hand, by Theorem 14.20, there is also a contraction $f : X \to Y$ of C to a point, with Y a normal algebraic space. Then by the last part of Theorem 14.20, there is an isomorphism $u : Y \to Y'$ such that $u \circ f = f'$. In particular, Y is a normal projective surface. □

Corollary 14.22. *Every normal complete algebraic surface Y over the algebraic closure $\Bbbk$ of a finite field is projective. Moreover, for every singular point $y \in Y$, the divisor class group of the $\mathfrak{m}_y$-adic completion $\hat{\mathcal{O}}_{Y,y}$ of the local ring $\mathcal{O}_{Y,y}$ is a torsion group.*

Proof. Let $f : X \to Y$ be the minimal desingularization of Y. If $y_1, \ldots, y_s \in Y$ are all the singular points of Y, let $C = f^{-1}(\{y_1, \ldots, y_s\})$. Then C is a curve on X whose intersection matrix is negative definite. By Theorem 1.28, X is a projective surface. Then, by Theorem 14.21, the curve C can be contracted to points on a normal projective surface Y', which is necessarily isomorphic to Y.

For the second part, make the base change via the morphism from $\mathrm{Spec}(\hat{\mathcal{O}}_{Y,y})$ to Y induced by $\mathcal{O}_{Y,y} \to \hat{\mathcal{O}}_{Y,y}$. Set $X' = X \times_Y \mathrm{Spec}(\hat{\mathcal{O}}_{Y,y})$. Then X' is a nonsingular scheme (see (4.2)); therefore we may assume that $Y = \mathrm{Spec}(\hat{\mathcal{O}}_{Y,y})$. Set $U = X \setminus C$, where $C = C_1 \cup \cdots \cup C_n = f^{-1}(y)$. By Step 1 of the proof of Theorem 14.21 and the proof of Theorem 4.6 we have $\mathrm{Pic}(X) = \mathrm{Pic}(E)$. The diagram in the proof of Theorem 4.6 yields the exact sequence

$$0 \to \mathrm{Pic}^0(E) \to \mathrm{Pic}(U) \to F \to 0,$$

where $F \subseteq \mathbb{Z}^n/\delta(H)$ is a finite group. On the other hand, by Step 2 of the proof of Theorem 14.21, $\mathrm{Pic}^0(E)$ is a torsion group (here we use the hypothesis that $\Bbbk$ is the algebraic closure of a finite field). It follows that $\mathrm{Pic}(U)$ is a torsion group, whence the conclusion because $\mathrm{Pic}(U)$ is identified with the divisor class group of $\hat{\mathcal{O}}_{Y,y}$. $\square$

Another contractability criterion that will be used in the sequel is the following.

Theorem 14.23. *In the hypotheses of Theorem 14.20, assume also that for every divisor $Z > 0$ with support in C the map $H^1(\mathcal{O}_X) \to H^1(\mathcal{O}_Z)$ is surjective, and if $\mathrm{char}(\Bbbk) > 0$, that $H^2(\mathcal{O}_X) = 0$. Then the curve C can be contracted projectively, that is, there is a birational morphism $f : X \to Y$ such that Y is a normal projective surface, $f(C)$ is a point of Y, and the restriction $f|_{X \setminus C} : X \setminus C \to Y \setminus f(C)$ is a biregular isomorphism.*

Proof. In the notation of Step 2 in the proof of Theorem 14.21, it suffices to find a very ample divisor H on X with the following properties:

(1) The unique positive divisor $Z = \sum_{i=1}^{n} b_i C_i$ determined by the conditions $(Z \cdot C_i) = -(H \cdot C_i)$, $\forall i = 1, \ldots, n$, is integral and $Z \geq E$, where E is the divisor given by Step 1 of Theorem 14.21 (this step works for every algebraically closed field $\Bbbk$).

(2) $H^1(X, \mathcal{O}_X(H)) = H^2(X, \mathcal{O}_X(H)) = 0$.

(3) $\mathcal{O}_X(H + Z) \otimes \mathcal{O}_Z \cong \mathcal{O}_Z$.

Property (1) is easy to arrange, because we can take for H a sufficiently large and divisible multiple of an arbitrary very ample divisor on X. The delicate point is to check (3). Start with an arbitrary very ample divisor H satisfying (1). Since $((H + Z) \cdot C_i) = 0$, $\forall i = 1, \ldots, n$, $\mathcal{O}_X(H + Z) \otimes \mathcal{O}_Z \in \mathrm{Pic}^0(Z)$.

If $H^1(\mathcal{O}_Z) = 0$ we can proceed as in the proof of Theorem 3.9. Assume therefore that $H^1(\mathcal{O}_Z) \neq 0$, that is, we are not in the case of rational contractions in the sense of Theorem 3.9. Consider then the morphism of Picard schemes $\alpha : \underline{\mathrm{Pic}}^0(X) \to \underline{\mathrm{Pic}}^0(Z)$. The tangent space map $T_{\underline{\mathrm{Pic}}^0(X),0} \to T_{\underline{\mathrm{Pic}}^0(Z),0}$ associated to α is identified to the restriction map $H^1(\mathcal{O}_X) \to H^1(\mathcal{O}_Z)$, which by hypothesis is surjective. By a result of Cartier, the Picard scheme $\underline{\mathrm{Pic}}^0(X)$ is always reduced (i.e., $\underline{\mathrm{Pic}}^0(X)$ is an Abelian variety) if $\mathrm{char}(\Bbbk) = 0$. If instead $\mathrm{char}(\Bbbk) > 0$, $\underline{\mathrm{Pic}}^0(X)$ is also an Abelian variety because in that case we have the additional hypothesis that $H^2(\mathcal{O}_X) = 0$ (see [Mum1, Lectures 25 and 27]).

Then the surjectivity of the tangent space map $T_{\underline{\mathrm{Pic}}^0(X),0} \to T_{\underline{\mathrm{Pic}}^0(Z),0}$ implies the surjectivity of the morphism $\alpha : \underline{\mathrm{Pic}}^0(X) \to \underline{\mathrm{Pic}}^0(Z)$.

Set $L = \mathcal{O}_X(H+Z) \otimes \mathcal{O}_Z$. It follows that there is a numerically trivial line bundle $M \in \mathrm{Pic}^0(X)$ such that $M|_Z \cong L$. Let H' be a divisor on X such that $\mathcal{O}_X(H') = \mathcal{O}_X(H) \otimes M^{-1}$. The line bundles $\mathcal{O}_X(H)$ and $\mathcal{O}_X(H')$ are numerically equivalent, so that H ample implies H' ample and $(H \cdot C_i) = (H' \cdot C_i)$, $\forall i = 1, \ldots, n$. Moreover, by construction, $\mathcal{O}_X(H') \otimes \mathcal{O}_Z \cong \mathcal{O}_X(H) \otimes M^{-1} \otimes \mathcal{O}_Z \cong L \otimes L^{-1} \cong \mathcal{O}_Z$.

In other words, the ample line bundle H' satisfies conditions (1) and (3). Finally, we have to check that we can choose a very ample line bundle H' which satisfies all three conditions (1), (2), and (3). If H' satisfies (1) and (3), then so does aH' for every $a > 0$. Indeed, we have to check that $\mathcal{O}_X(aH' + aZ) \otimes \mathcal{O}_{aZ} \cong \mathcal{O}_{aZ}$. But since $aZ \geq Z \geq E$, by Step 1 in the proof of Theorem 14.21, the existence of this isomorphism is equivalent to $\mathcal{O}_X(aH' + aZ) \otimes \mathcal{O}_E \cong \mathcal{O}_E$, and the existence of the latter isomorphism comes from $\mathcal{O}_X(H'+Z) \otimes \mathcal{O}_Z \cong \mathcal{O}_Z$. So, to check that condition (2) can be arranged, we may use a well-known result of Serre, which says that $H^i(\mathcal{O}_X(aH')) = 0$ for $i \geq 1$ and $a \gg 0$. □

Remark. If $\mathrm{char}(\Bbbk) > 0$ then the hypothesis $H^2(\mathcal{O}_X) = 0$ in Theorem 14.23 can be weakened to the following: if there is a divisor $Z > 0$ with support in C such that $H^1(\mathcal{O}_Z) \neq 0$, then the Bockstein operators on X (see [Mum1, Lecture 27]) are all zero. Therefore Theorem 14.23 can be considered as a generalization of Artin's Theorem 3.9.

14.24. Intersection of Weil divisors. Let X be a normal complete two-dimensional algebraic space (e.g., a normal complete algebraic surface). The concept of desingularization of a normal two-dimensional algebraic space can be defined as in (3.16), and the theorem of Zariski–Abhyankar concerning the existence of desingularizations extends to this case. In other words, there is a birational morphism $f : X' \to X$ from a surface X' (i.e., a nonsingular projective surface X'). Set $Z = \{x \in X \mid \dim f^{-1}(x) > 0\}$. Then Z is a finite set (containing the singular locus of X) and, by Zariski's Main Theorem, the restriction $f|_{X' \setminus f^{-1}(Z)} : X' \setminus f^{-1}(Z) \to X \setminus Z$ is an isomorphism. Moreover $f^{-1}(Z)$ is a curve on X' and the intersection matrix of its components $E_1, \ldots, E_n$ is negative definite. Let $D = \sum_{i=1}^{m} a_i C_i$ be a Weil divisor on X, with $a_i \in \mathbb{Z}$ and $C_1, \ldots, C_m$ pairwise distinct integral curves on X. According to [Mum5], one can define the inverse image $f^*(D)$ in the following way. Define first the proper transform D' of D (via f) by $D' \overset{\text{def}}{=} \sum_{i=1}^{m} a_i C_i'$, where C_i' is the closure in X' of $f^{-1}(C_i \setminus Z)$, that is, C_i' is the strict transform of C_i via f. Then we define

$$f^*(D) = D' + \sum_{j=1}^{n} b_j E_j,$$

where the coefficients b_j are the rational numbers uniquely determined by the conditions

$$(D' \cdot E_h) + \sum_{j=1}^{n} b_j (E_h \cdot E_j) = 0, \quad \forall h = 1, \ldots, n;$$

these b_j exist and are unique because the intersection matrix

$$I(E_1, \ldots, E_n) = \|(E_i \cdot E_j)\|_{i,j=1,\ldots,n}$$

is negative definite. Note that in general $f^*(D)$ is only a $\mathbb{Q}$-divisor, even if D is an integral Weil divisor on X. However, if D is a Cartier divisor, $f^*(D)$ is an integral divisor and coincides with the usual inverse image of Cartier divisors. For every two Weil divisors D_1 and D_2 on X, Mumford (loc. cit.) defines the intersection number by

$$(D_1 \cdot D_2) \overset{\text{def}}{=} (f^*(D_1) \cdot f^*(D_2)),$$

where the term on the right-hand side comes from the usual intersection number of two integral divisors on the surface X' by extending it to $\mathbb{Q}$-divisors. In particular, $(D_1 \cdot D_2)$ may be a rational number. This number is an integer if at least one of D_1 and D_2 is a Cartier divisor on X. The remarkable feature of this definition is that $(D_1 \cdot D_2)$ is independent of the choice of the desingularization. This fact is not difficult to prove and is left as an exercise to the reader. Of course, the definition of $f^*(D_1)$ and $(D_1 \cdot D_2)$ extends immediately to the case when D_1 and D_2 are Weil $\mathbb{Q}$-divisors.

14.25. In what follows we shall denote by (X, D) a pair consisting of a surface X and a divisor $D \in \text{Div}(X)$ such that $\kappa(X, D) = 2$. Then by Theorem 14.14 there is a Zariski decomposition $D = P + N$, with $P \in \text{Div}_{\mathbb{Q}}(X)$ nef (the positive part of D) and $N \in \text{Div}_{\mathbb{Q}}(X)$ effective (the negative part of D), satisfying the conditions of Theorem 14.14. By Lemma 14.18 the condition $\kappa(X, D) = 2$ is equivalent to $(P^2) > 0$. Let $\mathcal{A}$ be the set of all integral curves F of X such that $(P \cdot F) = 0$. By the Hodge index theorem (Corollary 2.4), $\mathcal{A}$ is a finite set $\{C_1, \ldots, C_n\}$ and the intersection matrix $I(C_1, \ldots, C_n)$ is negative definite. Set $C = C_1 \cup \cdots \cup C_n$. Now we can apply Theorem 14.20 to deduce that there exists a canonical morphism of algebraic spaces $f : X \to Y$, with Y a normal two-dimensional algebraic space, such that $f(C)$ is a finite set of points of Y and the restriction map $f|_{X \setminus C} : X \setminus C \to Y \setminus f(C)$ is a biregular isomorphism. Set $D' = f_*(D)$. The direct image of a divisor $D = \sum_i a_i E_i$ (with E_i pairwise distinct integral curves on X) is defined in the usual way: $f_*(D) = \sum_i a_i f_*(E_i)$, where $f_*(E_i) = f(E_i)$ if $f(E_i)$ is a curve, and $f_*(E_i) = 0$ if $f(E_i)$ is a point; cf. [Ful*]. In this way we get a Weil divisor D' on the two-dimensional algebraic space Y. Note that Y is a projective variety if $\Bbbk$ is the algebraic closure of a finite field (Theorem 14.21), or if the hypotheses of Theorem 14.23 are satisfied.

(14.25.1) Definition. *In the hypotheses of (14.25), (Y, f, D') is called* the model *of the pair (X, D).*

Proposition 14.26. *Let (Y, f, D') be the model of the pair (X, D), where $D \in \mathrm{Div}(X)$ is a divisor on the surface X such that $\kappa(X, D) = 2$. Then $(D'^2) > 0$ and $(D' \cdot F) > 0$ for every integral curve F on Y. Moreover, $f^*(D')$ is the positive part P in the Zariski decomposition $D = P + N$ of the divisor D.*

Proof. Since $D' = f_*(D)$, from the definition of $f^*(D')$ we get that $D = f^*(D') + N'$, where N' is a $\mathbb{Q}$-divisor whose support is contained in C. Since $(P \cdot C_i) = (f^*(D') \cdot C_i) = 0$, we get $(N \cdot C_i) = (N' \cdot C_i)$, $\forall i = 1, \ldots, n$. But $\mathrm{Supp}(N)$ and $\mathrm{Supp}(N')$ are both contained in C, and the matrix $I(C_1, \ldots, C_n)$ is negative definite; thus these equalities imply $N = N'$ (and hence $P = f^*(D')$). From the definition of the intersection number we have $(D'^2) = (f^*(D')^2) = (P^2)$, and the last number is > 0 because $\kappa(X, D) = 2$ (cf. Corollary 4.18). On the other hand, for every integral curve F on Y, $f^*(F) = F' + N''$, where F' is the proper transform of C via f and $\mathrm{Supp}(N'') \subseteq C$. Then $(D' \cdot F) = (P \cdot (F' + N'')) = (P \cdot F') > 0$, because F' does not belong to $\mathcal{A}$. $\qquad\square$

Proposition 14.27. *Let Y be a complete normal two-dimensional algebraic space, and let D' be a Weil $\mathbb{Q}$-divisor on Y such that $(D'^2) > 0$ and $(D' \cdot C) > 0$ for every integral curve C on Y. Let $f : X \to Y$ be the minimal desingularization of Y. Then for every divisor $D \in \mathrm{Div}(X)$ of the form $D = f^*(D') + N$, with $N \geq 0$ and $\mathrm{Supp}(N)$ contained in the exceptional fibers of f, one has $\kappa(X, D) = 2$ and (Y, f, D') is the model of (X, D).*

Proof. One sees immediately that $D = f^*(D') + N$ is the Zariski decomposition of D. Moreover, since $(f^*(D')^2) = (D'^2) > 0$, by Corollary 14.18, $\kappa(X, D) = 2$. The other statements are clear. $\qquad\square$

Theorem 14.28. *Let (X, D) be a pair as in Proposition 14.26, and let (Y, f, D') be the model of (X, D). Then the $\Bbbk$-algebra $R(X, D)$ is finitely generated if and only if the divisor D' is $\mathbb{Q}$-Cartier, that is, if there is an* *integer $m > 0$ such that mD' in an integral Cartier divisor. In particular, if the $\Bbbk$-algebra $R(X, D)$ is finitely generated then $Y \cong \mathrm{Proj}(R(X, D))$ and Y is a normal projective surface.*

Proof. Assume first that $D_1 \overset{\text{def}}{=} mD'$ is an integral Cartier divisor for some $m > 0$. Then by Proposition 14.26 $(D_1^2) > 0$ and $(D_1 \cdot C) > 0$ for every integral curve C on Y. Since D_1 is Cartier one can apply the generalized version (for algebraic spaces) of the Nakai–Moishezon criterion for ampleness (see, e.g., [BU*] if $\Bbbk = \mathbb{C}$) to conclude that D_1 is an ample divisor on Y. In particular, Y is a normal projective surface. (Note that if Y is a normal complete surface then the Nakai–Moishezon criterion 1.22 suffices.) Then for some $s > 0$ the linear system $|sD_1| = |smD'|$ has no base points. Since $f^*(D') = P$, it follows that the linear system $|smP|$ has no base points, and therefore by Zariski's theorem 14.19(ii), the $\Bbbk$-algebra $R(X, D)$ is finitely generated.

Conversely, assume that the $\Bbbk$-algebra $R(X, D)$ is finitely generated, and let $D = P + N$ be the Zariski decomposition of D (with P the positive part of D). By Theorem 14.19(ii) there is an integer $m > 0$ such that mP is integral and the linear system $|mP|$ has no base points. Let $u : X \to \mathbb{P}^d$ be the morphism associated to $|mP|$, so that $u^*(\mathcal{O}_{\mathbb{P}^d}(1)) = \mathcal{O}_X(mP)$. In the notation of (14.24), since $(P \cdot C_i) = 0$, $u(C_i)$ is a point for all $i = 1, \ldots, n$. Since f is the morphism that blows the curves C_i down to normal points, it follows that the morphism u factors as $u = v \circ f$, where $v : Y \to \mathbb{P}^d$ is the morphism induced by u. If we set $L = v^*(\mathcal{O}_{\mathbb{P}^d}(1))$, it follows that $f^*(L) = \mathcal{O}_X(mP)$. Moreover we have

$$f'^*(L|_{Y \setminus f(C)}) \cong \mathcal{O}_X(mP)|_{X \setminus C} \cong \mathcal{O}_X(mD)|_{X \setminus C} \cong f'^*(\mathcal{O}_Y(mD')|_{Y \setminus f(C)}),$$

where $f' = f|_{X \setminus C}$. Taking into account that f' is an isomorphism, it follows that $L|_{Y \setminus f(C)} \cong \mathcal{O}_Y(mD')|_{Y \setminus f(C)}$. Since Y is normal of dimension 2, $f(C)$ is a finite set of points, and the sheaves L and $\mathcal{O}_Y(mD')$ are both reflexive, this latter isomorphism yields an isomorphism $L \cong \mathcal{O}_Y(mD')$. This proves that the divisor mD' is Cartier, because L is invertible.

There only remains to prove that $Y \cong \mathrm{Proj}(R(X, D))$ if $R(X, D)$ is finitely generated. This means that mD' is Cartier for some $m > 0$, whence mD' is an ample Cartier divisor. From the definition of ampleness [EGA II, (4.5.2)] it follows then that $Y \cong \mathrm{Proj}(S)$, with $S = \oplus_{i=0}^{\infty} H^0(Y, \mathcal{O}_Y(imD'))$ (because Y is complete). On the other hand, from the projection formula we have $f_*(f^*(\mathcal{O}_Y(imD'))) \cong \mathcal{O}_Y(imD')$, $\forall i \geq 0$ ($\mathcal{O}_Y(imD')$ is an invertible sheaf on Y). It follows that mP is an integral divisor, and by Proposition 14.26, $f_*(\mathcal{O}_X(imP)) \cong \mathcal{O}_Y(imD')$, $\forall i \geq 0$, where P is the positive part in the Zariski decomposition of D. In other words, $S \cong R(X, P)^{(m)}$ (as $\Bbbk$-algebras), whence $Y \cong \mathrm{Proj}(R(X, P)^{(m)}) \cong \mathrm{Proj}(R(X, P))$, by [EGA II, (2.4.7)]. Finally, by Lemma 14.17, $R(X, P) = R(X, D)$. $\square$

Remark. If in Theorem 14.28 the $\Bbbk$-algebra $R(X,D)$ is not finitely generated, one can prove that $\mathrm{Proj}(R(X,D))$ is isomorphic (as a scheme) to the complement in Y of those singular points of Y around which the divisor D' is not $\mathbb{Q}$-Cartier (see, e.g., [Rus*]). In particular, $\mathrm{Proj}(R(X,D))$ is always a normal algebraic surface whether or not $R(X,D)$ is finitely generated (this reproves and refines an old result of Zariski, see [Zar6*]), and this surface is complete if and only if $R(X,D)$ is a finitely generated $\Bbbk$-algebra. See also [Mum6*] for more interpretation from the point of view of Hilbert's fourteenth problem.

Corollary 14.29. *Let (X,D) be a pair as in Proposition 14.26, and let (Y,f,D') be the model of (X,D). Assume that at least one of the following two conditions is satisfied:*

(i) *Every singularity of Y is rational; or*

(ii) *The ground field $\Bbbk$ is the algebraic closure of a finite field.*

Then the graded $\Bbbk$-algebra $R(X,D)$ is finitely generated.

Proof. In case (i), from Theorem 4.6 it follows that the divisor class group of every singularity y of Y is finite. In case (ii) the divisor class group of the local ring $\mathcal{O}_{Y,y}$ is a subgroup of the divisor class group of $\hat{\mathcal{O}}_{Y,y}$, and the latter is a torsion group by Corollary 14.22. Therefore in both cases the divisor class group of $\mathcal{O}_{Y,y}$ is a torsion group. The conclusion follows from Theorem 14.28. □

Corollary 14.30. *In the hypotheses of Corollary 14.29, the following three conditions are equivalent:*

(i) *The divisor D is nef.*

(ii) *The linear system $|mD|$ has no fixed components for some $m > 0$.*

(iii) *The linear system $|mD|$ has no base points for some $m > 0$.*

Proof. The equivalence [(ii) $\Longleftrightarrow$ (iii)] holds without any special assumptions and is a consequence of Theorem 9.17. The implication [(iii) $\Longrightarrow$ (i)] is obvious. In order to prove [(i) $\Longrightarrow$ (iii)] we can apply Corollary 14.29 to deduce that the $\Bbbk$-algebra $R(X,D)$ is finitely generated. Then by Theorem 14.19 D satisfies (iii), since by (i) D is nef. □

14.31. The Canonical Ring. Let X be a surface, and take $D = K$, a canonical divisor of X. We know that the canonical ring $R(X) = R(X,K)$ is a birational invariant. Therefore, without loss of generality, we may assume that X is minimal. By Theorem 14.19, $R(X)$ is always a finitely generated $\Bbbk$-algebra if the canonical dimension of X is ≤ 1. Therefore we may assume that X is a minimal surface of general type. Then K is nef and $(K^2) > 0$. Let (Y,f,D') be the model of (X,K) in the sense of Definition 14.25. The normal surface Y is obtained by blowing down all integral curves C such that $(K \cdot C) = 0$. Since $(K^2) > 0$, the Hodge index theorem implies $(C^2) < 0$, and therefore the genus formula yields $p_a(C) = 0$ and $(C^2) = -2$. Then by Theorem 3.31 Y is a normal projective surface with at most rational double points as singularities and $D' = K_Y$ is a canonical divisor of Y. Then by Theorem 14.28, the canonical ring $R(X)$ is a finitely generated k-algebra. This fact has already been proved directly in Theorem 9.1.

14.32. The anticanonical ring. Let X be a surface, and take $D = -K$, an anticanonical divisor of X. Set $R^{-1}(X) \overset{\text{def}}{=} R(X, -K)$. The k-algebra $R^{-1}(X)$ is called the *anticanonical ring* of the surface X. $R^{-1}(X)$ is no longer a birational invariant. As far as the problem of finite generation of $R^{-1}(X)$ is concerned, by Theorem 14.19 we may assume that $\kappa^{-1}(X) \overset{\text{def}}{=} \kappa(X, -K) = 2$. (We call $\kappa^{-1}(X)$ the *anticanonical dimension* of X.) Then it follows immediately that the canonical dimension of X is $-\infty$, and therefore, by Theorem 13.2, that X is a ruled surface. Let $g = h^1(X, \mathcal{O}_X)$ (the irregularity of X).

Let $-K = P + N$ be the Zariski decomposition of $-K$. Since $\kappa^{-1}(X) = 2$, $(P^2) > 0$ (Corollary 14.18). The model (Y, f, D') of $(X, -K)$ is called the *anticanonical model* of X. Let $C_1, \ldots, C_n$ denote all the integral curves of X such that $(P \cdot C_i) = 0$, and set $C \overset{\text{def}}{=} C_1 \cup \cdots \cup C_n$. Observe that D' is an anticanonical Weil divisor $-K_Y$ of Y. Indeed, since $f|_{X \setminus C}$ is an isomorphism $X \setminus C \cong Y \setminus f(C)$, $f_*(K)|_{Y \setminus f(C)}$ is a canonical divisor on the algebraic surface $Y \setminus f(C)$, whence $f_*(K) = K_Y$ because $f(C)$ is a finite set and Y is normal of dimension 2. So the anticanonical model of X is $(Y, f, -K_Y)$.

Let $Z > 0$ be an effective divisor with $\text{Supp}(Z) \subseteq C$. Then $K + Z = -P - N + Z$, whence $((K + Z) \cdot P) = -(P^2) < 0$. Since P is nef this implies

$$H^0(\mathcal{O}_X(m(K + Z))) = 0, \quad \forall m > 0 \text{ and } \forall Z > 0 \text{ with } \text{Supp}(Z) \subseteq C.$$
$$(14.32.1)$$

Then the exact sequence

$$0 \to \mathcal{O}_X(K) \to \mathcal{O}_X(K + Z) \to \omega_Z \to 0$$

yields the exact sequence

$$0 = H^0(\mathcal{O}_X(K + Z)) \to H^0(\omega_Z) \to H^1(\mathcal{O}_X(K)),$$

which by duality on Z gives that

the map $H^1(\mathcal{O}_X) \to H^1(\mathcal{O}_Z)$ is surjective; in particular, $h^1(\mathcal{O}_Z) \le g$.
$$(14.32.2)$$

Since X is ruled, $H^2(\mathcal{O}_X) = 0$. Therefore, by Theorem 14.23, Y is a projective surface.

We distinguish two cases:

(i) The surface X is *rational*, that is, $g = 0$.

In this case it follows that $h^1(\mathcal{O}_Z) = 0$ for all $Z > 0$ with support contained in C. Hence every singular point of Y is rational (see Lemma 3.8 and Theorem 3.9). Then every Weil divisor on Y is $\mathbb{Q}$-Cartier. Indeed, Y being normal there are finitely many singularities, and from Theorem 4.6 the divisor class group of every such singularity is finite (because the singularity is rational). Then one has the following result.

Theorem 14.33. *The anticanonical ring $R^{-1}(X)$ of any rational surface X over $\Bbbk$ is a finitely generated $\Bbbk$-algebra. If $(Y, f, -K_Y)$ is the anticanonical model of a rational surface X over $\Bbbk$ with $\kappa^{-1}(X) = 2$, then $Y \cong \operatorname{Proj}(R^{-1}(X))$ is a normal projective surface with at most rational singularities, $H^1(\mathcal{O}_Y) = H^2(\mathcal{O}_Y) = 0$, and $-K_Y$ is an ample $\mathbb{Q}$-Cartier divisor.*

Conversely, let Y be a normal projective surface with at most rational singularities over $\Bbbk$ such that $-K_Y$ is an ample $\mathbb{Q}$-Cartier divisor, and let $f : X \to Y$ be the minimal desingularization of X. Then X is a rational surface with $\kappa^{-1}(X) = 2$ and $(Y, f, -K_Y)$ is the anticanonical model of Y.

Proof. The statement concerning the anticanonical ring follows from Theorem 14.19, (i) if $\kappa^{-1}(X) \leq 1$, and from what we have said and Corollary 14.29 if $\kappa^{-1}(X) = 2$. On the other hand, by Proposition 14.26, $(K_Y^2) > 0$ and $(K_Y \cdot C) < 0$ for every integral curve C on Y. Recalling that $-K_Y$ is $\mathbb{Q}$-Cartier, by the Nakai–Moishezon criterion of ampleness, $-mK_Y$ is an ample Cartier divisor for some $m > 0$.

Conversely, let $f : X \to Y$ be the minimal desingularization of Y. By the definition of the inverse image of Weil divisors, we have $f^*(K_Y) = K_X + \Delta$, with $\Delta \in \operatorname{Div}_{\mathbb{Q}}(X)$ and $\operatorname{Supp}(\Delta) \subseteq C$, where C is the union of all exceptional fibers of f. Let $C_1, \ldots, C_n$ be the integral components of C. Then we have $(K_X \cdot C_i) + (\Delta \cdot C_i) = 0$, $\forall i = 1, \ldots, n$. Since f is the minimal desingularization of Y, $(K_X \cdot C_i) \geq 0$, whence $(\Delta \cdot C_i) \leq 0$, $\forall i = 1, \ldots, n$. Then by Lemma 14.9 it follows that $\Delta \geq 0$. In other words, $-K_X = f^*(-K_Y) + \Delta$, with $\Delta \geq 0$ and $\operatorname{Supp}(\Delta) \subseteq C$. Then from Proposition 14.27 it follows that $\kappa^{-1}(X) = 2$ and $(Y, f, -K_Y)$ is the anticanonical model of the surface X.

There only remains to prove that X is a rational surface. Since $\kappa^{-1}(X) = 2$, we have already observed that X is a ruled surface. Assume that X is not rational, that is, $g \geq 1$. Let $\pi : X \to B$ be the canonical ruled fibration of X, with B a curve of genus g. Then X dominates a minimal model X_0; denote by $\pi_0 : X_0 \to B$ the projection induced by π. By Theorem 12.2 we know that $X_0 \cong \mathbb{P}(E)$, with E a rank 2 vector bundle on B, which we can assume to be normalized in the sense of [Har1, pp. 372–373]. Set $e = -\deg(E)$. Let C_0 be the minimal section of π_0 determined by this normalization of E (loc. cit.). Then $(C_0^2) = -e$ and every divisor $D \in \operatorname{Div}(X_0)$ is linearly equivalent to $mC_0 + \pi_0^*(D')$ for some integer m and some $D' \in \operatorname{Div}(B)$.

Such a divisor is ample if and only if $m > 0$ and either $\deg(D') > me$ (if $e > 0$) or $\deg(D') > \frac{1}{2}me$ (if $e \leq 0$) (cf. [Har1, pp. 380, 382]). Using this, it is a simple exercise to show that D is pseudo-effective if and only if $m \geq 0$, and either $\deg(D') \geq 0$ (if $e > 0$) or $\deg(D') \geq \frac{1}{2}me$ (if $e \leq 0$). As a consequence, the anticanonical divisor of X_0 (which by [Har1, p. 373] is given by $-K_{X_0} = 2C_0 - \pi_0^*(K_B + \det(E)))$ is pseudo-effective if $g > 0$ only in the following cases: $[e \geq 2g - 2]$ or $[g = 1$ and $e = -1, 0]$.

Since X dominates X', $\kappa^{-1}(X) = 2$ immediately implies $\kappa^{-1}(X_0) = 2$. The possibility $g = 1$ with $e = -1, 0$ is ruled out because in this case $-K_{X_0}$ is nef and $(K_{X_0}^2) = 0$. Therefore $g \geq 1$ and $e \geq 2g - 2$. Then a simple calculation shows that the Zariski decomposition of $-K_{X_0}$ is given by $-K_{X_0} = P_0 + N_0$, with

$$P_0 = \left(1 - \frac{2g - 2}{e}\right) C_0 - \pi_0^*(K_B + \det(E)).$$

In particular,

$$(P_0^2) = e\left(1 - \frac{2g - 2}{e}\right)^2.$$

Since $\kappa^{-1}(X_0) = 2$ if and only if $(P_0^2) > 0$, it follows that $e > 2g - 2$, and hence E is of the form $E = \mathcal{O}_B \oplus \det(E)$ ([Har1, p. 376]). In particular, $(C_0^2) = -e < 0$ and $C_0 \cong B$, whence $p_a(C_0) = g \geq 1$. Let C be the strict transform of C_0 via the birational morphism $X \to X_0$. Since C dominates C_0 and $(C_0^2) < 0$, it follows that $(C^2) < 0$ and $p_a(C) > 0$. Then the genus' formula implies $(-K \cdot C) < 0$. If $-K = P + N$ is the Zariski decomposition of $-K$, it follows that $(P \cdot C) + (N \cdot C) < 0$. Since P is nef, we infer that $(N \cdot C) < 0$, and since N is effective, C is a component of $\mathrm{Supp}(N)$. In particular, $(P \cdot C) = 0$. But $p_a(C) > 0$ contradicts the fact that the surface Y has only rational singularities. In other words the assumption that $g > 0$ is absurd, whence X is a rational surface. $\square$

The second case to be analyzed is the following:

(ii) The surface X is *ruled and nonrational*. Then one has the following result.

Theorem 14.34. *Assume that* $\mathrm{char}(\Bbbk) = 0$. *Let* X *be a nonrational ruled surface over* $\Bbbk$ *with* $\kappa^{-1}(X) = 2$, *and let* $\pi : X \to B$ *be the canonical ruled fibration, with* B *a nonsingular curve of genus* $g \geq 1$. *Then the anticanonical model* $(Y, f, -K_Y)$ *has the following properties:*

(i) Y *is a normal projective surface such that* $(K_Y^2) > 0$ *and* $(K_Y \cdot D) < 0$ *for every integral curve* D *on* Y. *Moreover,* $H^1(\mathcal{O}_Y) = H^2(\mathcal{O}_Y) = 0$.

(ii) Y *has precisely one nonrational singularity* $y \in Y$, *and (possibly) finitely many rational singularities. The geometric genus of* (Y, y) *is* g, *that is,* $\dim_{\mathbb{C}} R^1 f_*(\mathcal{O}_X)_y = g$.

(iii) *The exceptional fibers of f over the rational singularities of Y are contained in the degenerated fibers of π, and the exceptional fiber of f over the nonrational singularity y consists of a section of π and (possibly) some irreducible components of the degenerated fibers of π.*

(iv) *The anticanonical ring $R^{-1}(X)$ is a finitely generated $\Bbbk$-algebra if and only if the nonrational singularity (Y, y) is $\mathbb{Q}$-Gorenstein, that is, if there is a positive integer $r > 0$ such that rK_Y is a Cartier divisor at y. (By (i) and the Nakai–Moishezon criterion, this last condition amounts to the fact that $-K_Y$ is an ample $\mathbb{Q}$-Cartier divisor.)*

Conversely, let Y be a normal two-dimensional algebraic space over $\Bbbk$ with at least one nonrational singularity, such that $(K_Y^2) > 0$ and $(K_Y \cdot D) < 0$ for every integral curve D on Y. Let $f : X \to Y$ be the minimal desingularization of Y. Then X is a nonrational ruled surface with $\kappa^{-1}(X) = 2$, and $(Y, f, -K_Y)$ is the anticanonical model of X. In particular, Y is a (normal) projective surface such that $H^1(\mathcal{O}_Y) = H^2(\mathcal{O}_Y) = 0$.

Proof. We know already that Y is a projective surface. We are keeping the notation of (14.32). By the Zariski–Grothendieck theorem on holomorphic functions [EGA III, (4.2.1)] we have $R^1 f_*(\mathcal{O}_X) = \operatorname{proj\,lim}_Z H^1(\mathcal{O}_Z)$ (where Z runs over all effective divisors on X with support contained in C), and since the maps $H^1(\mathcal{O}_Z) \to H^1(\mathcal{O}_{Z'})$ (with $Z \geq Z'$) are all surjective, by (14.32.2), we get

$$\dim_{\mathbb{C}} R^1 f_*(\mathcal{O}_X) \leq g, \qquad (14.32.3)$$

that is, the sum of the geometric genera of the singular points of Y is less than or equal to g. (By definition, the geometric genus of a singular point $y \in Y$ is $\dim_{\mathbb{C}} R^1 f_*(\mathcal{O}_X)_y$.)

The argument for the rationality of X in the proof of Theorem 14.33 shows that Y cannot have only rational singularities. Therefore Y has at least one nonrational singularity. We shall use the following two lemmas.

Lemma 14.35. *Let X be a nonrational ruled surface, and let $\pi : X \to B$ be the canonical ruled fibration, with B a nonsingular curve of genus $g \geq 1$. Let A be a connected curve on X of irreducible components $A_1, \ldots, A_s$. If the intersection matrix $I(A_1, \ldots, A_s)$ is negative definite and A is contained in a degenerated fiber of π, then $H^1(\mathcal{O}_Z) = 0$ for all positive divisors Z with $\operatorname{Supp}(Z) \subseteq A$.*

Lemma 14.36. *Assume that $\operatorname{char}(\Bbbk) = 0$. Let X be a ruled surface of genus (or irregularity) 1. Let E be an elliptic curve on X such that $|m(K + E)| = \varnothing, \forall m \geq 1$. Then E is a section of the canonical ruled fibration $\pi : X \to B$.*

For the time being we take these two lemmas for granted, and continue the proof of the theorem. From Lemma 14.35 we infer that there exists at least one component, say, C_1, which is not contained in any fiber of π. Then C_1 dominates B (via $\pi|_{C_1}$), whence $p_a(C_1) \geq p_a(B) = g$, so by (14.32.2), $p_a(C_1) = g$. If $g \geq 2$ then Hurwitz's formula [Har1, p. 301] implies that C_1 is a section of π. If $g = 1$ we deduce also that C_1 is a section of π (in Lemma 14.36 take $E = C_1$ and use (14.32.1)).

Now we claim that C_1 is the only curve among $\{C_1, \ldots, C_n\}$ that is not contained in any fiber of π. Indeed, assume that there is another curve, say, C_2, not contained in any fiber of π. As before, C_2 is a section of π, and in particular, $p_a(C_2) = g$. If C_1 and C_2 belong to different connected components of C we get a contradiction using (14.32.2). Assume therefore that C_1 and C_2 belong to the same connected component of C (i.e., to the same exceptional fiber of f). We have $p_a(C_1 + C_2) = p_a(C_1) + p_a(C_2) + (C_1 \cdot C_2) - 1$, which is $\geq 2g$ if $(C_1 \cdot C_2) > 0$ and $\geq 2g - 1$ otherwise. Since $p_a(C_1 + C_2) = h^1(\mathcal{O}_{C_1+C_2})$ we get the contradiction that $h^1(\mathcal{O}_{C_1+C_2}) > g$ except for the case $g = 1$ and $(C_1 \cdot C_2) = 0$. In this latter case, there exists a chain of s different curves $C_{i_1}, \ldots, C_{i_s}$ ($s \geq 2$) such that $C_{i_1} = C_1$, $C_{i_s} = C_2$, and $(C_{i_j} \cdot C_{i_{j+1}}) > 0$, $\forall j = 1, \ldots, s - 1$. Then, using induction on j, $p_a(C_{i_1} + \cdots + C_{i_j}) \geq 1$, $\forall j = 1, \ldots, s$. In particular,

$$
\begin{aligned}
p_a(C_{i_1} + \cdots + C_{i_s}) &= p_a(C_{i_1} + \cdots + C_{i_{s-1}}) + p_a(C_{i_s}) \\
&\quad + ((C_{i_1} + \cdots + C_{i_{s-1}}) \cdot C_{i_s}) - 1 \\
&\geq p_a(C_{i_1} + \cdots + C_{i_s}) + p_a(C_{i_s}) \\
&\geq 1 + 1 = 2.
\end{aligned}
$$

Since the curve $Z \stackrel{\text{def}}{=} C_{i_1} + \cdots + C_{i_s}$ is reduced, this implies $h^1(\mathcal{O}_Z) \geq 2$. This contradicts (14.32.2) once again, because $g = 1$.

Therefore we have shown that there is precisely one component C_1 of C that is a section of π, and $C_2, \ldots, C_n$ are all contained in the degenerated fibers of π. Recalling Lemma 14.35, this proves parts (ii) and (iii) of the theorem.

Consider now the exact sequence arising from the Leray spectral sequence of f:

$$0 \to H^1(\mathcal{O}_Y) \to H^1(\mathcal{O}_X) \stackrel{\epsilon}{\to} R^1 f_*(\mathcal{O}_X) \to H^2(\mathcal{O}_Y) \to H^2(\mathcal{O}_X). \quad (14.32.4)$$

Identifying $R^1 f_*(\mathcal{O}_X)$ to $\mathrm{proj}\lim_Z H^1(\mathcal{O}_Z)$, the map ϵ in this exact sequence is induced by the restriction maps $\epsilon_Z : H^1(\mathcal{O}_X) \to H^1(\mathcal{O}_Z)$. Taking $Z = C_1$ we infer that ϵ_{C_1} is an isomorphism (because C_1 is a section of π and the canonical map $\pi^* : H^1(\mathcal{O}_B) \to H^1(\mathcal{O}_X)$ is an isomorphism). Since $\dim_{\mathbb{C}} R^1 f_*(\mathcal{O}_X) = g$ and the maps $H^1(\mathcal{O}_Z) \to H^1(\mathcal{O}_{Z'})$ of this inverse limit are surjective, we infer that ϵ is an isomorphism. In particular, the exact sequence (14.32.4) implies that $H^1(\mathcal{O}_Y) = H^2(\mathcal{O}_Y) = 0$ ($H^2(\mathcal{O}_X) = 0$ because X is ruled). By Proposition 14.26, this proves part (i).

Part (iv) follows directly from Theorem 14.28.

Now we prove the converse (the last part of the theorem). By Proposition 14.27 it follows that $\kappa^{-1}(X) = 2$ (and in particular X is a ruled surface), and that $(Y, f, -K_Y)$ is the anticanonical model of X. Because Y has a nonrational singularity, X cannot be a rational surface, by Theorem 14.33. The rest follows from (i). In this way Theorem 14.34 is proved modulo Lemmas 14.35 and 14.36, which we will prove below.   □

Proof of Lemma 14.35. Let D be the degenerated fiber of π containing A. We claim that $H^1(\mathcal{O}_D) = 0$. To see this, observe that by the base-change theorems [Har1, p. 290] the canonical map $R^1\pi_*(\mathcal{O}_X) \otimes \Bbbk(b) \to H^1(\mathcal{O}_D)$ is an isomorphism. Therefore our assertion will follow if we prove that $R^1\pi_*(\mathcal{O}_X) = 0$. To prove this last fact, apply Theorem 12.2 to deduce that the morphism π factors as $\pi = \pi' \circ g$, where $g : X \to X' \overset{\text{def}}{=} \mathbb{P}(E)$ is a birational morphism onto a geometrically ruled surface $\mathbb{P}(E)$ with E a rank 2 vector bundle on B, and $\pi' : \mathbb{P}(E) \to B$ is the canonical projection of $\mathbb{P}(E)$. Moreover, $R^p g_*(\mathcal{O}_X) = 0$, $\forall p > 0$, $g_*(\mathcal{O}_X) = \mathcal{O}_{X'}$, and $R^q\pi'_*(\mathcal{O}_{X'}) = 0$, $\forall q > 0$. Then everything follows from the degenerated Leray spectral sequence of $\pi = \pi' \circ g$.

Therefore we have proved that $H^1(\mathcal{O}_D) = 0$. Using this and the exact sequence

$$0 \to \mathcal{O}_X(-mD) \otimes \mathcal{O}_D = \mathcal{O}_D \to \mathcal{O}_{(m+1)D} \to \mathcal{O}_{mD} \to 0, \qquad (m \geq 1)$$

we get by induction on m that $H^1(\mathcal{O}_{mD}) = 0$, $\forall m \geq 1$. Now, if Z is an effective divisor with support in A, there is a positive integer m such that $mD \geq Z$, and hence we get an exact sequence of the form

$$0 \to L \to \mathcal{O}_{mD} \to \mathcal{O}_Z \to 0,$$

with $\mathrm{Supp}(L) \subseteq D$. Taking cohomology and using the equalities $H^2(L) = 0$ and $H^1(\mathcal{O}_{mD}) = 0$, we get $H^1(\mathcal{O}_Z) = 0$.   □

Proof of Lemma 14.36. Assume that $m \overset{\text{def}}{=} (E \cdot F) \geq 2$, where F is a general fiber of π, and denote by $p : E \to B$ the restriction $\pi|_E$, and by $\tilde{X}$ the fiber product $X \times_B E$ (via p). Since E and B are elliptic curves, p is a finite étale morphism of degree m, so that the projection $p' : \tilde{X} \to X$ is also finite and étale of degree m. It follows that $\tilde{E} \overset{\text{def}}{=} p'^*(E)$ is a reduced curve consisting of m different sections of the ruled surface $\pi' : \tilde{X} \to E$. Clearly, $\tilde{K} + \tilde{E} = p'^*(K + E)$, where $\tilde{K}$ is a canonical divisor of $\tilde{X}$. The hypothesis says that $\kappa(X, K + E) = -\infty$. But since p' is a finite surjective morphism, by a general property, $\kappa(\tilde{X}, \tilde{K} + \tilde{E}) = \kappa(X, K + E)$ (see [Iit1*, Theorem 4]), so that $\kappa(\tilde{X}, \tilde{K} + \tilde{E}) = -\infty$. In particular, $|\tilde{K} + \tilde{E}| = \varnothing$. But this last fact contradicts Proposition 13.5(i) if $m \geq 2$ (because $\tilde{E}$ consists of m irreducible components, each of them isomorphic to E).   □

Remark. The hypothesis that $\mathrm{char}(\Bbbk) = 0$ is needed in Theorem 14.34 because the proof uses Lemma 14.36—and the proof of Lemma 14.36 makes essential use of the assumption that $\Bbbk$ has characteristic zero.

14.37. Examples of Rational Anticanonical Models. Let X be a surface obtained by blowing up m (possibly infinitely near) points $P_1, \ldots, P_m$ of the projective plane $\mathbb{P}^2$. If $m \leq 8$, it is a simple exercise to check that $\kappa^{-1}(X) = 2$. By Theorem 14.33 the anticanonical model $(Y, f, -K_Y)$ has the property that Y is a normal projective surface whose singularities are all rational. We distinguish the following situations:

(i) Assume that the points $P_1, \ldots, P_m$ are in general position, which means the following: no three of them are on the same line, no six of them are on the same conic, and (if $m = 8$) there is no cubic passing through seven of them and passing doubly through the eighth. Then it is a well-known fact that X is a (classical) Del Pezzo surface, that is, the anticanonical system is ample (see, e.g., [Dem*, exposé II] or [Man*]). Therefore the anticanonical model of X in this case is X itself.

(ii) Assume now that the points are in almost general position (see [Dem*, exposé III]). This is a natural condition which includes the following: no four points are on the same line, and no seven points are on the same irreducible conic. One proves that the points $P_1, \ldots, P_m$ are in almost general position if and only if the anticanonical system $|-K|$ of X is nef (and, of course, $(K^2) > 0$ because $m \leq 8$) (see [Dem*, exposé III, Theorem 1]). If $P_1, \ldots, P_m$ are in almost general position, then the anticanonical model $(Y, f, -K_Y)$ has the property that Y has at most rational double points as singularities.

(iii) If $P_1, \ldots, P_m$ $(m \leq 8)$ are not in almost general position then the anticanonical model $(Y, f, -K_Y)$ has the property that Y has at least one rational singularity of multiplicity ≥ 3.

As another example, take $X = \mathbb{F}_n = \mathbb{P}(\mathcal{O}_{\mathbb{P}^1} \oplus \mathcal{O}_{\mathbb{P}^1}(-n))$, $\forall n \geq 0$. Since the surfaces $\mathbb{F}_0 = \mathbb{P}^1 \times \mathbb{P}^1$ and $\mathbb{F}_1$ have the anticanonical system ample (i.e., they are classical Del Pezzo surfaces), we may assume $n \geq 2$. Let D_0 be the unique section of $\pi_n : \mathbb{F}_n \to \mathbb{P}^1$ with $(D_0^2) = -n$. Then the anticanonical divisor of $\mathbb{F}_n$ is $-K = 2D_0 + \pi_n^*(\mathcal{O}_{\mathbb{P}^1}(n+2))$, The positive part P in the Zariski decomposition of $-K$ is $P = (1 + \frac{2}{n})D_0 + \pi_n^*(\mathcal{O}_{\mathbb{P}^1}(n+2))$. One sees immediately that $(P^2) > 0$ and $(P \cdot D_0) = 0$, and since D_0 is the only irreducible curve on $\mathbb{F}_n$ with negative self-intersection (see (12.6)), it follows that the anticanonical model $(Y_n, f_n, -K_{Y_n})$ of $\mathbb{F}_n$ has the property that the surface Y_n is obtained by blowing the curve D_0 down to a point. Therefore the surface Y_n is isomorphic to the projective cone (in $\mathbb{P}^{n+1}$) over the rational normal curve of degree n in $\mathbb{P}^n$. The vertex of this cone is a rational singularity of multiplicity n.

We refer the reader to [Sak3*] for more examples and a more thorough study of the anticanonical models of rational surfaces (Theorem 14.33 is taken from that paper).

14.38. Examples of Nonrational Anticanonical Models. Let B be a nonsingular projective curve of genus $g \geq 1$, $L \in \mathrm{Pic}(B)$ a line bundle of degree $e > 2g - 2$, and set $X \stackrel{\mathrm{def}}{=} \mathbb{P}(\mathcal{O}_B \oplus L^{-1})$. Then X is a geometrically ruled surface over B; let $\pi : X \to B$ be the canonical projection. The canonical surjection $\mathcal{O}_B \oplus L^{-1} \to L^{-1}$ yields the section C_0 of π. Then $(C_0^2) = -e$, and if $L = \mathcal{O}_B(D)$, with $D \in \mathrm{Div}(B)$, then the anticanonical divisor $-K$ of X is given by $-K = 2C_0 + \pi^*(-K_B + D)$, where K_B is the canonical divisor of B. We already remarked in the proof of Theorem 14.33 that $\kappa^{-1}(X) = 2$ and the positive part P of the Zariski decomposition $-K = P + N$ is given by $P = (1 - \frac{2g-2}{e})C_0 + \pi^*(-K_B + D)$. Then it follows easily that C_0 is the only irreducible curve C on X such that $(P \cdot C) = 0$. Therefore the normal projective surface Y of the anticanonical model $(Y, f, -K_Y)$ is obtained by blowing down the curve C_0 to a point. The surface Y thus obtained can be easily described as follows. By [EGA II, 8], $Y = \mathrm{Proj}(R(B, L)[T])$, where $R(B, L) \stackrel{\mathrm{def}}{=} \oplus_{i=0}^{\infty} H^0(B, L^i)$, T is an indeterminate over $R(B, L)$, and $R(B, L)[T]$ is graded by the condition $\deg(aT^j) = i + j$, $\forall a \in H^0(B, L^i)$. In other words, Y is isomorphic to the projective cone over the polarized curve (B, L) (where the polarization of B is given by the ample line bundle L).

This example is interesting because it yields examples of ruled surfaces X with $\kappa^{-1}(X) = 2$ for which the anticanonical ring is not finitely generated. Indeed, we have the following.

Lemma 14.39. *Let $L \in \mathrm{Pic}(B)$ be a line bundle of degree $e > 0$ on a nonsingular projective curve B of genus $g \geq 2$, such that the line bundle (of degree zero) $L^{2g-2} \otimes \mathcal{O}_B(-eK_B)$ is not a torsion element in $\mathrm{Pic}^0(B)$ (there are many such examples when $\Bbbk = \mathbb{C}$). Denote by $Y = \mathrm{Proj}(R(B, L)[T])$ the projective cone over the polarized curve (B, L), that is, the normal surface obtained by blowing down the curve C_0 on the surface $X \stackrel{\mathrm{def}}{=} \mathbb{P}(\mathcal{O}_B \oplus L^{-1})$, as in (14.38). Then K_Y is not a $\mathbb{Q}$-Cartier divisor on Y.*

In particular, if $e > 2g - 2$ then the anticanonical ring $R^{-1}(X)$ of the surface X (with $\kappa^{-1}(X) = 2$ if $e > 2g - 2$) is not a finitely generated $\Bbbk$-algebra.

Proof. We have $N_{C_0} = \mathcal{O}_X(C_0) \otimes \mathcal{O}_{C_0} \cong L^{-1}$. Assume that there exists an integer $m > 0$ such that $-mK_Y$ is a Cartier divisor on Y. Then $f^*(\mathcal{O}_Y(-mK_Y)) \cong \mathcal{O}_X(-mK + sC_0)$, with $s \in \mathbb{Z}$. If y is the vertex of the cone Y, then $C_0 \cong f^{-1}(y)$. Restricting to $C_0 \cong B$, using the fact that $f^*(\mathcal{O}_Y(-mK_Y))$ is trivial along C_0, and taking into account the adjunction formula, we get $\mathcal{O}_B(-mK_B) \otimes L^{m+s} \cong \mathcal{O}_B$. But this contradicts the fact that $L^{2g-2} \otimes \mathcal{O}_B(-eK_B)$ is not a torsion element in $\mathrm{Pic}^0(B)$. The assertion about $R^{-1}(X)$ follows from the first part of the lemma and Theorem 14.34(iv). $\qquad\square$

By Theorem 14.28 we know that if the graded $\Bbbk$-algebra $R(X, D)$ associated to a pair (X, D) with $\kappa(X, D) = 2$ is finitely generated then Y is a projective surface, where (Y, f, D') is the model of (X, D). Lemma 14.39 shows that in general the projectivity of Y is not sufficient to ensure that the $\Bbbk$-algebra $R(X, D)$ is finitely generated.

14.40. Canonical models of Gorenstein surfaces. Let Z be a normal complete Gorenstein two-dimensional algebraic space over $\Bbbk$, and let $u :$ $X \to Z$ be the minimal desingularization of Z. Since Z is Gorenstein, any canonical divisor K_Z on Z is Cartier. Set $u^*(K_Z) = K + \Delta$, where $K = K_X$ is a canonical divisor on X and $\Delta \in \mathrm{Div}(X)$ is a positive divisor whose support is contained in the exceptional fibers of u. By Theorem 4.17, $\Delta \geq 0$, and $\Delta = 0$ if and only if Y has at most rational double points. Moreover, if $\Delta > 0$ then $\omega_\Delta \cong \mathcal{O}_\Delta$ and $\mathrm{Supp}(\Delta)$ coincides with the union of the exceptional fibers of u over all singularities of Z that are not rational double points.

We are interested in the canonical ring $R(Z, K_Z)$ of Z. By the projection formula, $u_*(u^*(\mathcal{O}_Y(iK_Z))) \cong \mathcal{O}_Z(iK_Z)$, $\forall i \geq 0$ (recall that $\omega_Z = \mathcal{O}_Z(K_Z)$ is invertible because Z is Gorenstein), whence $R(Z, K_Z) \cong R(X, K+\Delta)$. In other words, we have to consider the pair (X, D), with $\kappa(X, D) = 2$, where $D = K + \Delta$. (If $\kappa(X, D) \leq 1$, then $R(Z, K_Z)$ is automatically finitely generated—but we shall not be interested in that case.) Let (Y, f, D') be the model of (X, D), and let $D = P + N$ be the Zariski decomposition of D. We claim that D' coincides with the canonical divisor K_Y of Y. Indeed, on the one hand, $f_*(K) = K_Y$ (we have already observed this). On the other hand, it suffices to prove that f contracts every integral component C of an exceptional fiber of u. We have to prove that $(P \cdot C) = 0$. This follows from $(D \cdot C) = (K + \Delta \cdot C) = (u^*(K_Z) \cdot C) = 0$, whence $(P \cdot C) + (N \cdot C) = 0$. Since P is nef, $(P \cdot C) \geq 0$. Observe that we cannot have $(P \cdot C) > 0$ because this would imply $(N \cdot C) < 0$, whence C would be a component of N (because N is effective), contradicting the fact that every component of N is orthogonal to P. Thus the claim is proved.

Because f blows down all the exceptional fibers of u and $u_*(\mathcal{O}_X) = \mathcal{O}_Z$ (this equality is a consequence of the normality of Y), there is a unique birational morphism $f_Z : Z \to Y$ such that $f_Z \circ u = f$. Since $u_*(D) = K_Z$ and $f_*(D) = K_Y$, it follows that $(f_Z)_*(K_Z) = K_Y$.

Definition 14.41. In the notation of (14.40), (Y, f_Z, K_Y) is called the *canonical model* of Z. By abuse of language, sometimes we shall say that the surface Y is the canonical model of Z.

If E is an exceptional curve of the first kind on Z that does not pass through any singular point of Z, and if $v : Z \to Z'$ is the morphism that blows down E to a (nonsingular) point (cf. Castelnuovo's criterion of contractability, Theorem 3.30), then clearly $R(Z, K_Z) = R(Z', K_{Z'})$. Then we can blow down all the exceptional curves of the first kind not passing through the singular points of Z (there are only finitely many such curves), without changing the canonical ring. This shows that, without loss of generality, we may assume that there are no exceptional curves of the first kind on Z not passing through any singular point of Z.

On the other hand, let $u' : X' \to Z$ be the minimal desingularization of all rational double points of Z. Then there is a unique birational morphism $v : X \to X'$ such that $u = u' \circ v$. Since by Theorem 3.15 $u^*(\omega_Z) \cong \omega_{X'}$, we infer that $R(Z, K_Z) \cong R(X', K_{X'})$. This shows that, again without loss of generality, we may assume that Z has no rational double points as singularities.

With these two important reductions, we claim that the divisor $D = K + \Delta$ is nef.

To see this, assume that there is an integral curve E on X such that $(D \cdot E) = (K \cdot E) + (\Delta \cdot E) < 0$. Then E cannot be a component of $\mathrm{Supp}(\Delta)$, because this would imply $((K + \Delta) \cdot E) = 0$. Thus $(\Delta \cdot E) \geq 0$ because $\Delta \geq 0$. Therefore $(K \cdot E) < 0$. In the notation and assumptions of (14.40), the inequality $(D \cdot E) < 0$ implies $(P \cdot E) = 0$, and therefore, from the Hodge index theorem, $(E^2) < 0$ (because $(P^2) > 0$). It follows that E is an exceptional curve of the first kind on X. Moreover, $(\Delta \cdot E) = 0$ (otherwise $((K + \Delta) \cdot E) \geq 0$, because $(K \cdot E) = -1$), that is, E does not meet $\mathrm{Supp}(\Delta)$; that is, E does not meet any exceptional fiber of u. Then $u(E)$ is an exceptional curve of the first kind on Z, not passing through any singular point of Z. But by the reductions we made, there are no such curves. Therefore the claim that D is nef is proved.

Now we analyze the integral curves E such that $(D \cdot E) = ((K + \Delta) \cdot E) = 0$. We already know that every component of $\mathrm{Supp}(\Delta)$ has this property. Assume that E is not contained in $\mathrm{Supp}(\Delta)$; in particular, $(\Delta \cdot E) \geq 0$. From the Hodge index theorem, $(E^2) < 0$ (because by the claim, $K + \Delta = P$ and $(P^2) > 0$). If $(K \cdot E) \geq 0$, from $(D \cdot E) = 0$ it follows that $(K \cdot E) = (\Delta \cdot E) = 0$. In particular E does not meet $\mathrm{Supp}(\Delta)$. From the genus formula it follows that $p_a(E) = 0$ and $(E^2) = -2$. Therefore the curves of this type give rise to rational double points of Z. By our reduction there are no rational double points on Z, and so, there are no such curves. If $(K \cdot E) < 0$, since we also have $(E^2) < 0$, it follows that E is an exceptional curve of the first kind. In particular, $(K \cdot E) = -1$, whence $(\Delta \cdot E) = 1$. Such curves may exist on X.

Let E be an exceptional curve of the first kind on X such that $(\Delta \cdot E) = 1$, and let $v : X \to X'$ be the morphism blowing E down to a (smooth) point. Set $\Delta' = v_*(\Delta)$. Since $(\Delta \cdot E) = 1$, the restriction $v|_\Delta$ defines an isomorphism $\Delta \cong \Delta'$. Hence from $\omega_\Delta \cong \mathcal{O}_\Delta$ we get $\omega_{\Delta'} \cong \mathcal{O}_{\Delta'}$. Obviously the morphism $f : X \to Y$ factors as $f = f' \circ v$ (because $(D \cdot E) = 0$). Then $v_*(K + \Delta) = K' + \Delta'$, where K' is the canonical class of X'. It follows easily that $K + \Delta = v^*(K' + \Delta')$, whence $K' + \Delta'$ is nef (because $K + \Delta$ is nef) and $\kappa(X', K' + \Delta') = 2$. If X' contains no exceptional curves of the first kind E' such that $(\Delta' \cdot E') = 1$, the only integral curves C'

such that $(\Delta' \cdot C') = 1$ are the irreducible components of Δ'. If X' still contains exceptional curves of the first kind E' such that $(\Delta' \cdot E') = 1$, we continue this procedure until we reach a decomposition $f = f_0 \circ g$, with $f_0 : X_0 \to Y$ a birational morphism and X_0 a surface that contains no exceptional curves of the first kind E_0 such that $(\Delta_0 \cdot E_0) = 1$, where $\Delta_0 = g_*(\Delta)$ and $\omega_{\Delta_0} \cong \mathcal{O}_{\Delta_0}$. Then the union of all exceptional fibers of g is $\mathrm{Supp}(\Delta_0)$. In these circumstances we can apply Theorem 4.17 to deduce that all singularities of Y are Gorenstein (note that it might happen that Δ_0 contains exceptional curves of the first kind, but Theorem 4.17 can still be applied because the hypothesis that the desingularization $f_0 : X_0 \to Y$ was minimal was used only to deduce that Δ_0 is effective; but in our case we know that Δ is effective, and so $\Delta_0 = g_*(\Delta)$ is effective too).

Since Y has only Gorenstein singularities, K_Y is a Cartier divisor. Then we can apply Theorem 14.28 to deduce the following result.

Theorem 14.42. *Let Z be a normal complete Gorenstein two-dimensional algebraic space. Then the canonical ring $R(Z) = R(Z, K_Z)$ is a finitely generated $\Bbbk$-algebra. In particular, the canonical model Y of Z is isomorphic to $\mathrm{Proj}(R(Z))$, and hence Y is a normal projective surface.*

Remark. The canonical ring $R(Z) = R(Z, K_Z)$ of a non-Gorenstein normal complete two-dimensional algebraic space Z is not always finitely generated. To give an example, in Lemma 14.39 take e such that $0 < e < 2g-2$, $\Bbbk = \mathbb{C}$, and $L^{2g-2} \otimes \mathcal{O}_B(-eK_B)$ is not a torsion element in $\mathrm{Pic}^0(B)$. Then it is easy to see that $0 < e < 2g - 2$ implies $\kappa(X, K + \Delta) = 2$ (in this case Δ, defined by $f^*(K_Y) = K + \Delta$, is only a $\mathbb{Q}$-divisor), and that (Y, f, K_Y) (with $f : X \to Y$ the canonical morphism that blows the curve C_0 down to the vertex of the cone Y) is the model of $(X, K + \Delta)$. From Lemma 14.39 and Theorem 14.28 it follows that the $\mathbb{C}$-algebra $R(X, K + \Delta)$ is not finitely generated. On the other hand, one can prove that $R(X, K + \Delta) \cong R(Y, K_Y)$. This isomorphism is a consequence of the equalities $K_Y = f_*(f^*(K_Y)) = f_*(K + \Delta)$, which by a generalized projection formula [Sak4*, (2.1)] implies that $\mathcal{O}_Y(K_Y) \cong f_*(\mathcal{O}_X(K + \Delta))$.

EXERCISES

Exercise 14.1. Let $f : X' \to X$ be a birational morphism of complete (possibly singular) surfaces, and let D and D' be two $\mathbb{Q}$-divisors on X and on X', respectively.

(i) If X' is nonsingular and X is normal, show that $(f^*(D) \cdot D') = (D \cdot f_*(D'))$.

Now assume that X and X' are both nonsingular. Show that:

(ii) if D is nef then $f^*(D)$ is also nef;

(iii) if D' is pseudo-effective then $f_*(D')$ is also pseudo-effective;

(iv) if D is pseudo-effective then $f^*(D)$ is also pseudo-effective;

(v) if D' is nef then $f^*(f_*(D')) = D' + \Delta$, with $\Delta \in \mathrm{Div}(X')$ effective;

(vi) if D' is nef then $f_*(D')$ is also nef. If in addition $(D'^2) > 0$, then $(f_*(D')^2) > 0$.

Exercise 14.2. Let C be a nonsingular projective curve of genus $g \geq 2$, and denote by X the product $C \times C$ and by E the diagonal of $C \times C$. Prove that the curve E can be contracted projectively to a normal Gorenstein singularity. [*Hint:* One possible way in characteristic not equal to 2 is to consider the image of the morphism $g : C \times C \to J(C)$ into the Jacobian of C given by $g(x, y) = x - y$. Show that the image of g is a (singular) projective surface S. Obviously $g(E)$ is a point of S. To get the projective contraction take the normalization Y of S in the field of rational functions of X (via g). To prove the fact that the singularity obtained by contraction is Gorenstein use Exercise 4.6.]

Exercise 14.3. Let C be a nonsingular projective curve of genus $g \geq 2$, and denote by X the product $C \times C$ and by E the diagonal of $C \times C$. Prove that the divisor $D = K_X + 2E$ is nef, $\kappa(X, D) = 2$, and the graded k-algebra $R(X, D)$ is finitely generated, where K_X is a canonical divisor of X. [*Hint:* The fact that D is nef is easy. Observe that $(D^2) > 0, (D \cdot E) = 0$, and $(D \cdot F) > 0$ for every irreducible curve F of X other than E. Then use Theorem 14.28 and Exercise 14.2 or Theorem 14.42.]

Exercise 14.4. In Exercise 14.2 let Y be the Gorenstein projective surface obtained by contracting the diagonal E of $X = C \times C$. Show that the canonical class of Y is ample.

Exercise 14.5. Consider the elliptic curve E with $(E^2) = -1$ on the surface X from example 3.1. Show that there are nef divisors D on X such that $(D^2) > 0$, $(D \cdot E) = 0$, and $(D \cdot C) > 0$ for every irreducible curve C on X other than E. Then prove that for every such divisor D the k-algebra $R(X, D)$ is not finitely generated.

Exercise 14.6. With the notation of Theorem 14.28, let $\{y_1, \dots, y_n\}$ be all the singular points of the model (Y, f, D') around which D' is not a Cartier divisor. Prove that $\mathrm{Proj}(R(X, D))$ is canonically isomorphic to $Y \setminus \{y_1, \dots, y_n\}$. In particular, $\mathrm{Proj}(R(X, D))$ is always a quasi-projective variety (even though $R(X, D)$ might not be finitely generated). Moreover, $\mathrm{Proj}(R(X, D))$ is projective if and only if the k-algebra $R(X, D)$ is finitely generated. (Cf. [Zar6*], A. Constantinescu (unpublished), or [Rus*].)

Exercise 14.7. Generalize Exercise 12.7 to the case when the elliptic curve Y is an ample Cartier divisor on a normal projective surface X (with $\mathrm{char}(\mathbb{k}) = 0$), by showing that there are only three possibilities:

(a) X is a surface with at most rational double points as singularities and $-Y$ is a canonical divisor of X (these are called *singular Del Pezzo surfaces*, classified in [Dem*]; see also (14.37)), or

(b) X is a geometrically ruled surface $\mathbb{P}(E)$ over Y and Y is embedded in X as a section of the canonical projection $\mathbb{P}(E) \to Y$, or

(c) X is a cone over the polarized curve (Y, N_Y), with N_Y the normal bundle of Y in X.

Exercise 14.8. Let $X = \mathbb{P}(\mathcal{O}_B \oplus L^{-1})$ be the geometrically ruled surface over the nonsingular projective curve B of genus $g \geq 1$ considered in (14.38). Assume that $e = \deg(L) > 4g - 4$ and let d be a positive integer such that $d < e - 4g + 4$. Show that the surface X' obtained by blowing up d arbitrary points of X has anticanonical dimension 2.

Bibliographic References. The Zariski decomposition of divisors is presented after [Zar1] and [Fuj*]. Theorem 14.21 is proved in [Art1], and Theorem 14.23 is inspired by [Bre*]. Theorem 14.28 (which is a reinterpretation of Theorem 14.19 of Zariski in the spirit of [Mum6*]) and Corollary 14.29 were observed by the author in [Băd3*] and (in some special cases) by Sakai. Theorems 14.33 and 14.42 are due to Sakai [Sak3*, Sak1*], and Theorem 14.34 to Bădescu [Băd2*].

15
Appendix: Further Reading

In this appendix we discuss briefly a few themes on algebraic surfaces in which some relevant progress has been made in the last twenty years.

15.1. Further Monographs on Surfaces. Besides the classical monograph of Enriques [Enr], we want to call the reader's attention to the following monographs on the classification of surfaces published after 1980: Barth–Peters–Van de Ven [BPV*] (in which the Enriques–Kodaira classification of all compact complex surfaces is presented), Kurke [Kur2*], and Reid [Rei1*] (where part of the classification is in the spirit of Mori's theory [Mor*]). We also mention the monographs Cossec–Dolgachev [CD*], devoted to Enriques surfaces, and Barthel–Hirzebruch–Höffner [BHH*]. Furthermore, there are several surveys serving as an introduction to the classification of complex surfaces, for example, Catanese [Cat1*, Cat4*] and Peters [Pet*].

15.2. Mori's Theory and Classification of Surfaces. Using his theory of extremal rays (developed as a first step in the classification of minimal models of projective threefolds), Mori proved that a minimal surface X whose canonical class K is not nef is isomorphic to either $\mathbb{P}^2$ or a $\mathbb{P}^1$-bundle over a curve (see [Mor*, Rei1*]). However, it is not clear whether these methods could yield a proof of the implication [X minimal with canonical dimension $-\infty$] $\Rightarrow$ [K is not nef]. In other words, Mori's methods do not seem to yield a proof of Enriques' Theorem 13.2.

15.3. Reider's Method. In the 1980s Reider [Rdr*] and Serrano [Srn*] discovered new methods to study linear systems on surfaces. Reider's results (sometimes called Reider's method) are particularly spectacular because they simplify and improve the results of Bombieri [Bom] on pluricanonical maps. Another nice application of Reider's method is to prove important results concerning the so-called adjunction mapping (see [SV*, Ion*]). Finally the ideas of Reider and Serrano have also been applied successfully to singular surfaces by Sakai [Sak5*].

15.4. Enriques' Classification of Singular and Open Surfaces. Enriques' classification of surfaces has been extended to compact normal two-dimensional Moishezon spaces by Sakai [Sak1*, Sak4*] and to open (i.e., non-projective) surfaces by Kawamata [Kaw*].

15.5. Moduli and Geography of Surfaces of General Type. For a minimal surface of general type over $\mathbb{C}$, there are certain classical inequalities among the numerical invariants of X. In 1977, Miyaoka [Miy*] and S. T. Yau [Yau*] independently proved the important inequality $c_1^2(X) \leq 3c_2(X)$. This inequality had been conjectured by Van de Ven [VdV2*]. For some earlier inequalities see [VdV1*, VdV2*] ($c_1^2(X) \leq 8c_2(X)$) and [Bog*, Gie2*, Rei2*] ($c_1^2(X) \leq 4c_2(X)$).

One of the fundamental problems of the geography of surfaces is to determine which invariants can occur. We refer the reader to [Cat2*] and [Per*] (and the bibliography therein). On the other hand, Gieseker [Gie1*] proved that there is a coarse moduli space $\mathcal{M}_{(K^2),c_2}$ parametrizing the isomorphism classes of surfaces of general type X with $c_1^2(X) = (K^2)$ and $c_2(X) = c_2$. Then the problem is to describe $\mathcal{M}_{(K^2),c_2}$ (which is a quasi-projective variety). Important results in this direction have been obtained by Catanese and his students; see, for example, [Cat3*] and [Mnt*].

References

[Abh]   S. Abhyankar, On the field of definition of a nonsingular transform of an algebraic surface. *Ann. of Math.* **65** (1957), 268–281.

[AK]   S. Altman and S. Kleiman, *Introduction to Grothendieck Duality Theory.* Springer Lect. Notes Math. **146**, 1970.

[ACGH*]   E. Arbarello, M. Cornalba, Ph. Griffiths, and J. Harris, *Geometry of Algebraic Curves, I.* Grundlehren **267**, Springer-Verlag, New York, 1985.

[Art1]   M. Artin, Some numerical criteria for contractability of curves on an algebraic surface. *Amer. J. Math.* **84** (1962), 485–496.

[Art2]   M. Artin, On isolated rational singularities of surfaces. *Amer. J. Math.* **88** (1966), 129–136.

[Art3]   M. Artin, Algebraization of formal moduli: II. Existence of modifications. *Ann. of Math.* **91** (1970), 88–136.

[Ati]   M. F. Atiyah, Vector bundles over an elliptic curve. *Proc. London Math. Soc.* **27** (1957), 414–452.

[BPV*]   W. Barth, C. Peters, and A. Van de Ven, *Compact Complex Surfaces.* Springer-Verlag, Berlin–Heidelberg–New York, 1984.

[BHH*]   G. Barthel, F. Hirzebruch, and Th. Höfer, *Geraden-konfigurationen und Algebraische Flächen.* Aspekte der Mathematik, Vieweg, 1987

248   References

[Băd1]   L. Bădescu, Applications of the Grothendieck duality theory to the study of normal isolated singularities. *Revue Roumaine Math. Pures Appl.* **24** (1979), 673–689.

[Băd2*]   L. Bădescu, Anticanonical models of ruled surfaces. *Ann. Univ. Ferrara* **29** (1983), 165–177.

[Băd3*]   L. Bădescu, The graded algebra associated to a divisor on a smooth projective surface (in Romanian). In *Analiză Complexă: Aspecte Clasice şi Moderne.* Editura Ştiinţifică şi Enciclopedică, Bucureşti, 1988. 295–337.

[BBI*]   L. Bădescu, M. Beltrametti, and P. Ionescu, Almost-lines and quasi-lines on projective manifolds. *Proc. Symp. Complex Analysis and Algebraic Geometry, held in memory of M. Schneider, 1998.* De Gruyter, 2000, pp. 1–27.

[BU*]   C. Bănică and K. Ueno, On the Hilbert–Samuel polynomial in complex-analytic geometry. *J. Math. Kyoto Univ.* **20** (1980), 381–389.

[Bea]   A. Beauville, *Surfaces Algébriques Complexes.* Astérisque **54** (1978). Soc. Math. France.

[Bog*]   F. A. Bogomolov, Holomorphic tensors and vector bundles on projective varieties. *Izv. Akad. Nauk USSR* **13** (1979), 499–555.

[Bom]   E. Bombieri, Pluricanonical models of surfaces of general type. *Inst. Hautes Études Sci. Publ. Math.* **42** (1972), 171–220.

[BH]   E. Bombieri and D. Husemoller, Classification and embeddings of surfaces. *Proc. Symp. Pure Math.* **29** (1975), 329–420.

[BM1]   E. Bombieri and D. Mumford, Enriques' classification of surfaces in characteristic p: II. In *Complex Analysis and Algebraic Geometry (dedicated to K. Kodaira).* Iwanami Shoten Publ., Tokyo, Cambridge Univ. (Part I), 1977.

[BM2]   E. Bombieri and D. Mumford, Enriques' classification of surfaces in characteristic p: III. *Invent. Math.* **35** (1976), 197–232.

[Bor]   A. Borel, *Linear Algebraic Groups.* Benjamin, New York–Amsterdam, 1969.

[BS]   A. Borel and J.-P. Serre, Le théorème de Riemann–Roch. *Bull. Soc. Math. France* **86** (1958), 97–136.

[Bbk]   N. Bourbaki, *Algèbre Commutative, Chap. VII.* Hermann, Paris, 1965.

[Bre*]   L. Brenton, Some algebraicity criteria for singular surfaces. *Invent. Math.* **41** (1977), 129–147.

[Buc1]   I. Bucur, *Seminar de Geometrie Algebrică* (in Romanian). Rome, 1970. Unpublished manuscript.

[Buc2]   I. Bucur, *Capitole Speciale de Algebră* (in Romanian). Editura Academiei, Bucureşti, 1980. English translation: *Selected Topics in Algebra and Its Interrelations with Logic, Number Theory and Algebraic Geometry*, D. Reidel Publishing Co. and Editura Academiei Române, 1984.

[CE]   H. Cartan and S. Eilenberg, *Homological Algebra*. Princeton University Press, 1956.

[Cat1*]   F. Catanese, Superficie complesse compatte. In *Atti del Convegno GNSAGA del CNR, Valetto, Torino, 1984*, 1986, 7–58.

[Cat2*]   F. Catanese, On the moduli spaces of surfaces of general type. *J. Diff. Geom.* **19** (1984), 483–515.

[Cat3*]   F. Catanese, Moduli of surfaces of general type. In *Algebraic Geometry—Open Problems, Ravello, 1982*, Springer Lect. Notes Math. **997**, 1983, 90–112.

[Cat4*]   F. Catanese, Old and new results on algebraic surfaces. In *First European Congress of Mathematics, Paris, July 6–11, 1992, Vol. I*. Progress in Mathematics **119**, Birkhäuser, 1994, 445–490.

[CG]   C. H. Clemens and P. Griffiths, The intermediate Jacobian of the cubic threefold. *Ann. of Math.* **95** (1972), 281–356.

[CD*]   F. Cossec and I. Dolgachev, *Enriques Surfaces I*. Birkhäuser, 1989.

[Alm*]   J. d'Almeida, *L'Enseignement Math.* **41** (1995), 135–139.

[Del]   P. Deligne et al., *Cohomologie Étale (SGA $4\frac{1}{2}$)*. Springer Lect. Notes Math. **569**, 1977.

[DK]   P. Deligne and N. Katz, *Groupe de monodromie en géométrie algébrique*. SGA VII-2, éxposé XIX, Springer Lect. Notes Math. **340**, 1973.

[Dem*]   M. Demazure, Surfaces de Del Pezzo II–V. In *Séminaire sur les surfaces algébriques*, Springer Lect. Notes Math. **777**, 1980, 23–69.

[DuV]   P. Du Val, On isolated singularities which do not affect the condition of adjunction I. *Proc. Camb. Phil. Soc.* **30** (1933/34), 483–491.

[EGA]   This is another name for [GD].

[Enr]   F. Enriques, *Superficie Algebriche.* Zanichelli, Bologna, 1949.

[FAC]   This is another name for [Ser1].

[Fuj*]   T. Fujita, On Zariski problem. *Proc. Japan. Acad.* **55** (1979), 106–110.

[Ful*]   W. Fulton, *Intersection Theory.* Springer-Verlag, New York, 1984.

[Gie1*]   D. Gieseker, Global moduli for surfaces of general type. *Invent. Math.* **43** (1977), 233–282.

[Gie2*]   D. Gieseker, On a theorem of Bogomolov on Chern classes of stable bundles. *Amer. J. Math.* **101** (1979), 77–85.

[Gra*]   H. Grauert, Über Modifikazionen und excepzionelle analytische Mengen. *Math. Ann.* **146** (1962), 331–368.

[GH*]   Ph. Griffiths and J. Harris, *Principles of Algebraic Geometry.* Wiley Interscience, New York, 1978.

[Gro1]   A. Grothendieck, Sur une note de Mattuck–Tate. *J. Reine Angew. Mathematik* **200** (1958), 208–215.

[Gro2]   A. Grothendieck, La théorie des classes de Chern. *Bull. Soc. Math. France* **86** (1958), 137–154.

[Gro3]   A. Grothendieck, *Fondements de la Géométrie Algébrique (Extraits du Séminaire Bourbaki, 1957–1962).* Secrét. Paris, 1962.

[Gro4]   A. Grothendieck, *Local Cohomology.* Springer Lect. Notes Math. **41**, 1967.

[Gro5]   A. Grothendieck, La classification des fibrés holomorphes sur la sphère de Riemann. *Amer. J. Math.* **79** (1957), 121–138.

[Gro6]   A. Grothendieck, *Séminaire de Géométrie Algébrique (SGA II).* Inst. Hautes Études Sci., Paris, 1960.

[Gro7]   A. Grothendieck, *Revêtements Étales et Groupe Fondamental (SGA I).* Springer Lect. Notes Math. **224**, 1971.

[Gro8]   A. Grothendieck et al., *Théorie des Intersections et Théorème de Riemann–Roch (SGA 6).* Springer Lect. Notes Math. **225**, 1971.

[GD]   A. Grothendieck and J. Dieudonné, *Éléments de Géométrie Algébrique I–IV (EGA I–IV)*. *Inst. Hautes Études Sci. Publ. Math.* **4** (1960), **8** (1961), **11** (1961), and **28** (1966).

[Har1]   R. Hartshorne, *Algebraic Geometry*. Graduate Texts in Mathematics **52**, Springer-Verlag, New York, 1977.

[Har2]   R. Hartshorne, *Residues and Duality*. Springer Lect. Notes Math. **20**, 1966.

[Har3]   R. Hartshorne, *Ample Subvarieties of Algebraic Varieties*. Springer Lect. Notes Math. **156**, 1970.

[Har4]   R. Hartshorne, Curves with high self-intersection on an algebraic surface. *Inst. Hautes Études Sci. Publ. Math.* **36** (1969), 111–126.

[Hir]   H. Hironaka, Resolution of singularities of an algebraic variety over a field of characteristic zero. *Ann. of Math.* **79** (1964), 109–326.

[Igu]   J.-I. Igusa, Betti and Picard numbers of an abstract algebraic surface. *Proc. Nat. Acad. Sci. USA* **46** (1960), 724–726.

[Iit1*] S. Iitaka, On *D*-dimension of algebraic varieties. *J. Math. Soc. Japan* **23** (1971), 356–373.

[Iit2*] S. Iitaka, *An Introduction to Birational Geometry of Algebraic Varieties*. Springer Verlag, Berlin–Heidelberg–New York, 1982.

[Ion*]   P. Ionescu, Ample and very ample divisors on surfaces. *Rev. Roumaine Math. Pures Appl.* **33** (1988), 349–358.

[Isk]   V. A. Iskovskih, Minimal models of rational surfaces over arbitrary fields (in Russian). *Izvestia Akad. Nauk USSR* **43** (1979), 19–43.

[IM]   V. A. Iskovskih and Yu. I. Manin, Three-dimensional quartics and counter-examples to the Lüroth problem (in Russian). *Mat. Sbornik* **86** (1971), 140–166.

[Kaw*]   Y. Kawamata, *On the Classification of Non-Complete Algebraic Surfaces*. Springer Lect. Notes Math. **732**, 1979, 215–232.

[Kle]   S. Kleiman, Towards a numerical theory of ampleness. *Ann. of Math.* **84** (1966), 293–344.

[Knö]   F. W. Knöller, 2-dimensionale Singularitäten und Differentialformen. *Math. Ann.* **206** (1973), 205–213.

[Knu*]   D. Knudson, *Algebraic Spaces*. Springer Lect. Notes Math. **203**, 1971.

[Kod1]   K. Kodaira, On compact analytic surfaces I, II, III. *Ann. of Math.* **71, 77, 78** (1960,1963), 111–152, 563–626, and 1–40.

[Kod2]   K. Kodaira, On the structure of compact complex analytic surfaces I, II, III, IV. *Amer. J. Math.* **86, 88** (1964, 1966), 751–798, 682–721, 55–83, and 1048–1066.

[Kur1]   H. Kurke, On Castelnuovo's criterion for rational surfaces. *Proc. Internat. Symposium Algebraic Geometry.* Kyoto, 1977, 557–563. Kinokuniya Book Store.

[Kur2*]   H. Kurke, *Vorlesungen über algebraische Flächen.* Teubner-Texte zur Math., Band **43**, Teubner, Leipzig, 1982.

[Lau]   H. Laufer, On rational singularities. *Amer. J. Math.* **94** (1972), 597–608.

[Lip]   J. Lipman, Rational singularities with applications to algebraic surfaces and unique factorization. *Inst. Hautes Études Sci. Publ. Math.* **36** (1969), 195–297.

[Mnt*]   M. Manetti, *Degenerations of Algebraic Surfaces and Applications to Moduli Problems* (tesi di perfezionamento). Scuola Normale Superiore, Pisa, 1996.

[Man*]   Yu. I. Manin, *Cubic Forms: Algebra, Geometry, Arithmetic.* North-Holland, Amsterdam–London, 1974.

[Miy*]   Y. Miyaoka, On the Chern numbers of surfaces of general type. *Invent. Math.* **42** (1977), 225–237.

[Moi*]   B.G. Moishezon, Algebraic analogue of compact complex spaces with sufficiently large field of meromorphic functions I, II. *Izv. Akad Nauk SSSR* **33** (1969), 174–238 and 323–367.

[Mor*]   S. Mori, Threefolds whose canonical bundles are not numerically effective. *Ann. of Math.* **116** (1982), 133–176.

[Mum1]   D. Mumford, *Lectures on Curves on an Algebraic Surface.* Annals of Mathematics Studies **59**, 1966. Princeton.

[Mum2]   D. Mumford, *Abelian Varieties.* Oxford Univ. Press, Oxford, 1970.

[Mum3]   D. Mumford, The canonical ring of an algebraic surface. *Ann. of Math.* **76** (1962), 612–615.

[Mum4]   D. Mumford, Enriques' classification of surfaces in characteristic p: I. In *Global analysis (dedicated to K. Kodaira)*. 325–339. Princeton Univ. Press, 1969.

[Mum5]   D. Mumford, The topology of normal singularities of an algebraic surface and a criterion for simplicity. *Inst. Hautes Études Sci. Publ. Math.* **9** (1961), 229–246.

[Mum6*]   D. Mumford, Hilbert's fourteenth problem–the finite generation of subrings such as rings of invariants. In *Mathematical Developments Arising from Hilbert's Problems*, Proc. Symposia in Pure Math. **28**, Part 2, Amer. Math. Soc., Providence, Rhode Island, 1976, 431–444.

[Nag]   M. Nagata, On rational surfaces I. *Mem. Coll. Sci. Kyoto Univ.*, Ser. A, **32** (1960), 635–639.

[Ogu]   A. Ogus, Zariski's theorem on several linear systems. *Proc. Amer. Math. Soc.* **37** (1973), 59–62.

[Oor]   F. Oort, A construction of generalized Jacobian varieties by group extensions. *Math. Ann.* **147** (1962), 277–286.

[Per*]   U. Persson, An Introduction to the Geography of Surfaces of General Type. *Proc. Symp. Pure Math.* **46**(1), Amer. Math. Soc., Providence, Rhode Island, 1987, 195–218.

[Pet*]   C. Peters, Introduction to the theory of compact complex surfaces. *Canadian Math. Soc. Conf. Proc.*, **12**, 1992.

[Ram]   C. P. Ramanujam, Remarks on the Kodaira vanishing theorem. *J. Indian Math. Soc.* **36** (1972), 41–51.

[Ray]   M. Raynaud, Spécialization du foncteur de Picard. *Inst. Hautes Études Sci. Publ. Math.* **38** (1970), 27–76.

[Rei1*]   M. Reid, *Chapters on Algebraic Surfaces*. IAS/Park City Math. Series, **3**, 1997, 5–159.

[Rei2*]   M. Reid, Bogomolov's theorem $c_1^2 \leq 4c_2$. In *Proc. Internat. Symposium Algebraic Geometry, Kyoto, 1977*. Kinokuniya, Tokyo, 1977, 623–642.

[Rei3*]   M. Reid, Elliptic Gorenstein singularities of surfaces. Unpublished manuscript, 1975.

[Rdr*]   I. Reider, Vector bundles of rank 2 and linear systems on algebraic surfaces. *Ann. of Math.* **127** (1988), 309–316.

254   References

[Rus*]   F. Russo, On the complement of a nef and big divisor on an algebraic variety. *Math. Proc. Camb. Phil. Soc.* **120** (1996), 411–422.

[Sak1*]   F. Sakai, Enriques' classification of normal Gorenstein surfaces. *Amer. J. Math.* **104** (1981), 1233–1241.

[Sak2*]   F. Sakai, *D*-dimensions of algebraic surfaces and numerically effective divisors. *Compositio Math.* **48** (1983), 101–118.

[Sak3*]   F. Sakai, Anticanonical models of rational surfaces. *Math. Ann.* **269** (1984), 389–410.

[Sak4*]   F. Sakai, Weil divisors on normal surfaces. *Duke Math. J.* **51** (1984), 877–888.

[Sak5*]   F. Sakai, Reider-Serrano method on normal surfaces. Springer Lect. Notes Math. **1417**, 1990, 301–319.

[Srn*]   F. Serrano, The adjunction mapping and hyperelliptic divisors on a surface. *J. Reine Angew. Mathematik* **381** (1987), 90–109.

[Ser1]   J.-P. Serre, Faisceaux algébriques cohérents (FAC). *Ann. of Math.* **61** (1955), 197–278.

[Ser2]   J.-P. Serre, Géométrie algébrique et géométrie analytique. *Ann. Inst. Fourier* **6** (1956), 1–42.

[Ser3]   J.-P. Serre, *Groupes Algébriques et Corps de Classes.* Hermann, Paris, 1959.

[Ser4]   J.-P. Serre, Morphismes universels et variétés d'Albanese. *Séminaire Chevalley*, E.N.S. 1958/59, Exposé 10.

[Ser5]   J.-P. Serre, *Cours d'arithmétique*, Paris, 1970.

[SGA $4\frac{1}{2}$] This is another name for [Del].

[Sha1]   I. R. Shafarevich, *Basic Algebraic Geometry.* Grundlehren **213**, Springer-Verlag, Heidelberg, 1974. (Second edition, 1994)

[Sha2]   I. R. Shafarevich, *Lectures on Minimal Models and Birational Transformations of Two-Dimensional Schemes.* Tata Institute, Bombay, 1966.

[Sha3]   I. R. Shafarevich et al., *Algebraic Surfaces.* A.M.S., Providence, Rhode Island, 1967.

[SV*]   A. Sommese and A. Van de Ven, On the adjunction mapping. *Math. Ann.* **278** (1987), 593–603.

[Tat]   J. Tate, Algebraic formulas in arbitrary characteristic. Appendix 1 in S. Lang's *Elliptic Functions*. Addison Wesley Publ. Comp., 1973.

[VdV1*]   A. Van de Ven, On the Chern numbers of certain complex and almost complex manfolds. *Proc. Nat. Acad. Sci. USA* **55** (1966), 1624–1627.

[VdV2*]   A. Van de Ven, On the Chern numbers of surfaces of general type. *Invent. Math.* **36** (1976), 285–293.

[Wei*]   A. Weil, *Sur les courbes algébriques et les variétés qui s'en déduisent*. Hermann, Paris, 1948.

[Yau*]   S. T. Yau, Calabi's conjecture and some new results in algebraic geometry. *Proc. Nat. Acad. Sci. USA* **74** (1977), 1798–1799.

[Zar1]   O. Zariski, The theorem of Riemann–Roch for high multiples of a divisor on an algebraic surface. *Ann. of Math.* **76** (1962), 560–612.

[Zar2]   O. Zariski, The problem of minimal models in the theory of algebraic surfaces. *Amer. J. Math.* **80** (1958), 146–184.

[Zar3]   O. Zariski, On Castelnuovo's criterion of rationality $p_a = p_2 = 0$ of an algebraic surface. *Illinois J. Math.* **2** (1958), 303–315.

[Zar4]   O. Zariski, The reduction of singularities of an algebraic surface. *Ann. of Math.* **40** (1939), 639–689.

[Zar5]   O. Zariski, A simplified proof for the reduction of singularities of an algebraic surface. *Ann. of Math.* **43** (1942), 583–593.

[Zar6*]   O. Zariski, Interprétation algébro-géométrique du 14-ième problème de Hilbert. *Bull. Sci. Math.* **78** (1954), 155–164.

Index

adjunction terminates, 195
Albanese variety, 73
algebraic equivalence of
    invertible sheaves, 70
ampleness
    Nakai–Moishezon criterion,
    9
anticanonical ring of a surface,
    232
arithmetic genus of a curve, 24
Artin, M., 31, 35, 37, 41, 213,
    223, 224

Bagnera–DeFranchis list, 160
Betti numbers, 69

canonical dimension of a surface,
    76
canonical model of a surface of
    general type, 126, 240
canonical ring of a surface, 76,
    123, 231

Castelnuovo, G., vii, 43, 81, 195,
    207, 212, 214, 240
contractability
    Castelnuovo's criterion, 43
    M. Artin's criterion, 31, 35
curve
    arithmetic genus, 24
    exceptional of the first kind,
    55
    of canonical type, 92
      indecomposable, 92

D-dimension, 216
desingularization of a normal
    singularity, 35, 58
divisor(s)
    nef, 217
    numerical equivalence, 70
    pseudo-effective, 217
    Zariski decomposition,
    221
dualizing sheaf, 33
Dynkin diagrams, 44, 50

elementary transformation

generalized, of a
    geometrically ruled
    surface over $\mathbb{P}^1$, 186
elementary transformation of a
    geometrically ruled
    surface, 181
(quasi)elliptic fibration, 91
    canonical class, 100
Enriques, F., vii, 13, 143, 153,
    195, 212, 245, 246
exceptional curve of the first
    kind, 55
exceptional fiber, 100

fiber
    exceptional, 100
    generic irreducibility, 87, 89
    multiple, 100, 117, 156
fibration over a curve, 90
fundamental cycle, 36

generalized elementary
    transformation, 186
geometric genus, 76
Gorenstein singularity, 34, 64

Hodge index theorem, 18

Igusa–Severi inequality, 75, 114,
    141, 202
indecomposable curve of
    canonical type, 92
intersection number, 3
intersection theory of Weil
    divisors, 227
invertible sheaves
    algebraic equivalence, 70
    numerical equivalence, 70

Laufer–Ramanujam vanishing
    theorem, 53

m-graded module, 129
    polyfinite, 129, 130
m-graded ring, 129

minimal desingularization of a
    surface singularity, 55
minimal model
    existence, 81
minimal models
    of nonrational ruled
        surfaces, 181
    of rational surfaces, 186
model of a pair (X, D) with
    $\kappa(X, D) = 2$, 228
multiple fiber, 100, 117, 156

Nagata's example, 23
Nagata, M., 23, 186, 224
Nakai–Moishezon criterion, 9
nef divisor, 217
Noether–Tsen criterion, 165
normal singularity
    desingularization, 35, 58
numerical equivalence of
    invertible sheaves, 70

pencil of curves on a surface, 165
Picard scheme of a surface, 74
Picard variety, 74
plurigenera, 76
polyfinite m-graded module, 130
pseudo-effective divisor, 217
    Zariski decomposition, 221

rational double point, 43, 50
rational singularity, 36
Riemann–Roch theorem
    for vector bundles on a
        curve, 174
ruled surface, 165

singularity
    duality theorem, 59
    Gorenstein, 34, 64
    minimal desingularization,
        55
    rational double point, 43, 50
singularity, rational, 36
splitting of a vector bundle on a
    curve, 174

surface, 17, 23, 69
    Abelian, 151
    anticanonical ring, 232
    Betti numbers, 69
    canonical dimension, 76
        classification according
            to, 126, 129, 195
    canonical ring, 76, 123, 231
    Chern classes, 69
    class (a), (b), (c), or (d), 112
    Del Pezzo, 192, 213
    Enriques, 143
        examples, 145
        properties, 143–145
    geometric genus, 76
    geometrically ruled, 172
        elementary
            transformation, 181
    Godeaux, 127
    (quasi)hyperelliptic, 115
        properties, 113, 115
    K3, 138
        examples, 139
        properties, 138, 139

    Kummer, 139
    of general type, 126
        canonical model, 126, 240
    Picard scheme, 74
    plurigenera, 76
    rational
        Castelnuovo's criterion,
            195
        minimal models, 186
    ruled, 165
        Enriques' criterion, 195
        minimal models, 181
        Noether–Tsen criterion,
            165

unirational variety, 212

variety
    Albanese, 73
    Picard, 74
    unirational, 212

Zariski decomposition, 221
Zariski–Goodman theorem, 12
Zariski–Mumford theorem, 129

Universitext   *(continued)*

Luecking/Rubel: Complex Analysis: A Functional Analysis Approach
MacLane/Moerdijk: Sheaves in Geometry and Logic
Marcus: Number Fields
McCarthy: Introduction to Arithmetical Functions
Meyer: Essential Mathematics for Applied Fields
Mines/Richman/Ruitenburg: A Course in Constructive Algebra
Moise: Introductory Problems Course in Analysis and Topology
Morris: Introduction to Game Theory
Poizat: A Course In Model Theory: An Introduction to Contemporary Mathematical Logic
Polster: A Geometrical Picture Book
Porter/Woods: Extensions and Absolutes of Hausdorff Spaces
Radjavi/Rosenthal: Simultaneous Triangularization
Ramsay/Richtmyer: Introduction to Hyperbolic Geometry
Reisel: Elementary Theory of Metric Spaces
Ribenboim: Classical Theory of Algebraic Numbers
Rickart: Natural Function Algebras
Rotman: Galois Theory
Rubel/Colliander: Entire and Meromorphic Functions
Sagan: Space-Filling Curves
Samelson: Notes on Lie Algebras
Schiff: Normal Families
Shapiro: Composition Operators and Classical Function Theory
Simonnet: Measures and Probability
Smith: Power Series From a Computational Point of View
Smith/Kahanpää/Kekäläinen/Traves: An Invitation to Algebraic Geometry
Smoryński: Self-Reference and Modal Logic
Stillwell: Geometry of Surfaces
Stroock: An Introduction to the Theory of Large Deviations
Sunder: An Invitation to von Neumann Algebras
Tondeur: Foliations on Riemannian Manifolds
Wong: Weyl Transforms
Zhang: Matrix Theory: Basic Results and Techniques
Zong: Sphere Packings
Zong: Strange Phenomena in Convex and Discrete Geometry